A Course in Fuzzy Systems and Control

Li-Xin Wang

For book and bookstore information

http://www.prenhall.com
gopher to gopher.prenhall.com

Prentice Hall PTR, Upper Saddle River, NJ 07458

Library of Congress Cataloging-in-Publication Data
Wang, Li-Xin, 1962-
    A course in fuzzy systems and control / Li-Xin Wang.
       p. cm.
    Includes bibliographical references and index.
    ISBN 0-13-540882-2 (hard cover)
    1. Neural networks (Computer science) 2. Fuzzy Systems.
I. Title.
QA76.87.W35 1997
629.8'3--dc20   96-17721
                                              CIP

Editorial/Production Supervision: *Joe Czerwinski*
Buyer: *Alexis R. Heydt*
Publisher: *Bernard Goodwin*

©1997 by Prentice Hall PTR
Prentice-Hall, Inc.
A Division of Simon and Schuster
Upper Saddle River, NJ 07458

The publisher offers discounts on this book when ordered
in bulk quantities. For more information, contact:

    Corporate Sales Department
    Prentice Hall PTR
    One Lake Street
    Upper Saddle River, NJ 07458

    Phone: 800-382-3419
    Fax: 201-236-7141
    E-mail: corpsales@prenhall.com

Printed in the United States of America

10 9 8 7 6 5 4 3 2 1

ISBN: 0-13-540882-2

Prentice-Hall International (UK) Limited, *London*
Prentice-Hall of Australia Pty. Limited, *Sydney*
Prentice-Hall of Canada Inc., *Toronto*
Prentice-Hall Hispanoamericana, S.A., *Mexico*
Prentice-Hall of India Pte. Ltd., *New Delhi*
Prentice-Hall of Japan, Inc., *Tokyo*
Simon & Schuster Asia Pte. Ltd., *Singapore*
Editora Prentice-Hall do Brasil, Ltda., *Rio de Janeiro*

*To the pioneers of fuzzy theory
and those who are working
rigorously in this fuzzy world*

Contents

V   Adaptive Fuzzy Control   289

Preface

The field of fuzzy systems and control has been making rapid progress in recent years. Motivated by the practical success of fuzzy control in consumer products and industrial process control, there has been an increasing amount of work on the rigorous theoretical studies of fuzzy systems and fuzzy control. Researchers are trying to explain why the practical results are good, systematize the existing approaches, and develop more powerful ones. As a result of these efforts, the whole picture of fuzzy systems and fuzzy control theory is becoming clearer. Although there are many books on fuzzy theory, most of them are either research monographs that concentrate on special topics, or collections of papers, or books on fuzzy mathematics. We desperately need a real textbook on fuzzy systems and control that provides the skeleton of the field and summarizes the fundamentals.

This book, which is based on a course developed at the Hong Kong University of Science and Technology, is intended as a textbook for graduate and senior students, and as a self-study book for practicing engineers. When writing this book, we required that it be:

- **Well-Structured:** This book is not intended as a collection of existing results on fuzzy systems and fuzzy control; rather, we first establish the structure that a reasonable theory of fuzzy systems and fuzzy control should follow, and then fill in the details. For example, when studying fuzzy control systems, we should consider the stability, optimality, and robustness of the systems, and classify the approaches according to whether the plant is linear, nonlinear, or modeled by fuzzy systems. Fortunately, the major existing results fit very well into this structure and therefore are covered in detail in this book. Because the field is not mature, as compared with other mainstream fields, there are holes in the structure for which no results exist. For these topics, we either provide our preliminary approaches, or point out that the problems are open.

- **Clear and Precise:** Clear and logical presentation is crucial for any book, especially for a book associated with the word "fuzzy." Fuzzy theory itself is precise; the "fuzziness" appears in the phenomena that fuzzy theory tries to study. Once a fuzzy description (for example, "hot day") is formulated in terms of fuzzy theory, nothing will be fuzzy anymore. We pay special attention to the use of precise language to introduce the concepts, to develop the approaches, and to justify the conclusions.

- **Practical:** We recall that the driving force for fuzzy systems and control is practical applications. Most approaches in this book are tested for problems that have practical significance. In fact, a main objective of the book is to teach students and practicing engineers how to use the fuzzy systems approach to solving engineering problems in control, signal processing, and communications.

- **Rich and Rigorous:** This book should be intelligently challenging for students. In addition to the emphasis on practicality, many theoretical results are given (which, of course, have practical relevance and importance). All the theorems and lemmas are proven in a mathematically rigorous fashion, and some effort may have to be taken for an average student to comprehend the details.

- **Easy to Use as Textbook:** To facilitate its use as a textbook, this book is written in such a style that each chapter is designed for a one and one-half hour lecture. Sometimes, three chapters may be covered by two lectures, or vice versa, depending upon the emphasis of the instructor and the background of the students. Each chapter contains some exercises and mini-projects that form an integrated part of the text.

The book is divided into six parts. Part I (Chapters 2-6) introduces the fundamental concepts and principles in the general field of fuzzy theory that are particularly useful in fuzzy systems and fuzzy control. Part II (Chapters 7-11) studies the fuzzy systems in detail. The operations inside the fuzzy systems are carefully analyzed and certain properties of the fuzzy systems (for example, approximation capability and accuracy) are studied. Part III (Chapters 12-15) introduces four methods for designing fuzzy systems from sensory measurements, and all these methods are tested for a number of control, signal processing, or communication problems. Part IV (Chapters 16-22) and Part V (Chapters 23-26) parts concentrate on fuzzy control, where Part IV studies nonadaptive fuzzy control and Part V studies adaptive fuzzy control. Finally, Part VI (Chapters 27-31) reviews a number of topics that are not included in the main structure of the book, but are important and strongly relevant to fuzzy systems and fuzzy control.

The book can be studied in many ways, according to the particular interests of the instructor or the reader. Chapters 1-15 cover the general materials that can be applied to a variety of engineering problems. Chapters 16-26 are more specialized in control problems. If the course is not intended as a control course, then some materials in Chapters 16-26 may be omitted, and the time saved may be used for a more detailed coverage of Chapters 1-15 and 27-31. On the other hand, if it is a control course, then Chapters 16-26 should be studied in detail. The book also can be used, together with a book on neural networks, for a course on neural networks and fuzzy systems. In this case, Chapters 1-15 and selected topics from Chapters 16-31 may be used for the fuzzy system half of the course. If a practicing engineer wants to learn fuzzy systems and fuzzy control quickly, then the proofs of the theorems and lemmas may be skipped.

This book has benefited from the review of many colleagues, students, and friends. First of all, I would like thank my advisors, Lotfi Zadeh and Jerry Mendel, for their continued encouragement. I would like to thank Karl Åström for sending his student, Mikael Johansson, to help me prepare the manuscript during the summer of 1995. Discussions with Kevin Passino, Frank Lewis, Jyh-Shing Jang, Hua Wang, Hideyuki Takagi, and other researchers in fuzzy theory have helped the organization of the materials. The book also benefited from the input of the students who took the course at HKUST.

Support for the author from the Hong Kong Research Grants Council was greatly appreciated.

Finally, I would like to express my gratitude to my department at HKUST for providing the excellent research and teaching environment. Especially, I would like to thank my colleagues Xiren Cao, Zexiang Li, Li Qiu, Erwei Bai, Justin Chuang, Philip Chan, and Kwan-Fai Cheung for their collaboration and critical remarks on various topics in fuzzy theory.

Li-Xin Wang
The Hong Kong University of Science and Technology

Chapter 1

Introduction

1.1 Why Fuzzy Systems?

According to the Oxford English Dictionary, the word "fuzzy" is defined as "blurred, indistinct; imprecisely defined; confused, vague." We ask the reader to disregard this definition and view the word "fuzzy" as a technical adjective. Specifically, fuzzy systems are systems to be precisely defined, and fuzzy control is a special kind of nonlinear control that also will be precisely defined. This is analogous to linear systems and control where the word "linear" is a technical adjective used to specify "systems and control;" the same is true for the word "fuzzy." Essentially, what we want to emphasize is that although the phenomena that the fuzzy systems theory characterizes may be fuzzy, the theory itself is precise.

In the literature, there are two kinds of justification for fuzzy systems theory:

- The real world is too complicated for precise descriptions to be obtained, therefore approximation (or fuzziness) must be introduced in order to obtain a reasonable, yet trackable, model.

- As we move into the information era, human knowledge becomes increasingly important. We need a theory to formulate human knowledge in a systematic manner and put it into engineering systems, together with other information like mathematical models and sensory measurements.

The first justification is correct, but does not characterize the unique nature of fuzzy systems theory. In fact, almost all theories in engineering characterize the real world in an approximate manner. For example, most real systems are nonlinear, but we put a great deal of effort in the study of linear systems. A good engineering theory should be precise to the extent that it characterizes the key features of the real world and, at the same time, is trackable for mathematical analysis. In this aspect, fuzzy systems theory does not differ from other engineering theories.

The second justification characterizes the unique feature of fuzzy systems theory and justifies the existence of fuzzy systems theory as an independent branch in engineering. As a general principle, a good engineering theory should be capable of making use of all available information effectively. For many practical systems, important information comes from two sources: one source is human experts who describe their knowledge about the system in natural languages; the other is sensory measurements and mathematical models that are derived according to physical laws. An important task, therefore, is to combine these two types of information into system designs. To achieve this combination, a key question is how to formulate human knowledge into a similar framework used to formulate sensory measurements and mathematical models. In other words, the key question is how to transform a human knowledge base into a mathematical formula. Essentially, what a fuzzy system does is to perform this transformation. In order to understand how this transformation is done, we must first know what fuzzy systems are.

1.2 What Are Fuzzy Systems?

Fuzzy systems are knowledge-based or rule-based systems. The heart of a fuzzy system is a knowledge base consisting of the so-called fuzzy IF-THEN rules. A fuzzy IF-THEN rule is an IF-THEN statement in which some words are characterized by continuous membership functions. For example, the following is a fuzzy IF-THEN rule:

$$IF\ the\ speed\ of\ a\ car\ is\ high,\ THEN\ apply\ less\ force\ to\ the\ accelerator \quad (1.1)$$

where the words "high" and "less" are characterized by the membership functions shown in Figs.1.1 and 1.2, respectively.[1] A fuzzy system is constructed from a collection of fuzzy IF-THEN rules. Let us consider two examples.

Example 1.1. Suppose we want to design a controller to automatically control the speed of a car. Conceptually, there are two approaches to designing such a controller: the first approach is to use conventional control theory, for example, designing a PID controller; the second approach is to emulate human drivers, that is, converting the rules used by human drivers into an automatic controller. We now consider the second approach. Roughly speaking, human drivers use the following three types of rules to drive a car in normal situations:

$$IF\ speed\ is\ low,\ THEN\ apply\ more\ force\ to\ the\ accelerator \quad (1.2)$$
$$IF\ speed\ is\ medium,\ THEN\ apply\ normal\ force\ to\ the\ accelerator \quad (1.3)$$
$$IF\ speed\ is\ high,\ THEN\ apply\ less\ force\ to\ the\ accelerator \quad (1.4)$$

where the words "low," "more," "medium," "normal," "high," and "less" are characterized by membership functions similar to those in Figs.1.1-1.2. Of course, more rules are needed in real situations. We can construct a fuzzy system based on these

---

[1]A detailed definition and analysis of membership functions will be given in Chapter 2. At this point, an intuitive understanding of the membership functions in Figs. 1.1 and 1.2 is sufficient.

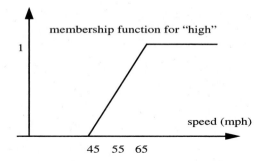

Figure 1.1. Membership function for "high," where the horizontal axis represents the speed of the car and the vertical axis represents the membership value for "high."

rules. Because the fuzzy system is used as a controller, it also is called a fuzzy controller. □

Example 1.2. In Example 1.1, the rules are control instructions, that is, they represent what a human driver does in typical situations. Another type of human knowledge is descriptions about the system. Suppose a person pumping up a balloon wished to know how much air he could add before it burst, then the relationship among some key variables would be very useful. With the balloon there are three key variables: the air inside the balloon, the amount it increases, and the surface tension. We can describe the relationship among these variables in the following fuzzy IF-THEN rules:

$$\text{IF the amount of air is small and it is increased slightly,} \atop \text{THEN the surface tension will increase slightly} \qquad (1.5)$$

$$\text{IF the amount of air is small and it is increased substantially,} \atop \text{THEN the surface tension will increase substantially} \qquad (1.6)$$

$$\text{IF the amount of air is large and it is increased slightly,} \atop \text{THEN the surface tension will increase moderately} \qquad (1.7)$$

$$\text{IF the amount of air is large and it is increased substantially,} \atop \text{THEN the surface tension will increase very substantially} \qquad (1.8)$$

where the words "small," "slightly," "substantially," etc., are characterized by membership functions similar to those in Figs.1.1 and 1.2. Combining these rules into a fuzzy system, we obtain a model for the balloon. □

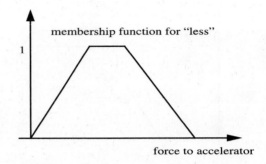

Figure 1.2. Membership function for "less," where the horizontal axis represents the force applied to the accelerator and the vertical axis represents the membership value for "less."

In summary, the starting point of constructing a fuzzy system is to obtain a collection of fuzzy IF-THEN rules from human experts or based on domain knowledge. The next step is to combine these rules into a single system. Different fuzzy systems use different principles for this combination. So the question is: what are the commonly used fuzzy systems?

There are three types of fuzzy systems that are commonly used in the literature: (i) pure fuzzy systems, (ii) Takagi-Sugeno-Kang (TSK) fuzzy systems, and (iii) fuzzy systems with fuzzifier and defuzzifier. We now briefly describe these three types of fuzzy systems.

The basic configuration of a pure fuzzy system is shown in Fig. 1.3. The *fuzzy rule base* represents the collection of fuzzy IF-THEN rules. For examples, for the car controller in Example 1.1, the fuzzy rule base consists of the three rules (1.2)-(1.4), and for the balloon model of Example 1.2, the fuzzy rule base consists of the four rules (1.5)-(1.8). The *fuzzy inference engine* combines these fuzzy IF-THEN rules into a mapping from fuzzy sets[2] in the input space $U \subset R^n$ to fuzzy sets in the output space $V \subset R$ based on fuzzy logic principles. If the dashed feedback line in Fig. 1.3 exists, the system becomes the so-called fuzzy dynamic system.

The main problem with the pure fuzzy system is that its inputs and outputs are

---

[2]The precise definition of fuzzy set is given in Chapter 2. At this point, it is sufficient to view a fuzzy set as a word like, for example, "high," which is characterized by the membership function shown in Fig.1.1.

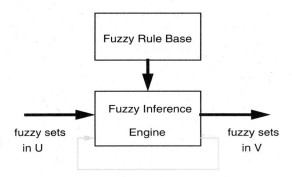

Figure 1.3. Basic configuration of pure fuzzy systems.

fuzzy sets (that is, words in natural languages), whereas in engineering systems the inputs and outputs are real-valued variables. To solve this problem, Takagi, Sugeno, and Kang (Takagi and Sugeno [1985] and Sugeno and Kang [1988]) proposed another fuzzy system whose inputs and outputs are real-valued variables.

Instead of considering the fuzzy IF-THEN rules in the form of (1.1), the Takagi-Sugeno-Kang (TSK) system uses rules in the following form:

$$IF \ the \ speed \ x \ of \ a \ car \ is \ high, \\ THEN \ the \ force \ to \ the \ accelerator \ is \ y = cx \tag{1.9}$$

where the word "high" has the same meaning as in (1.1), and c is a constant. Comparing (1.9) and (1.1) we see that the THEN part of the rule changes from a description using words in natural languages into a simple mathematical formula. This change makes it easier to combine the rules. In fact, the Takagi-Sugeno-Kang fuzzy system is a weighted average of the values in the THEN parts of the rules. The basic configuration of the Takagi-Sugeno-Kang fuzzy system is shown in Fig. 1.4.

The main problems with the Takagi-Sugeno-Kang fuzzy system are: (i) its THEN part is a mathematical formula and therefore may not provide a natural framework to represent human knowledge, and (ii) there is not much freedom left to apply different principles in fuzzy logic, so that the versatility of fuzzy systems is not well-represented in this framework. To solve these problems, we use the third type of fuzzy systems—fuzzy systems with fuzzifier and defuzzifier.

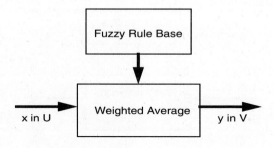

Figure 1.4. Basic configuration of Takagi-Sugeno-Kang fuzzy system.

In order to use pure fuzzy systems in engineering systems, a simple method is to add a fuzzifier, which transforms a real-valued variable into a fuzzy set, to the input, and a defuzzifier, which transforms a fuzzy set into a real-valued variable, to the output. The result is the fuzzy system with fuzzifier and defuzzifier, shown in Fig. 1.5. This fuzzy system overcomes the disadvantages of the pure fuzzy systems and the Takagi-Sugeno-Kang fuzzy systems. Unless otherwise specified, from now on when we refer fuzzy systems we mean fuzzy systems with fuzzifier and defuzzifier.

To conclude this section, we would like to emphasize a distinguished feature of fuzzy systems: on one hand, fuzzy systems are multi-input-single-output mappings from a real-valued vector to a real-valued scalar (a multi-output mapping can be decomposed into a collection of single-output mappings), and the precise mathematical formulas of these mappings can be obtained (see Chapter 9 for details); on the other hand, fuzzy systems are knowledge-based systems constructed from human knowledge in the form of fuzzy IF-THEN rules. *An important contribution of fuzzy systems theory is that it provides a systematic procedure for transforming a knowledge base into a nonlinear mapping.* Because of this transformation, we are able to use knowledge-based systems (fuzzy systems) in engineering applications (control, signal processing, or communications systems, etc.) in the same manner as we use mathematical models and sensory measurements. Consequently, the analysis and design of the resulting combined systems can be performed in a mathematically rigorous fashion. The goal of this text is to show how this transformation is done, and how the analysis and design are performed.

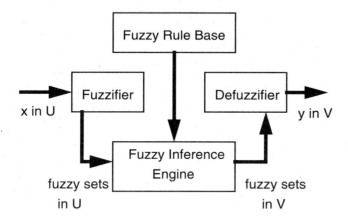

Figure 1.5. Basic configuration of fuzzy systems with fuzzifier and defuzzifier.

1.3   Where Are Fuzzy Systems Used and How?

Fuzzy systems have been applied to a wide variety of fields ranging from control, signal processing, communications, integrated circuit manufacturing, and expert systems to business, medicine, psychology, etc. However, the most significant applications have concentrated on control problems. Therefore, instead of listing the applications of fuzzy systems in the different fields, we concentrate on a number of control problems where fuzzy systems play a major role.

Fuzzy systems, as shown in Fig. 1.5, can be used either as open-loop controllers or closed-loop controllers, as shown in Figs. 1.6 and 1.7, respectively. When used as an open-loop controller, the fuzzy system usually sets up some control parameters and then the system operates according to these control parameters. Many applications of fuzzy systems in consumer electronics belong to this category. When used as a closed-loop controller, the fuzzy system measures the outputs of the process and takes control actions on the process continuously. Applications of fuzzy systems in industrial processes belong to this category. We now briefly describe how fuzzy systems are used in a number of consumer products and industrial systems.

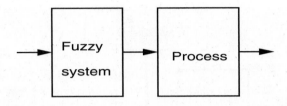

Figure 1.6. Fuzzy system as open-loop controller.

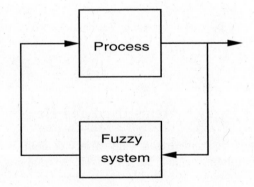

Figure 1.7. Fuzzy system as closed-loop controller.

1.3.1   Fuzzy Washing Machines

The fuzzy washing machines were the first major consumer products to use fuzzy systems. They were produced by Matsushita Electric Industrial Company in Japan around 1990. They use a fuzzy system to automatically set the proper cycle according to the kind and amount of dirt and the size of the load. More specifically, the fuzzy system used is a three-input-one-output system, where the three inputs are measurements of dirtiness, type of dirt, and load size, and the output is the correct cycle. Sensors supply the fuzzy system with the inputs. The optical sensor sends a beam of light through the water and measures how much of it reaches the other side. The dirtier the water, the less light crosses. The optical sensor also can tell whether the dirt is muddy or oily. Muddy dirt dissolves faster. So, if the light readings reach minimum quickly, the dirt is muddy. If the downswing is slower, it is oily. And if the curve slopes somewhere in between, the dirt is mixed. The machine also has a load sensor that registers the volume of clothes. Clearly, the more volume of the clothes, the more washing time is needed. The heuristics above were summarized in a number of fuzzy IF-THEN rules that were then used to construct the fuzzy system.

1.3.2   Digital Image Stabilizer

Anyone who has ever used a camcorder realizes that it is very difficult for a human hand to hold the camcorder without shaking it slightly and imparting an irksome quiver to the tape. Smoothing out this jitter would produce a new generation of camcorders and would have tremendous commercial value. Matsushita introduced what it calls a digital image stabilizer, based on fuzzy systems, which stabilizes the picture when the hand is shaking. The digital image stabilizer is a fuzzy system that is constructed based on the following heuristics:

$$\text{IF all the points in the picture are moving in the same direction,} \\ \text{THEN the hand is shaking} \tag{1.10}$$

$$\text{IF only some points in the picture are moving,} \\ \text{THEN the hand is not shaking} \tag{1.11}$$

More specifically, the stabilizer compares each current frame with the previous images in memory. If the whole appears to have shifted, then according to (1.10) the hand is shaking and the fuzzy system adjusts the frame to compensate. Otherwise, it leaves it alone. Thus, if a car crosses the field, only a portion of the image will change, so the camcorder does not try to compensate. In this way the picture remains steady, although the hand is shaking.

1.3.3   Fuzzy Systems in Cars

An automobile is a collection of many systems—engine, transmission, brake, suspension, steering, and more—and fuzzy systems have been applied to almost all of them. For example, Nissan has patented a fuzzy automatic transmission that saves fuel by 12 to 17 percent. It is based on the following observation. A normal transmission shifts whenever the car passes a certain speed, it therefore changes quite often and each shift consumes gas. However, human drivers not only shift less frequently, but also consider nonspeed factors. For example, if accelerating up a hill, they may delay the shift. Nissan's fuzzy automatic transmission device summarized these heuristics into a collection of fuzzy IF-THEN rules that were then used to construct a fuzzy system to guide the changes of gears.

Nissan also developed a fuzzy antilock braking system. The challenge here is to apply the greatest amount of pressure to the brake without causing it to lock. The Nissan system considers a number of heuristics, for example,

$$IF \; the \; car \; slows \; down \; very \; rapidly,$$
$$THEN \; the \; system \; assumes \; brake - lock \; and \; eases \; up \; on \; pressure \qquad (1.12)$$

In April 1992, Mitsubishi announced a fuzzy omnibus system that controls a car's automatic transmission, suspension, traction, four-wheel steering, four-wheel drive, and air conditioner. The fuzzy transmission downshifts on curves and also keeps the car from upshifting inappropriately on bends or when the driver releases the accelerator. The fuzzy suspension contains sensors in the front of the car that register vibration and height changes in the road and adjusts the suspension for a smoother ride. Fuzzy traction prevents excess speed on corners and improves the grip on slick roads by deciding whether they are level or sloped. Finally, fuzzy steering adjusts the response angle of the rear wheels according to road conditions and the car's speed, and fuzzy air conditioning monitors sunlight, temperature, and humidity to enhance the environment inside the car.

1.3.4 Fuzzy Control of a Cement Kiln

Cement is manufactured by finegrinding of cement clinker. The clinkers are produced in the cement kiln by heating a mixture of linestone, clay, and sand components. Because cement kilns exhibit time-varying nonlinear behavior and relatively few measurements are available, they are difficult to control using conventional control theory.

In the late 1970s, Holmblad and Østergaard of Denmark developed a fuzzy system to control the cement kiln. The fuzzy system (fuzzy controller) had four inputs and two outputs (which can be viewed as two fuzzy systems in the form of Fig. 1.5, which share the same inputs). The four inputs are: (i) oxygen percentage in exhausted gases, (ii) temperature of exhaust gases, (iii) kiln drive torque, and (iv) litre weight of clinker (indicating temperature level in the burning zone and quality of clinker). The two outputs are: (i) coal feed rate and (ii) air flow. A collection of fuzzy IF-THEN rules were constructed that describe how the outputs should be related to the inputs. For example, the following two rules were used:

$$IF \; the \; oxygen \; percentage \; is \; high \; and \; the \; temperature \; is \; low,$$
$$THEN \; increase \; air \; flow \qquad (1.13)$$

$$IF \; the \; oxygen \; percentage \; is \; high \; and \; the \; temperature \; is \; high,$$
$$THEN \; reduce \; the \; coal \; feed \; rate \; slightly \qquad (1.14)$$

The fuzzy controller was constructed by combining these rules into fuzzy systems. In June 1978, the fuzzy controller ran for six days in the cement kiln of F.L. Smidth & Company in Denmark—the first successful test of fuzzy control on a full-scale industrial process. The fuzzy controller showed a slight improvement over the results of the human operator and also cut fuel consumption. We will show more details about this system in Chapter 16.

1.3.5 Fuzzy Control of Subway Train

The most significant application of fuzzy systems to date may be the fuzzy control system for the Sendai subway in Japan. On a single north-south route of 13.6 kilometers and 16 stations, the train runs along very smoothly. The fuzzy control system considers four performance criteria simutaneously: safety, riding comfort, traceability to target speed, and accuracy of stopping gap. The fuzzy control system consists of two parts: the constant speed controller (it starts the train and keeps the speed below the safety limit), and the automatic stopping controller (it regulates the train speed in order to stop at the target position). The constant speed controller was constructed from rules such as:

$$For\ safety;\ IF\ the\ speed\ of\ train\ is\ approaching\ the\ limit\ speed, \\ THEN\ select\ the\ maximum\ brake\ notch \tag{1.15}$$

$$For\ riding\ comfort;\ IF\ the\ speed\ is\ in\ the\ allowed\ range, \\ THEN\ do\ not\ change\ the\ control\ notch \tag{1.16}$$

More rules were used in the real system for traceability and other factors. The automatic stopping controller was constructed from the rules like:

$$For\ riding\ comfort;\ IF\ the\ train\ will\ stop\ in\ the\ allowed\ zone, \\ THEN\ do\ not\ change\ the\ control\ notch \tag{1.17}$$

$$For\ riding\ comfort\ and\ safety;\ IF\ the\ train\ is\ in\ the\ allowed\ zone, \\ THEN\ change\ the\ control\ notch\ from\ acceleration\ to\ slight\ braking \tag{1.18}$$

Again, more rules were used in the real system to take care of the accuracy of stopping gap and other factors. By 1991, the Sendai subway had carried passengers for four years and was still one of the most advanced subway systems.

1.4 What Are the Major Research Fields in Fuzzy Theory?

By fuzzy theory we mean all the theories that use the basic concept of fuzzy set or continuous membership function. Fuzzy theory can be roughly classified according to Fig.1.8. There are five major branches: (i) fuzzy mathematics, where classical mathematical concepts are extended by replacing classical sets with fuzzy sets; (ii) fuzzy logic and artificial intelligence, where approximations to classical logic

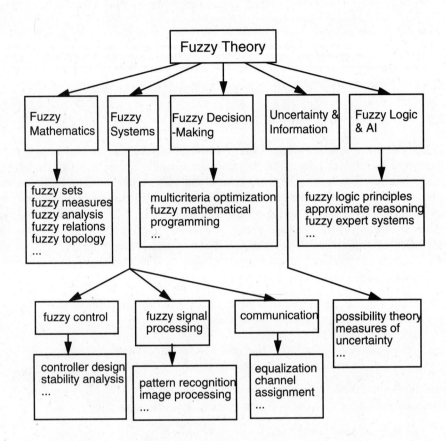

Figure 1.8. Classification of fuzzy theory.

are introduced and expert systems are developed based on fuzzy information and approximate reasoning; (iii) fuzzy systems, which include fuzzy control and fuzzy approaches in signal processing and communications; (iv) uncertainty and information, where different kinds of uncertainties are analyzed; and (v) fuzzy decision making, which considers optimalization problems with soft constraints.

Of course, these five branches are not independent and there are strong interconnections among them. For example, fuzzy control uses concepts from fuzzy mathematics and fuzzy logic.

From a practical point of view, the majority of applications of fuzzy theory has concentrated on fuzzy systems, especially fuzzy control, as we could see from the examples in Section 1.3. There also are some fuzzy expert systems that perform medical diagnoses and decision support (Terano, Asai and Sugeno [1994]). Because fuzzy theory is still in its infancy from both theoretical and practical points of view, we expect that more solid practical applications will appear as the field matures.

From Fig. 1.8 we see that fuzzy theory is a huge field that comprises a variety of research topics. In this text, we concentrate on fuzzy systems and fuzzy control. We first will study the basic concepts in fuzzy mathematics and fuzzy logic that are useful in fuzzy systems and control (Chapters 2-6), then we will study fuzzy systems and control in great detail (Chapters 7-26), and finally we will briefly review some topics in other fields of fuzzy theory (Chapters 27-31).

1.5 A Brief History of Fuzzy Theory and Applications

1.5.1 The 1960s: The Beginning of Fuzzy Theory

Fuzzy theory was initiated by Lotfi A. Zadeh in 1965 with his seminal paper "Fuzzy Sets" (Zadeh [1965]). Before working on fuzzy theory, Zadeh was a well-respected scholar in control theory. He developed the concept of "state," which forms the basis for modern control theory. In the early '60s, he thought that classical control theory had put too much emphasis on precision and therefore could not handle the complex systems. As early as 1962, he wrote that to handle biological systems "we need a radically different kind of mathematics, the mathematics of fuzzy or cloudy quantities which are not describable in terms of probability distributions" (Zadeh [1962]). Later, he formalized the ideas into the paper "Fuzzy Sets."

Since its birth, fuzzy theory has been sparking controversy. Some scholars, like Richard Bellman, endorsed the idea and began to work in this new field. Other scholars objected to the idea and viewed "fuzzification" as against basic scientific principles. The biggest challenge, however, came from mathematicians in statistics and probability who claimed that probability is sufficient to characterize uncertainty and any problems that fuzzy theory can solve can be solved equally well or better by probability theory (see Chapter 31). Because there were no real practical applications of fuzzy theory in the beginning, it was difficult to defend the field from a purely philosophical point of view. Almost all major research institutes in the world failed to view fuzzy theory as a serious research field.

Although fuzzy theory did not fall into the mainstream, there were still many researchers around the world dedicating themselves to this new field. In the late 1960s, many new fuzzy methods like fuzzy algorithms, fuzzy decision making, etc., were proposed.

1.5.2 The 1970s: Theory Continued to Grow and Real Applications Appeared

It is fair to say that the establishment of fuzzy theory as an independent field is largely due to the dedication and outstanding work of Zadeh. Most of the fundamental concepts in fuzzy theory were proposed by Zadeh in the late '60s and early '70s. After the introduction of fuzzy sets in 1965, he proposed the concepts of fuzzy algorithms in 1968 (Zadeh [1968]), fuzzy decision making in 1970 (Bellman and Zadeh [1970]), and fuzzy ordering in 1971 (Zadeh [1971b]). In 1973, he published another seminal paper, "Outline of a new approach to the analysis of complex systems and decision processes" (Zadeh [1973]), which established the foundation for fuzzy control. In this paper, he introduced the concept of linguistic variables and proposed to use fuzzy IF-THEN rules to formulate human knowledge.

A big event in the '70s was the birth of fuzzy controllers for real systems. In 1975, Mamdani and Assilian established the basic framework of fuzzy controller (which is essentially the fuzzy system in Fig.1.5) and applied the fuzzy controller to control a steam engine. Their results were published in another seminal paper in fuzzy theory "An experiment in linguistic synthesis with a fuzzy logic controller" (Mamdani and Assilian [1975]). They found that the fuzzy controller was very easy to construct and worked remarkably well. Later in 1978, Holmblad and Østergaard developed the first fuzzy controller for a full-scale industrial process—the fuzzy cement kiln controller (see Section 1.3).

Generally speaking, the foundations of fuzzy theory were established in the 1970s. With the introduction of many new concepts, the picture of fuzzy theory as a new field was becoming clear. Initial applications like the fuzzy steam engine controller and the fuzzy cement kiln controller also showed that the field was promising. Usually, the field should be founded by major resources and major research institutes should put some manpower on the topic. Unfortunately, this never happened. On the contrary, in the late '70s and early '80s, many researchers in fuzzy theory had to change their field because they could not find support to continue their work. This was especially true in the United States.

1.5.3   The 1980s: Massive Applications Made a Difference

In the early '80s, this field, from a theoretical point of view, progressed very slowly. Few new concepts and approaches were proposed during this period, simply because very few people were still working in the field. It was the application of fuzzy control that saved the field.

Japanese engineers, with their sensitivity to new technology, quickly found that fuzzy controllers were very easy to design and worked very well for many problems. Because fuzzy control does not require a mathematical model of the process, it could be applied to many systems where conventional control theory could not be used due to a lack of mathematical models. In 1980, Sugeno began to create Japan's first fuzzy application—control of a Fuji Electric water purification plant. In 1983, he began the pioneer work on a fuzzy robot, a self-parking car that was controlled by calling out commands (Sugeno and Nishida [1985]). In the early 1980s, Yasunobu and Miyamoto from Hitachi began to develop a fuzzy control system for the Sandai subway. They finished the project in 1987 and created the most advanced subway system on earth. This very impressive application of fuzzy control made a very big difference.

In July 1987, the Second Annual International Fuzzy Systems Association Conference was held in Tokyo. The conference began three days after the Sendai subway began operation, and attendees were amused with its dreamy ride. Also, in the conference Hirota displayed a fuzzy robot arm that played two-dimensional Ping-Pong in real time (Hirota, Arai and Hachisu [1989]), and Yamakawa demonstrated a fuzzy system that balanced an inverted pendulum (Yamakawa [1989]). Prior to this event, fuzzy theory was not well-known in Japan. After it, a wave of pro-fuzzy sentiment swept through the engineering, government, and business communities. By the early 1990s, a large number of fuzzy consumer products appeared in the market (see Section 1.3 for examples).

1.5.4   The 1990s: More Challenges Remain

The success of fuzzy systems in Japan surprised the mainstream researchers in the United States and in Europe. Some still criticize fuzzy theory, but many others have been changing their minds and giving fuzzy theory a chance to be taken seriously. In February 1992, the first IEEE International Conference on Fuzzy Systems was held in San Diego. This event symbolized the acceptance of fuzzy theory by the largest engineering organization—IEEE. In 1993, the IEEE Transactions on Fuzzy Systems was inaugurated.

From a theoretical point of view, fuzzy systems and control has advanced rapidly in the late 1980s and early 1990s. Although it is hard to say there is any breakthrough, solid progress has been made on some fundamental problems in fuzzy systems and control. For examples, neural network techniques have been used to determine membership functions in a systematic manner, and rigorous stability analysis of fuzzy control systems has appeared. Although the whole picture of fuzzy systems and control theory is becoming clearer, much work remains to be done. Most approaches and analyses are preliminary in nature. We believe that only when the top research institutes begin to put some serious man power on the research of fuzzy theory can the field make major progress.

1.6   Summary and Further Readings

In this chapter we have demonstrated the following:

- The goal of using fuzzy systems is to put human knowledge into engineering systems in a systematic, efficient, and analyzable order.

- The basic architectures of the commonly used fuzzy systems.

- The fuzzy IF-THEN rules used in certain industrial processes and consumer products.

- Classification and brief history of fuzzy theory and applications.

A very good non-technical introduction to fuzzy theory and applications is McNeill and Freiberger [1993]. It contains many interviews and describes the major events. Some historical remarks were made in Kruse, Gebhardt, and Klawonn [1994]. Klir and Yuan [1995] is perhaps the most comprehensive book on fuzzy sets and fuzzy logic. Earlier applications of fuzzy control were collected in Sugeno [1985] and more recent applications (mainly in Japan) were summarized in Terano, Asai, and Sugeno [1994].

1.7   Exercises

Exercise 1.1. Is the fuzzy washing machine an open-loop control system or a closed-loop control system? What about the fuzzy cement kiln control system? Explain your answer.

Exercise 1.2. List four to six applications of fuzzy theory to practical problems other than those in Section 1.3. Point out the references where you find these applications.

Exercise 1.3. Suppose we want to design a fuzzy system to balance the inverted pendulum shown in Fig. 1.9. Let the angle θ and its derivation $\dot{\theta}$ be the inputs to the fuzzy system and the force u applied to the cart be its output.

(a) Determine three to five fuzzy IF-THEN rules based on the common sense of how to balance the inverted pendulum.

(b) Suppose that the rules in (a) can successfully control a particular inverted pendulum system. Now if we want to use the rules to control another inverted pendulum system with different values of m_c, m, and l, what parts of the rules should change and what parts may remain the same.

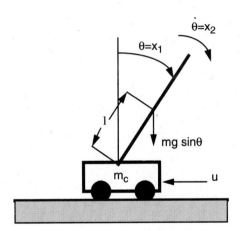

Figure 1.9. The inverted pendulum system.

Part I

The Mathematics of Fuzzy Systems and Control

Fuzzy mathematics provide the starting point and basic language for fuzzy systems and fuzzy control. Fuzzy mathematics by itself is a huge field, where fuzzy mathematical principles are developed by replacing the sets in classical mathematical theory with fuzzy sets. In this way, all the classical mathematical branches may be "fuzzified." We have seen the birth of fuzzy measure theory, fuzzy topology, fuzzy algebra, fuzzy analysis, etc. Understandably, only a small portion of fuzzy mathematics has found applications in engineering. In the next five chapters, we will study those concepts and principles in fuzzy mathematics that are useful in fuzzy systems and fuzzy control.

In Chapter 2, we will introduce the most fundamental concept in fuzzy theory —the concept of fuzzy set. In Chapter 3, set-theoretical operations on fuzzy sets such as complement, union, and intersection will be studied in detail. Chapter 4 will study fuzzy relations and introduce an important principle in fuzzy theory— the extension principle. Linguistic variables and fuzzy IF-THEN rules, which are essential to fuzzy systems and fuzzy control, will be precisely defined and studied in Chapter 5. Finally, Chapter 6 will focus on three basic principles in fuzzy logic that are useful in the fuzzy inference engine of fuzzy systems.

Chapter 2

Fuzzy Sets and Basic Operations on Fuzzy Sets

2.1 From Classical Sets to Fuzzy Sets

Let U be the *universe of discourse*, or *universal set*, which contains all the possible elements of concern in each particular context or application. Recall that a *classical (crisp) set* A, or simply a set A, in the universe of discourse U can be defined by listing all of its members (*the list method*) or by specifying the properties that must be satisfied by the members of the set (*the rule method*). The list method can be used only for finite sets and is therefore of limited use. The rule method is more general. In the rule method, a set A is represented as

$$A = \{x \in U | x \text{ meets some conditions}\} \tag{2.1}$$

There is yet a third method to define a set A—*the membership method*, which introduces a zero-one membership function (also called characteristic function, discrimination function, or indicator function) for A, denoted by $\mu_A(x)$, such that

$$\mu_A(x) = \begin{cases} 1 & if \quad x \in A \\ 0 & if \quad x \notin A \end{cases} \tag{2.2}$$

The set A is mathematically equivalent to its membership function $\mu_A(x)$ in the sense that knowing $\mu_A(x)$ is the same as knowing A itself.

Example 2.1. Consider the set of all cars in Berkeley; this is the universe of discourse U. We can define different sets in U according to the properties of cars. Fig. 2.1 shows two types of properties that can be used to define sets in U: (a) US cars or non-US cars, and (b) number of cylinders. For example, we can define a set A as all cars in U that have 4 cylinders, that is,

$$A = \{x \in U | x \text{ has 4 cylinders}\} \tag{2.3}$$

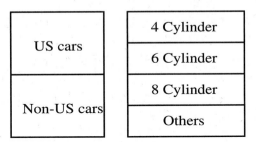

Figure 2.1. Partitioning of the set of all cars in Berkeley into subsets by: (a) US cars or non-US cars, and (b) number of cylinders.

or

$$\mu_A(x) = \begin{cases} 1 & if \quad x \in U \ and \ x \ has \ 4 \ cylinders \\ 0 & if \quad x \in U \ and \ x \ does \ not \ have \ 4 \ cylinders \end{cases} \tag{2.4}$$

If we want to define a set in U according to whether the car is a US car or a non-US car, we face a difficulty. One perspective is that a car is a US car if it carries the name of a USA auto manufacturer; otherwise it is a non-US car. However, many people feel that the distinction between a US car and a non-US car is not as crisp as it once was, because many of the components for what we consider to be US cars (for examples, Fords, GM's, Chryslers) are produced outside of the United States. Additionally, some "non-US" cars are manufactured in the USA. How to deal with this kind of problems? □

Essentially, the difficulty in Example 2.1 shows that *some sets do not have clear boundaries.* Classical set theory requires that a set must have a well-defined property, therefore it is unable to define the set like "all US cars in Berkeley." To overcome this limitation of classical set theory, the concept of fuzzy set was introduced. It turns out that this limitation is fundamental and a new theory is needed—this is the fuzzy set theory.

Definition 2.1. A *fuzzy set* in a universe of discourse U is characterized by a membership function $\mu_A(x)$ that takes values in the interval $[0, 1]$.

Therefore, a fuzzy set is a generalization of a classical set by allowing the membership function to take any values in the interval $[0, 1]$. In other words, the membership function of a classical set can only take two values—zero and one, whereas the membership function of a fuzzy set is a continuous function with range $[0, 1]$. We see from the definition that there is nothing "fuzzy" about a fuzzy set; it is simply a set with a continuous membership function.

A fuzzy set A in U may be represented as a set of ordered pairs of a generic element x and its membership value, that is,

$$A = \{(x, \mu_A(x)) | x \in U\} \tag{2.5}$$

When U is continuous (for example, $U = R$), A is commonly written as

$$A = \int_U \mu_A(x)/x \tag{2.6}$$

where the integral sign does not denote integration; it denotes the collection of all points $x \in U$ with the associated membership function $\mu_A(x)$. When U is discrete, A is commonly written as

$$A = \sum_U \mu_A(x)/x \tag{2.7}$$

where the summation sign does not represent arithmetic addition; it denotes the collection of all points $x \in U$ with the associated membership function $\mu_A(x)$.

We now return to Example 2.1 and see how to use the concept of fuzzy set to define US and non-US cars.

Example 2.1 (Cont'd). We can define the set "US cars in Berkeley," denoted by D, as a fuzzy set according to the percentage of the car's parts made in the USA. Specifically, D is defined by the membership function

$$\mu_D(x) = p(x) \tag{2.8}$$

where $p(x)$ is the percentage of the parts of car x made in the USA and it takes values from 0% to 100%. For example, if a particular car x_0 has 60% of its parts made in the USA, then we say that the car x_0 belongs to the fuzzy set D to the degree of 0.6.

Similarly, we can define the set "non-US cars in Berkeley," denoted by F, as a fuzzy set with the membership function

$$\mu_F(x) = 1 - p(x) \tag{2.9}$$

where $p(x)$ is the same as in (2.8). Thus, if a particular car x_0 has 60% of its parts made in the USA, then we say the car x_0 belongs to the fuzzy set F to the degree of 1-0.6=0.4. Fig. 2.2 shows (2.8) and (2.9). Clearly, an element can belong to different fuzzy sets to the same or different degrees. □

We now consider another example of fuzzy sets and from it draw some remarks.

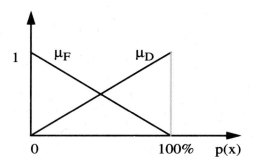

Figure 2.2. Membership functions for US (μ_D) and non-US (μ_F) cars based on the percentage of parts of the car made in the USA ($p(x)$).

Example 2.2. Let Z be a fuzzy set named "numbers close to zero." Then a possible membership function for Z is

$$\mu_Z(x) = e^{-x^2} \qquad (2.10)$$

where $x \in R$. This is a Gaussian function with mean equal to zero and standard derivation equal to one. According to this membership function, the numbers 0 and 2 belong to the fuzzy set Z to the degrees of $e^0 = 1$ and e^{-4}, respectively.

We also may define the membership function for Z as

$$\mu_Z(x) = \begin{cases} 0 & if \quad x < -1 \\ x+1 & if \quad -1 \le x < 0 \\ 1-x & if \quad 0 \le x < 1 \\ 0 & if \quad 1 \le x \end{cases} \qquad (2.11)$$

According to this membership function, the numbers 0 and 2 belong to the fuzzy set Z to the degrees of 1 and 0, respectively. (2.10) and (2.11) are plotted graphically in Figs. 2.3 and 2.4, respectively. We can choose many other membership functions to characterize "numbers close to zero." □

From Example 2.2 we can draw three important remarks on fuzzy sets:

- The properties that a fuzzy set is used to characterize are usually fuzzy, for example, "numbers close to zero" is not a precise description. Therefore, we

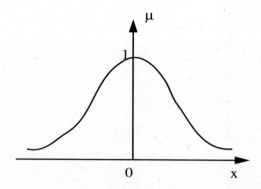

Figure 2.3. A possible membership function to characterize "numbers close to zero."

may use different membership functions to characterize the same description.
However, the membership functions themselves are not fuzzy—they are precise
mathematical functions. Once a fuzzy property is represented by a membership function, for example, once "numbers close to zero" is represented by the
membership function (2.10) or (2.11), nothing will be fuzzy anymore. Thus,
by characterizing a fuzzy description with a membership function, we essentially *defuzzify* the fuzzy description. A common misunderstanding of fuzzy
set theory is that fuzzy set theory tries to fuzzify the world. We see, on the
contrary, that fuzzy sets are used to defuzzify the world.

- Following the previous remark is an important question: how to determine the
  membership functions? Because there are a variety of choices of membership
  functions, how to choose one from these alternatives? Conceptually, there are
  two approaches to determining a membership function. The first approach
  is to use the knowledge of human experts, that is, ask the domain experts
  to specify the membership functions. Because fuzzy sets are often used to
  formulate human knowledge, the membership functions represent a part of
  human knowledge. Usually, this approach can only give a rough formula of the
  membership function; fine-tuning is required. In the second approach, we use
  data collected from various sensors to determine the membership functions.
  Specifically, we first specify the structures of the membership functions and
  then fine-tune the parameters of the membership functions based on the data.
  Both approaches, especially the second approach, will be studied in detail in

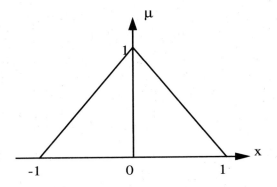

Figure 2.4. Another possible membership function to characterize "numbers close to zero."

later chapters.

- Finally, it should be emphasized that although (2.10) and (2.11) are used to characterize the same description "numbers close to zero," they are different fuzzy sets. Hence, rigorously speaking, we should use different labels to represent the fuzzy sets (2.10) and (2.11); for example, we should use $\mu_{Z_1}(x)$ in (2.10) and $\mu_{Z_2}(x)$ in (2.11). A fuzzy set has a one-to-one correspondence with its membership function. That is, when we say a fuzzy set, there must be a unique membership function associated with it; conversely, when we give a membership function, it represents a fuzzy set. Fuzzy sets and their membership functions are equivalent in this sense.

Let us consider two more examples of fuzzy sets, one in continuous domain and the other in discrete domain; they are classical examples from Zadeh's seminal paper (Zadeh [1965]).

Example 2.3. Let U be the interval $[0, 100]$ representing the age of ordinary humans. Then we may define fuzzy sets "young" and "old" as (using the integral notation (2.6))

$$young = \int_0^{25} 1/x + \int_{25}^{100} (1 + (\frac{x - 25}{5})^2)^{-1}/x \qquad (2.12)$$

$$old = \int_{50}^{100} (1 + (\frac{x - 50}{5})^{-2})^{-1}/x \qquad (2.13)$$

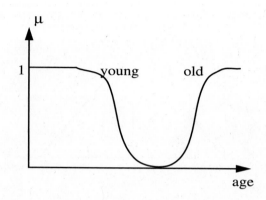

Figure 2.5. Diagrammatic representation of "young" and "old."

See Fig. 2.5. □

Example 2.4. Let U be the integers from 1 to 10, that is, $U = \{1, 2, ..., 10\}$. Then the fuzzy set "several" may be defined as (using the summation notation (2.7))

$$several = 0.5/3 + 0.8/4 + 1/5 + 1/6 + 0.8/7 + 0.5/8 \qquad (2.14)$$

That is, 5 and 6 belong to the fuzzy set "several" with degree 1, 4 and 7 with degree 0.8, 3 and 8 with degree 0.5, and 1,2,9 and 10 with degree 0. See Fig. 2.6. □

2.2   Basic Concepts Associated with Fuzzy Set

We now introduce some basic concepts and terminology associated with a fuzzy set. Many of them are extensions of the basic concepts of a classical (crisp) set, but some are unique to the fuzzy set framework.

Definition 2.2. The concepts of support, fuzzy singleton, center, crossover point, height, normal fuzzy set, α-cut, convex fuzzy set, and projections are defined as follows.

The *support* of a fuzzy set A in the universe of discourse U is a crisp set that contains all the elements of U that have nonzero membership values in A, that is,

$$supp(A) = \{x \in U | \mu_A(x) > 0\} \qquad (2.15)$$

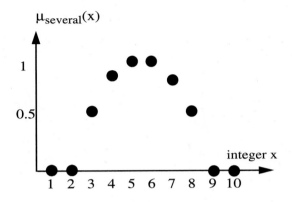

Figure 2.6. Membership function for fuzzy set "several."

where $supp(A)$ denotes the support of fuzzy set A. For example, the support of fuzzy set "several" in Fig. 2.6 is the set of integers $\{3, 4, 5, 6, 7, 8\}$. If the support of a fuzzy set is empty, it is called an *empty fuzzy set*. A *fuzzy singleton* is a fuzzy set whose support is a single point in U.

The *center* of a fuzzy set is defined as follows: if the mean value of all points at which the membership function of the fuzzy set achieves its maximum value is finite, then define this mean value as the center of the fuzzy set; if the mean value equals positive (negative) infinite, then the center is defined as the smallest (largest) among all points that achieve the maximum membership value. Fig. 2.7 shows the centers of some typical fuzzy sets. The *crossover point* of a fuzzy set is the point in U whose membership value in A equals 0.5.

The *height* of a fuzzy set is the largest membership value attained by any point. For example, the heights of all the fuzzy sets in Figs.2.2-2.4 equal one. If the height of a fuzzy set equals one, it is called a *normal fuzzy set*. All the fuzzy sets in Figs. 2.2-2.4 are therefore normal fuzzy sets.

An α-*cut* of a fuzzy set A is a crisp set A_α that contains all the elements in U that have membership values in A greater than or equal to α, that is,

$$A_\alpha = \{x \in U | \mu_A(x) \geq \alpha\} \tag{2.16}$$

For example, for $\alpha = 0.3$, the α-cut of the fuzzy set (2.11) (Fig. 2.4) is the crisp set $[-0.7, 0.7]$, and for $\alpha = 0.9$, it is $[-0.1, 0.1]$.

When the universe of discourse U is the n-dimensional Euclidean space R^n, the

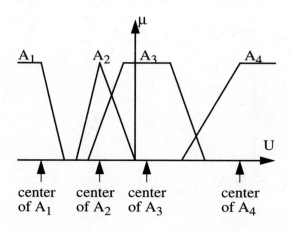

Figure 2.7. Centers of some typical fuzzy sets.

concept of set convexity can be generalized to fuzzy set. A fuzzy set A is said to be *convex* if and only if its α-cut A_α is a convex set for any α in the interval $(0, 1]$. The following lemma gives an equivalent definition of a convex fuzzy set.

Lemma 2.1. A fuzzy set A in R^n is convex if and only if

$$\mu_A[\lambda x_1 + (1 - \lambda)x_2] \geq min[\mu_A(x_1), \mu_A(x_2)] \qquad (2.17)$$

for all $x_1, x_2 \in R^n$ and all $\lambda \in [0, 1]$.

Proof: First, suppose that A is convex and we prove the truth of (2.17). Let x_1 and x_2 be arbitrary points in R^n and without loss of generality we assume $\mu_A(x_1) \leq \mu_A(x_2)$. If $\mu_A(x_1) = 0$, then (2.17) is trivially true, so we let $\mu_A(x_1) = \alpha > 0$. Since by assumption the α-cut A_α is convex and $x_1, x_2 \in A_\alpha$ (since $\mu_A(x_2) \geq \mu_A(x_1) = \alpha$), we have $\lambda x_1 + (1 - \lambda)x_2 \in A_\alpha$ for all $\lambda \in [0, 1]$. Hence, $\mu_A[\lambda x_1 + (1 - \lambda)x_2] \geq \alpha = \mu_A(x_1) = min[\mu_A(x_1), \mu_A(x_2)]$.

Conversely, suppose (2.17) is true and we prove that A is convex. Let α be an arbitrary point in $(0, 1]$. If A_α is empty, then it is convex (empty sets are convex by definition). If A_α is nonempty, then there exists $x_1 \in R^n$ such that $\mu_A(x_1) = \alpha$ (by the definition of A_α). Let x_2 be an arbitrary element in A_α, then $\mu_A(x_2) \geq \alpha = \mu_A(x_1)$. Since (2.17) is true by assumption, we have $\mu_A[\lambda x_1 + (1 - \lambda)x_2] \geq min[\mu_A(x_1), \mu_A(x_2)] = \mu_A(x_1) = \alpha$ for all $\lambda \in [0, 1]$, which means that $\lambda x_1 + (1 - \lambda)x_2 \in A_\alpha$. So A_α is a convex set. Since α is an arbitrary point in $(0, 1]$, the convexity of A_α implies the convexity of A. $\square$

Let A be a fuzzy set in R^n with membership function $\mu_A(x) = \mu_A(x_1, ..., x_n)$ and H be a hyperplane in R^n defined by $H = \{x \in R^n | x_1 = 0\}$ (for notational simplicity, we consider this special case of hyperplane; generalization to general hyperplanes is straightforward). The *projection* of A on H is a fuzzy set A_H in R^{n-1} defined by

$$\mu_{A_H}(x_2, ..., x_n) = \sup_{x_1 \in R} \mu_A(x_1, ..., x_n) \qquad (2.18)$$

where $\sup_{x_1 \in R} \mu_A(x_1, ..., x_n)$ denotes the maximum value of the function $\mu_A(x_1, ..., x_n)$ when x_1 takes values in R.

2.3 Operations on Fuzzy Sets

The basic concepts introduced in Sections 2.1 and 2.2 concern only a single fuzzy set. In this section, we study the basic operations on fuzzy sets. In the sequel, we assume that A and B are fuzzy sets defined in the same universe of discourse U.

Definition 2.3. The equality, containment, complement, union, and intersection of two fuzzy sets A and B are defined as follows.

We say A and B are *equal* if and only if $\mu_A(x) = \mu_B(x)$ for all $x \in U$. We say B *contains* A, denoted by $A \subset B$, if and only if $\mu_A(x) \leq \mu_B(x)$ for all $x \in U$. The *complement* of A is a fuzzy set $\bar{A}$ in U whose membership function is defined as

$$\mu_{\bar{A}}(x) = 1 - \mu_A(x) \qquad (2.19)$$

The *union* of A and B is a fuzzy set in U, denoted by $A \cup B$, whose membership function is defined as

$$\mu_{A \cup B}(x) = max[\mu_A(x), \mu_B(x)] \qquad (2.20)$$

The *intersection* of A and B is a fuzzy set $A \cap B$ in U with membership function

$$\mu_{A \cap B}(x) = min[\mu_A(x), \mu_B(x)] \qquad (2.21)$$

The reader may wonder why we use "max" for union and "min" for intersection; we now give an intuitive explanation. An intuitively appealing way of defining the union is the following: the union of A and B is the smallest fuzzy set containing both A and B. More precisely, if C is any fuzzy set that contains both A and B, then it also contains the union of A and B. To show that this intuitively appealing definition is equivalent to (2.20), we note, first, that $A \cup B$ as defined by (2.20) contains both A and B because $max[\mu_A, \mu_B] \geq \mu_A$ and $max[\mu_A, \mu_B] \geq \mu_B$. Furthermore, if C is any fuzzy set containing both A and B, then $\mu_C \geq \mu_A$ and $\mu_C \geq \mu_B$. Therefore, $\mu_C \geq max[\mu_A, \mu_B] = \mu_{A \cup B}$, which means that $A \cup B$ as defined by (2.20) is the smallest fuzzy set containing both A and B. The intersection as defined by (2.21) can be justified in the same manner.

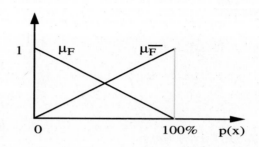

Figure 2.8. The membership functions for $\bar{F}$ and F.

Example 2.5. Consider the two fuzzy sets D and F defined by (2.8) and (2.9) (see also Fig. 2.2). The complement of F, $\bar{F}$, is the fuzzy set defined by

$$\mu_{\bar{F}}(x) = 1 - \mu_F(x) = 1 - p(x) \qquad (2.22)$$

which is shown in Fig. 2.8. Comparing (2.22) with (2.9) we see that $\bar{F} = D$. This makes sense because if a car is not a non-US car (which is what the complement of F means intuitively), then it should be a US car; or more accurately, the less a car is a non-US car, the more the car is a US car. The union of F and D is the fuzzy set $F \cup D$ defined by

$$\mu_{F \cup D}(x) = max[\mu_F, \mu_D] = \begin{cases} \mu_F(x) & if \quad 0 \le p(x) \le 0.5 \\ \mu_D(x) & if \quad 0.5 \le p(x) \le 1 \end{cases} \qquad (2.23)$$

which is plotted in Fig. 2.9. The intersection of F and D is the fuzzy set $F \cap D$ defined by

$$\mu_{F \cap D}(x) = min[\mu_F, \mu_D] = \begin{cases} \mu_D(x) & if \quad 0 \le p(x) \le 0.5 \\ \mu_F(x) & if \quad 0.5 \le p(x) \le 1 \end{cases} \qquad (2.24)$$

which is plotted in Fig. 2.10. $\square$

With the operations of complement, union and intersection defined as in (2.19), (2.20) and (2.21), many of the basic identities (not all!) which hold for classical sets can be extended to fuzzy sets. As an example, let us consider the following lemma.

Lemma 2.2. The De Morgan's Laws are true for fuzzy sets. That is, suppose A and B are fuzzy sets, then

$$\overline{A \cup B} = \bar{A} \cap \bar{B} \qquad (2.25)$$

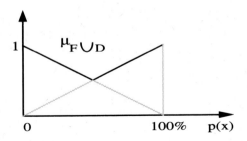

Figure 2.9. The membership function for $F \cup D$, where F and D are defined in Fig. 2.2.

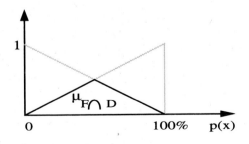

Figure 2.10. The membership function for $F \cap D$, where F and D are defined in Fig. 2.2.

and

$$\overline{A \cap B} = \bar{A} \cup \bar{B} \tag{2.26}$$

Proof: We only prove (2.25); (2.26) can be proven in the same way and is left as an exercise. First, we show that the following identity is true:

$$1 - max[\mu_A, \mu_B] = min[1 - \mu_A, 1 - \mu_B] \tag{2.27}$$

To show this we consider the two possible cases: $\mu_A \geq \mu_B$ and $\mu_A < \mu_B$. If $\mu_A \geq \mu_B$, then $1 - \mu_A \leq 1 - \mu_B$ and $1 - max[\mu_A, \mu_B] = 1 - \mu_A = min[1 - \mu_A, 1 - \mu_B]$, which is (2.27). If $\mu_A < \mu_B$, then $1 - \mu_A > 1 - \mu_B$ and $1 - max[\mu_A, \mu_B] = 1 - \mu_B = min[1 - \mu_A, 1 - \mu_B]$, which is again (2.27). Hence, (2.27) is true. From the definitions (2.19)-(2.21) and the definition of the equality of two fuzzy sets, we see that (2.27) implies (2.25). $\square$

2.4   Summary and Further Readings

In this chapter we have demonstrated the following:

- The definitions of fuzzy set, basic concepts associated with a fuzzy set (support, α-cut, convexity, etc.) and basic operations (complement, union, intersection, etc.) of fuzzy sets.

- The intuitive meaning of membership functions and how to determine intuitively appealing membership functions for specific fuzzy descriptions.

- Performing operations on specific examples of fuzzy sets and proving basic properties concerning fuzzy sets and their operations.

Zadeh's original paper (Zadeh [1965]) is still the best source to learn fuzzy set and related concepts. The paper was extremely well-written and the reader is encouraged to read it. The basic operations and concepts associated with a fuzzy set were also introduced in Zadeh [1965].

2.5   Exercises

Exercise 2.1. Determine reasonable membership functions for "short persons," "tall persons," and "heavy persons."

Exercise 2.2. Model the following expressions as fuzzy sets: (a) hard-working students, (b) top students, and (c) smart students.

Exercise 2.3. Consider the fuzzy sets F, G and H defined in the interval $U = [0, 10]$ by the membership functions

$$\mu_F(x) = \frac{x}{x+2}, \ \mu_G(x) = 2^{-x}, \ \mu_H(x) = \frac{1}{1 + 10(x-2)^2} \tag{2.28}$$

Determine the mathematical formulas and graphs of membership functions of each of the following fuzzy sets:

(a) $\bar{F}, \bar{G}, \bar{H}$

(b) $F \cup G, F \cup H, G \cup H$

(c) $F \cap G, F \cap H, G \cap H$

(d) $F \cup G \cup H, F \cap G \cap H$

(e) $F \cap \bar{H}, \overline{\bar{G} \cap H}, \overline{F \cup H}$

Exercise 2.4. Determine the α-cuts of the fuzzy sets F, G and H in Exercise 2.3 for: (a) $\alpha = 0.2$, (b) $\alpha = 0.5$, (c) $\alpha = 0.9$, and (d) $\alpha = 1$.

Exercise 2.5. Let fuzzy set A be defined in the closed plane $U = [-1, 1] \times [-3, 3]$ with membership function

$$\mu_A(x_1, x_2) = e^{-(x_1^2 + x_2^2)} \tag{2.29}$$

Determine the projections of A on the hyperplanes $H_1 = \{x \in U | x_1 = 0\}$ and $H_2 = \{x \in U | x_2 = 0\}$, respectively.

Exercise 2.6. Show that the law of the excluded middle, $F \cup \bar{F} = U$, is not true if F is a fuzzy set.

Exercise 2.7. Prove the identity (2.26) in Lemma 2.2.

Exercise 2.8. Show that the intersection of two convex fuzzy sets is also a convex fuzzy set. What about the union?

Chapter 3

Further Operations on Fuzzy Sets

In Chapter 2 we introduced the following basic operators for complement, union, and intersection of fuzzy sets:

$$\mu_{\bar{A}}(x) = 1 - \mu_A(x) \tag{3.1}$$

$$\mu_{A \cup B}(x) = max[\mu_A(x), \mu_B(x)] \tag{3.2}$$

$$\mu_{A \cap B}(x) = min[\mu_A(x), \mu_B(x)] \tag{3.3}$$

We explained that the fuzzy set $A \cup B$ defined by (3.2) is the smallest fuzzy set containing both A and B, and the fuzzy set $A \cap B$ defined by (3.3) is the largest fuzzy set contained by both A and B. Therefore, (3.1)-(3.3) define only one type of operations on fuzzy sets. Other possibilities exist. For example, we may define $A \cup B$ as any fuzzy set containing both A and B (not necessarily the smallest fuzzy set). In this chapter, we study other types of operators for complement, union, and intersection of fuzzy sets.

Why do we need other types of operators? The main reason is that the operators (3.1)-(3.3) may not be satisfactory in some situations. For example, when we take the intersection of two fuzzy sets, we may want the larger fuzzy set to have an impact on the result. But if we use the *min* operator of (3.3), the larger fuzzy set will have no impact. Another reason is that from a theoretical point of view it is interesting to explore what types of operators are possible for fuzzy sets. We know that for nonfuzzy sets only one type of operation is possible for complement, union, or intersection. For fuzzy sets there are other possibilities. But what are they? What are the properties of these new operators? These are the questions we will try to answer in this chapter.

The new operators will be proposed on axiomatic bases. That is, we will start with a few axioms that complement, union, or intersection should satisfy in order to be qualified as these operations. Then, we will list some particular formulas that satisfy these axioms.

34

3.1 Fuzzy Complement

Let $c : [0, 1] \to [0, 1]$ be a mapping that transforms the membership function of fuzzy set A into the membership function of the complement of A, that is,

$$c[\mu_A(x)] = \mu_{\bar{A}}(x) \qquad (3.4)$$

In the case of (3.1), $c[\mu_A(x)] = 1 - \mu_A(x)$. In order for the function c to be qualified as a complement, it should satisfy at least the following two requirements:

Axiom c1. $c(0) = 1$ and $c(1) = 0$ (boundary condition).

Axiom c2. For all $a, b \in [0, 1]$, if $a < b$, then $c(a) \geq c(b)$ (nonincreasing condition), where (and throughout this chapter) a and b denote membership functions of some fuzzy sets, say, $a = \mu_A(x)$ and $b = \mu_B(x)$.

Axiom c1 shows that if an element belongs to a fuzzy set to degree zero (one), then it should belong to the complement of this fuzzy set to degree one (zero). Axiom c2 requires that an increase in membership value must result in a decrease or no change in membership value for the complement. Clearly, any violation of these two requirements will result in an operator that is unacceptable as complement.

Definition 3.1. Any function $c : [0, 1] \to [0, 1]$ that satisfies Axioms c1 and c2 is called a *fuzzy complement*.

One class of fuzzy complements is the *Sugeno class* (Sugeno [1977]) defined by

$$c_\lambda(a) = \frac{1 - a}{1 + \lambda a} \qquad (3.5)$$

where $\lambda \in (-1, \infty)$. For each value of the parameter λ, we obtain a particular fuzzy complement. It is a simple matter to check that the complement defined by (3.5) satisfies Axioms c1 and c2. Fig. 3.1 illustrates this class of fuzzy complements for different values of λ. Note that when $\lambda = 0$ it becomes the basic fuzzy complement (3.1).

Another type of fuzzy complement is the *Yager class* (Yager [1980]) defined by

$$c_w(a) = (1 - a^w)^{1/w} \qquad (3.6)$$

where $w \in (0, \infty)$. For each value of w, we obtain a particular fuzzy complement. It is easy to verify that (3.6) satisfies Axioms c1 and c2. Fig. 3.2 illustrates the Yager class of fuzzy complements for different values of w. When $w = 1$, (3.6) becomes (3.1).

3.2 Fuzzy Union—The S-Norms

Let $s : [0, 1] \times [0, 1] \to [0, 1]$ be a mapping that transforms the membership functions of fuzzy sets A and B into the membership function of the union of A and B, that

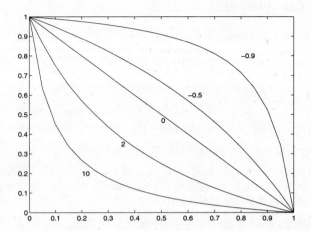

Figure 3.1. Sugeno class of fuzzy complements $c_\lambda(a)$ for different values of λ.

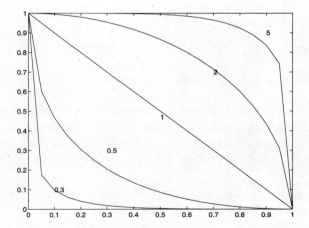

Figure 3.2. Yager class of fuzzy complements $c_w(a)$ for different values of w.

is,

$$s[\mu_A(x), \mu_B(x)] = \mu_{A \cup B}(x) \tag{3.7}$$

In the case of (3.2), $s[\mu_A(x), \mu_B(x)] = max[\mu_A(x), \mu_B(x)]$. In order for the function s to be qualified as an union, it must satisfied at least the following four requirements:

Axiom s1. $s(1,1) = 1, s(0,a) = s(a,0) = a$ (boundary condition).

Axiom s2. $s(a,b) = s(b,a)$ (commutative condition).

Axiom s3. If $a \leq a'$ and $b \leq b'$, then $s(a,b) \leq s(a',b')$ (nondecreasing condition).

Axiom s4. $s(s(a,b),c) = s(a,s(b,c))$ (associative condition).

Axiom s1 indicates what an union function should be in extreme cases. Axiom s2 insures that the order in which the fuzzy sets are combined has no influence on the result. Axiom s3 shows a natural requirement for union: an increase in membership values in the two fuzzy sets should result in an increase in membership value in the union of the two fuzzy sets. Axiom s4 allows us to extend the union operations to more than two fuzzy sets.

Definition 3.2. Any function $s : [0,1] \times [0,1] \rightarrow [0,1]$ that satisfies Axioms s1-s4 is called an *s-norm*.

It is a simple matter to prove that the basic fuzzy union *max* of (3.2) is a s-norm. We now list three particular classes of s-norms:

- Dombi class (Dombi [1982]):

$$s_\lambda(a,b) = \frac{1}{1 + [(\frac{1}{a} - 1)^{-\lambda} + (\frac{1}{b} - 1)^{-\lambda}]^{-1/\lambda}} \tag{3.8}$$

where the parameter $\lambda \in (0, \infty)$.

- Dubois-Prade class (Dubois and Prade [1980]):

$$s_\alpha(a,b) = \frac{a + b - ab - min(a,b,1-\alpha)}{max(1-a, 1-b, \alpha)} \tag{3.9}$$

where the parameter $\alpha \in [0,1]$.

- Yager class (Yager [1980]):

$$s_w(a,b) = min[1, (a^w + b^w)^{1/w}] \tag{3.10}$$

where the parameter $w \in (0, \infty)$.

With a particular choice of the parameters, (3.8)-(3.10) each defines a particular s-norm. It is straightforward to verify that (3.8)-(3.10) satisfy Axioms s1-s4. These s-norms were obtained by generalizing the union operation for classical sets from different perspectives.

Many other s-norms were proposed in the literature. We now list some of them below:

- Drastic sum:

$$s_{ds}(a,b) = \begin{cases} a \; if \; b = 0 \\ b \; if \; a = 0 \\ 1 \; otherwise \end{cases} \qquad (3.11)$$

- Einstein sum:

$$s_{es}(a,b) = \frac{a+b}{1+ab} \qquad (3.12)$$

- Algebraic sum:

$$s_{as}(a,b) = a + b - ab \qquad (3.13)$$

- Maximum: (3.2)

Why were so many s-norms proposed in the literature? The theoretical reason is that they become identical when the membership values are restricted to zero or one; that is, they are all extensions of nonfuzzy set union. The practical reason is that some s-norms may be more meaningful than others in some applications.

Example 3.1: Consider the fuzzy sets D and F defined in Example 2.1 of Chapter 2 ((2.8) and (2.9)). If we use the Yager s-norm (3.10) for fuzzy union, then the fuzzy set $D \cup F$ is computed as

$$\mu_{D \cup F}(x) = s_w[\mu_D(x), \mu_F(x)]$$
$$= min[1, ((p(x))^w + (1 - p(x))^w)^{1/w}] \qquad (3.14)$$

Fig. 3.3 illustrates this $\mu_{D \cup F}(x)$ for $w = 3$. If we use the algebraic sum (3.13) for the fuzzy union, the fuzzy set $D \cup F$ becomes

$$\mu_{D \cup F}(x) = s_{as}[\mu_D(x), \mu_F(x)]$$
$$= p(x) + (1 - p(x)) - p(x)(1 - p(x))$$
$$= 1 - p(x) + (p(x))^2 \qquad (3.15)$$

which is plotted in Fig. 3.4. □

Comparing Figs. 3.3 and 3.4 with Fig. 2.9, we see that the Yager s-norm and algebraic sum are larger than the maximum operator. In general, we can show that *maximum* (3.2) is the smallest s-norm and *drastic sum* (3.11) is the largest s-norm.

Theorem 3.1: For any s-norm s, that is, for any function $s : [0,1] \times [0,1] \to [0,1]$ that satisfies Axioms s1-s4, the following inequality holds:

$$max(a,b) \leq s(a,b) \leq s_{ds}(a,b) \qquad (3.16)$$

for any $a, b \in [0,1]$.

Proof: We first prove $max(a,b) \leq s(a,b)$. From the nondecreasing condition Axiom s3 and the boundary condition Axium s1, we obtain

$$s(a,b) \geq s(a,0) = a \qquad (3.17)$$

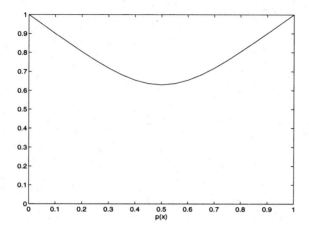

Figure 3.3. Membership function of $D \cup F$ using the Yager s-norm (3.10) with $w = 3$.

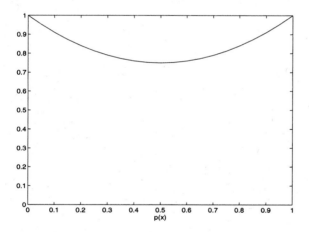

Figure 3.4. Membership function of $D \cup F$ using the algebraic sum (3.13).

Furthermore, the commutative condition Axiom s2 gives

$$s(a, b) = s(b, a) \geq s(b, 0) = b \qquad (3.18)$$

Combining (3.17) and (3.18) we have $s(a, b) \geq max(a, b)$.

Next we prove $s(a, b) \leq s_{ds}(a, b)$. If $b = 0$, then from Axiom s1 we have $s(a, b) =$

$s(a, 0) = a$, thus $s(a, b) = s_{ds}(a, b)$. By the commutative condition Axiom s2 we have $s(a, b) = s_{ds}(a, b)$ if $a = 0$. If $a \neq 0$ and $b \neq 0$, we have

$$s_{ds}(a, b) = 1 \geq s(a, b) \tag{3.19}$$

Thus $s(a, b) \leq s_{ds}(a, b)$ for all $a, b \in [0, 1]$. □

Finally, we prove an interesting property of the Dombi s-norm $s_\lambda(a, b)$ (3.8): $s_\lambda(a, b)$ converges to the basic fuzzy union $max(a, b)$ as the parameter λ goes to infinity and converges to the drastic sum $s_{ds}(a, b)$ as λ goes to zero. Therefore, the Dombi s-norm covers the whole spectrum of s-norms.

Lemma 3.1: Let $s_\lambda(a, b)$ be defined as in (3.8) and $s_{ds}(a, b)$ be defined as in (3.11), then

$$\lim_{\lambda \to \infty} s_\lambda(a, b) = max(a, b) \tag{3.20}$$

$$\lim_{\lambda \to 0} s_\lambda(a, b) = s_{ds}(a, b) \tag{3.21}$$

Proof: We first prove (3.20). If $a = b \neq 0$, then from (3.8) we have $\lim_{\lambda \to \infty} s_\lambda(a, b) = \lim_{\lambda \to \infty} [1/(1 + 2^{-1/\lambda}(\frac{1}{a} - 1))] = a = max(a, b)$. If $a = b = 0$, then $\lim_{\lambda \to \infty} s_\lambda(a, b) = \lim_{\lambda \to \infty} 1/(1 + 0^{-1/\lambda}) = 0 = max(a, b)$. If $a \neq b$, then without loss of generality (due to Axiom s2) we assume $a < b$. Let $z = [(\frac{1}{a} - 1)^{-\lambda} + (\frac{1}{b} - 1)^{-\lambda}]^{-1/\lambda}$, then using l'Hospital's rule, we have

$$\lim_{\lambda \to \infty} ln(z) = \lim_{\lambda \to \infty} - \frac{ln[(\frac{1}{a} - 1)^{-\lambda} + (\frac{1}{b} - 1)^{-\lambda}]}{\lambda}$$

$$= \lim_{\lambda \to \infty} \frac{(\frac{1}{a} - 1)^{-\lambda} ln(\frac{1}{a} - 1) + (\frac{1}{b} - 1)^{-\lambda} ln(\frac{1}{b} - 1)}{(\frac{1}{a} - 1)^{-\lambda} + (\frac{1}{b} - 1)^{-\lambda}}$$

$$= \lim_{\lambda \to \infty} \frac{[(\frac{1}{a} - 1)/(\frac{1}{b} - 1)]^{-\lambda} ln(\frac{1}{a} - 1) + ln(\frac{1}{b} - 1)}{[(\frac{1}{a} - 1)/(\frac{1}{b} - 1)]^{-\lambda} + 1}$$

$$= ln(\frac{1}{b} - 1) \tag{3.22}$$

Hence, $\lim_{\lambda \to \infty} z = \frac{1}{b} - 1$, and

$$\lim_{\lambda \to \infty} s_\lambda(a, b) = \lim_{\lambda \to \infty} \frac{1}{1 + z} = b = max(a, b) \tag{3.23}$$

Next we prove (3.21). If $a = 0$ and $b \neq 0$, we have $s_\lambda(a, b) = 1/[1 + (\frac{1}{b} - 1)^{-\lambda \frac{-1}{\lambda}}] = b = s_{ds}(a, b)$. By commutativity, we have $s_\lambda(a, b) = a = s_{ds}(a, b)$ if $b = 0$ and $a \neq 0$. If $a \neq 0$ and $b \neq 0$, we have $lim_{\lambda \to 0} s_\lambda(a, b) = \lim_{\lambda \to 0} 1/[1 + 2^{-1/\lambda}] = 1 = s_{ds}(a, b)$. Finally, if $a = b = 0$, we have $\lim_{\lambda \to 0} s_\lambda(a, b) = \lim_{\lambda \to 0} 1/[1 + 0^{-1/\lambda}] = 0 = s_{ds}(a, b)$. □

Similarly, it can be shown that the Yager s-norm (3.10) converges to the basic fuzzy union $max(a, b)$ as w goes to infinity and converges to the drastic sum $s_{ds}(a, b)$ as w goes to zero; the proof is left as an exercise.

3.3 Fuzzy Intersection—The T-Norms

Let $t : [0,1] \times [0,1] \to [0,1]$ be a function that transforms the membership functions of fuzzy sets A and B into the membership function of the intersection of A and B, that is,

$$t[\mu_A(x), \mu_B(x)] = \mu_{A \cap B}(x) \tag{3.24}$$

In the case of (3.3), $t[\mu_A(x), \mu_B(x)] = min[\mu_A(x), \mu_B(x)]$. In order for the function t to be qualified as an intersection, it must satisfy at least the following four requirements:

Axiom t1: $t(0,0) = 0; t(a,1) = t(1,a) = a$ (boundary condition).

Axiom t2: $t(a,b) = t(b,a)$ (commutativity).

Axiom t3: If $a \le a'$ and $b \le b'$, then $t(a,b) \le t(a',b')$ (nondecreasing).

Axiom t4: $t[t(a,b),c] = t[a,t(b,c)]$ (associativity).

These axioms can be justified in the same way as for Axioms s1-s4.

Definition 3.3. Any function $t : [0,1] \times [0,1] \to [0,1]$ that satisfies Axioms t1-t4 is called a *t-norm*.

We can verify that the basic fuzzy intersection *min* of (3.3) is a t-norm. For any t-norm, there is an s-norm associated with it and vice versa. Hence, associated with the s-norms of Dombi, Dubois-Prade and Yager classes ((3.8)-(3.10)), there are t-norms of Dombi, Dubois-Prade and Yager classes, which are defined as follows:

- Dombi class (Dombi [1982]):

$$t_\lambda(a,b) = \frac{1}{1 + [(\frac{1}{a} - 1)^\lambda + (\frac{1}{b} - 1)^\lambda]^{1/\lambda}} \tag{3.25}$$

  where $\lambda \in (0,\infty)$.

- Dubois-Prade class (Dubois and Prade [1980]):

$$t_\alpha(a,b) = \frac{ab}{max(a,b,\alpha)} \tag{3.26}$$

  where $\alpha \in [0,1]$.

- Yager class (Yager [1980]):

$$t_w(a,b) = 1 - min[1, ((1-a)^w + (1-b)^w)^{1/w}] \tag{3.27}$$

  where $w \in (0,\infty)$.

With a particular choice of the parameters, (3.25)-(3.27) each defines a particular t-norm. We can verify that (3.25)-(3.27) satisfy Axiom t1-t4. Associated with the particular s-norms (3.11)-(3.13) and (3.2), there are t-norms that are listed below:

- Drastic product:

$$t_{dp}(a,b) = \begin{cases} a \ if \ b = 1 \\ b \ if \ a = 1 \\ 0 \ otherwise \end{cases} \qquad (3.28)$$

- Einstein product:

$$t_{ep}(a,b) = \frac{ab}{2 - (a + b - ab)} \qquad (3.29)$$

- Algebraic product:

$$t_{ap}(a,b) = ab \qquad (3.30)$$

- Minimum: (3.3)

Example 3.2: Consider the fuzzy sets D and F defined in Example 2.1 of Chapter 2. If we use the Yager t-norm (3.27) for fuzzy intersection, then $D \cap F$ is obtained as

$$\mu_{D \cap F}(x) = t_w[\mu_D(x), \mu_F(x)]$$
$$= 1 - min[1, ((1 - p(x))^w + (p(x))^w)^{1/w}] \qquad (3.31)$$

Fig. 3.5 shows this $\mu_{D \cap F}(x)$ for $w = 3$. If we use the algebraic product (3.30) for fuzzy intersection, the fuzzy set $D \cap F$ becomes

$$\mu_{D \cap F}(x) = t_{ap}[\mu_D(x), \mu_F(x)]$$
$$= p(x)(1 - p(x)) \qquad (3.32)$$

which is plotted in Fig. 3.6. □

Comparing Figs. 3.5 and 3.6 with Fig. 2.10, we see that the Yager t-norm and algebraic product are smaller than the minimum operator. In general, we can show that minimum is the largest t-norm and drastic product is the smallest t-norm.

Theorem 3.2: For any t-norm t, that is, for any function $t : [0,1] \times [0,1] \to [0,1]$ that satisfies Axioms t1-t4, the following inequality holds:

$$t_{dp}(a,b) \leq t(a,b) \leq min(a,b) \qquad (3.33)$$

for any $a, b \in [0,1]$.

The proof of this theorem is very similar to that of Theorem 3.1 and is left as an exercise. Similar to Lemma 3.1, we can show that the Dombi t-norm $t_\lambda(a,b)$ of (3.25) converges to the basic fuzzy intersection $min(a,b)$ as λ goes to infinity and converges to the drastic product $t_{dp}(a,b)$ as λ goes to zero. Hence, the Dombi t-norm covers the whole range of t-norms.

Lemma 3.2: Let $t_\lambda(a,b)$ be defined as in (3.25), then

$$\lim_{\lambda \to \infty} t_\lambda(a,b) = min(a,b) \qquad (3.34)$$

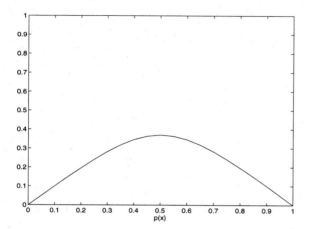

Figure 3.5. Membership function of $D \cap F$ using the Yager t-norm (3.27) with $w = 3$.

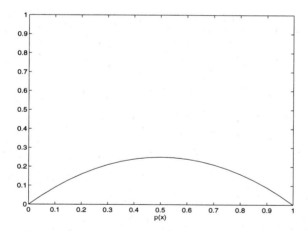

Figure 3.6. Membership function of $D \cap F$ using the algebraic product (3.30).

and

$$\lim_{\lambda \to 0} t_\lambda(a, b) = t_{dp}(a, b) \qquad (3.35)$$

This lemma can be proven in a similar way as for Lemma 3.1.

Comparing (3.8)-(3.13) with (3.25)-(3.30), respectively, we see that for each s-norm there is a t-norm associated with it. But what does this "associated" mean? It means that there exists a fuzzy complement such that the three together satisfy the DeMorgan's Law. Specifically, we say the s-norm $s(a, b)$, t-norm $t(a, b)$ and fuzzy complement $c(a)$ form *an associated class* if

$$c[s(a, b)] = t[c(a), c(b)] \tag{3.36}$$

Example 3.3: The Yager s-norm $s_w(a, b)$ of (3.10), Yager t-norm $t_w(a, b)$ of (3.27), and the basic fuzzy complement (3.1) form an associated class. To show this, we have from (3.1) and (3.10) that

$$c[s_w(a, b)] = 1 - min[1, (a^w + b^w)^{1/w}] \tag{3.37}$$

where $c(a)$ denotes the basic fuzzy complement (3.1). On the other hand, we have from (3.1) and (3.27) that

$$t_w[c(a), c(b)] = 1 - min[1, ((1 - 1 + a)^w + (1 - 1 + b)^w)^{1/w}] \tag{3.38}$$

From (3.37) and (3.38) we obtain (3.36). □

Example 3.4: The algebraic sum (3.13), algebraic product (3.30), and the basic fuzzy complement (3.1) form an associated class. To show this, we have from (3.1) and (3.13) that

$$c[s_{as}(a, b)] = 1 - a - b + ab \tag{3.39}$$

On the other hand, from (3.1) and (3.30) we have

$$t_{ap}[c(a), c(b)] = (1 - a)(1 - b) = 1 - a - b + ab \tag{3.40}$$

Hence, they satisfy the DeMorgan's Law (3.36). □

3.4 Averaging Operators

From Theorem 3.1 we see that for any membership values $a = \mu_A(x)$ and $b = \mu_B(x)$ of arbitrary fuzzy sets A and B, the membership value of their union $A \cup B$ (defined by any s-norm) lies in the interval $[max(a, b), s_{ds}(a, b)]$. Similarly, from Theorem 3.2 we have that the membership value of the intersection $A \cap B$ (defined by any t-norm) lies in the interval $[t_{dp}(a, b), min(a, b)]$. See Fig.3.7. Therefore, the union and intersection operators cannot cover the interval between $min(a, b)$ and $max(a, b)$. The operators that cover the interval $[min(a, b), max(a, b)]$ are called *averaging operators*. Similar to the s-norms and t-norms, an averaging operator, denoted by v, is a function from $[0, 1] \times [0, 1]$ to $[0, 1]$.

Many averaging operators were proposed in the literature. Here we list four of them:

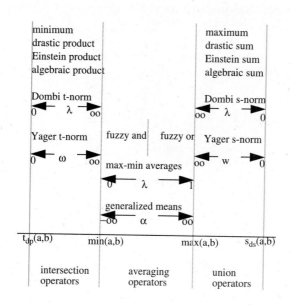

Figure 3.7. The full scope of fuzzy aggregation operators.

- Max-min averages:

$$v_\lambda(a, b) = \lambda max(a, b) + (1 - \lambda)min(a, b) \tag{3.41}$$

where $\lambda \in [0, 1]$.

- Generalized means:

$$v_\alpha(a, b) = (\frac{a^\alpha + b^\alpha}{2})^{1/\alpha} \tag{3.42}$$

where $\alpha \in R$ ($\alpha \neq 0$).

- "Fuzzy and":

$$v_p(a, b) = pmin(a, b) + \frac{(1 - p)(a + b)}{2} \tag{3.43}$$

where $p \in [0, 1]$.

- "Fuzzy or":

$$v_\gamma(a, b) = \gamma max(a, b) + \frac{(1 - \gamma)(a + b)}{2} \tag{3.44}$$

where $\gamma \in [0, 1]$.

Clearly, the max-min averages cover the whole interval $[min(a, b), max(a, b)]$ as the parameter λ changes from 0 to 1. The "fuzzy and" covers the range from $min(a, b)$ to $(a+b)/2$, and the "fuzzy or" covers the range from $(a+b)/2$ to $max(a, b)$. It also can be shown that the generalized means cover the whole range from $min(a, b)$ to $max(a, b)$ as α changes from $-\infty$ to ∞.

3.5 Summary and Further Readings

In this chapter we have demonstrated the following:

- The axiomatic definitions of fuzzy complements, s-norms (fuzzy unions), and t-norms (fuzzy intersections).

- Some specific classes of fuzzy complements, s-norms, t-norms, and averaging operators, and their properties.

- How to prove various properties of some particular fuzzy complements, s-norms, t-norms, and averaging operators.

The materials in this chapter were extracted from Klir and Yuan [1995] where more details on the operators can be found. Dubois and Prade [1985] provided a very good review of fuzzy union, fuzzy intersection, and averaging operators.

3.6 Exercises

Exercise 3.1. The *equilibrium* of a fuzzy complement c is defined as $a \in [0, 1]$ such that $c(a) = a$.

(a) Determine the equilibrium of the Yager fuzzy complement (3.6).

(b) Prove that every fuzzy complement has at most one equilibrium.

(c) Prove that a continuous fuzzy complement has a unique equilibrium.

Exercise 3.2. Show that the Yager s-norm (3.10) converges to the basic fuzzy union (3.2) as w goes to infinity and converges to the drastic sum (3.11) as w goes to zero.

Exercise 3.3. Let the fuzzy sets F and G be defined as in Exercise 2.3.

(a) Determine the membership functions for $F \cup G$ and $F \cap G$ using the Yager s-norm (3.10) and t-norm (3.27) with $w = 2$.

(b) Using (3.1) as fuzzy complement, algebraic sum (3.13) as fuzzy union, and algebraic product (3.30) as fuzzy intersection, compute the membership functions for $F \cap \bar{G}$, $\overline{F} \cap G$, and $\overline{F \cup G}$.

Exercise 3.4. Prove Theorem 3.2.

Exercise 3.5. A fuzzy complement c is said to be *involutive* if $c[c(a)] = a$ for all $a \in [0, 1]$.

(a)Show that the Sugeno fuzzy complement (3.5) and the Yager fuzzy complement (3.6) are involutive.

(b) Let c be an involutive fuzzy complement and t be any t-norm. Show that the operator $u : [0, 1] \times [0, 1] \to [0, 1]$ defined by

$$u(a, b) = c[t(c(a), c(b))] \tag{3.45}$$

is an s-norm.

(c) Prove that the c, t, and u in (b) form an associated class.

Exercise 3.6. Determine s-norm $s_x(a, b)$ such that $s_x(a, b)$, the minimum t-norm (3.3), and the Yager complement (3.6) with $w = 2$ form an associated class.

Exercise 3.7. Prove that the following triples form an associated class with respect to any fuzzy complement c: (a) (min, max, c), and (b) (t_{dp}, s_{ds}, c).

Exercise 3.8. Prove that the generalized means (3.42) become min and max operators as $\alpha \to -\infty$ and $\alpha \to \infty$, respectively.

Chapter 4

Fuzzy Relations and the Extension Principle

4.1 From Classical Relations to Fuzzy Relations

4.1.1 Relations

Let U and V be two arbitrary classical (nonfuzzy, crisp) sets. The *Cartesian product* of U and V, denoted by $U \times V$, is the nonfuzzy set of all ordered pairs (u, v) such that $u \in U$ and $v \in V$; that is,

$$U \times V = \{(u, v) | u \in U \ and \ v \in V\} \tag{4.1}$$

Note that the order in which U and V appears is important; that is, if $U \neq V$, then $U \times V \neq V \times U$. In general, the Cartesian product of arbitrary n nonfuzzy sets $U_1, U_2, ..., U_n$, denoted by $U_1 \times U_2 \times \cdots \times U_n$, is the nonfuzzy set of all n-tuples $(u_1, u_2, ..., u_n)$ such that $u_i \in U_i$ for $i \in \{1, 2, ..., \}$; that is,

$$U_1 \times U_2 \times \cdots \times U_n = \{(u_1, u_2, ..., u_n) | u_1 \in U_1, \ u_2 \in U_2, \cdots, \ u_n \in U_n\} \tag{4.2}$$

A (nonfuzzy) *relation* among (nonfuzzy) sets $U_1, U_2, ..., U_n$ is a subset of the Cartesian product $U_1 \times U_2 \times \cdots \times U_n$; that is, if we use $Q(U_1, U_2, ..., U_n)$ to denote a relation among $U_1, U_2, ..., U_n$, then

$$Q(U_1, U_2, ..., U_n) \subset U_1 \times U_2 \times \cdots \times U_n \tag{4.3}$$

As a special case, a *binary relation* between (nonfuzzy) sets U and V is a subset of the Cartesian product $U \times V$.

Example 4.1. Let $U = \{1, 2, 3\}$ and $V = \{2, 3, 4\}$. Then the cartesian product of U and V is the set $U \times V = \{(1, 2), (1, 3), (1, 4), (2, 2), (2, 3), (2, 4), (3, 2), (3, 3), (3, 4)\}$. A relation between U and V is a subset of $U \times V$. For example, let $Q(U, V)$ be a relation named "the first element is no smaller than the second element," then

$$Q(U, V) = \{(2, 2), (3, 2), (3, 3)\} \tag{4.4}$$

48

□

Because a relation is itself a set, all of the basic set operations can be applied to it without modification. Also, we can use the following membership function to represent a relation:

$$\mu_Q(u_1, u_2, ..., u_n) = \begin{cases} 1 \; if \; (u_1, u_2, ..., u_n) \in Q(U_1, U_2, ..., U_n) \\ 0 \; otherwise \end{cases} \tag{4.5}$$

For binary relation $Q(U, V)$ defined over $U \times V$ which contains finite elements, we often collect the values of the membership function μ_Q into a *relational matrix*; see the following example.

Example 4.1 (Cont'd). The relation $Q(U, V)$ of (4.4) can be represented by the following relational matrix:

$$\begin{array}{c} & V \\ & \begin{array}{ccc} 2 & 3 & 4 \end{array} \\ & \begin{array}{c} 1 \\ U \quad 2 \\ 3 \end{array} \begin{array}{ccc} 0 & 0 & 0 \\ 1 & 0 & 0 \\ 1 & 1 & 0 \end{array} \end{array} \tag{4.6}$$

□

A classical relation represents a crisp relationship among sets, that is, either there is such a relationship or not. For certain relationships, however, it is difficult to give a zero-one assessment; see the following example.

Example 4.2. Let $U = \{SanFrancisco, HongKong, Tokyo\}$ and $V = \{Boston, HongKong\}$. We want to define the relational concept "very far" between these two sets of cities. Clearly, classical relations are not useful because the concept "very far" is not well-defined in the framework of classical sets and relations. However, "very far" does mean something and we should find a numerical system to characterize it. If we use a number in the interval $[0, 1]$ to represent the degree of "very far," then the concept "very far" may be represented by the following (fuzzy) relational matrix:

$$\begin{array}{c} & V \\ & \begin{array}{cc} Boston & HK \end{array} \\ & \begin{array}{c} SF \\ U \quad HK \\ Tokyo \end{array} \begin{array}{cc} 0.3 & 0.9 \\ 1 & 0 \\ 0.95 & 0.1 \end{array} \end{array} \tag{4.7}$$

□

Example 4.2 shows that we need to generalize the concept of classical relation in order to formulate more relationships in the real world. The concept of fuzzy relation was thus introduced.

Definition 4.1. A *fuzzy relation* is a fuzzy set defined in the Cartesian product of crisp sets $U_1, U_2, ..., U_n$. With the representation scheme (2.5), a fuzzy relation

Q in $U_1 \times U_2 \times \cdots \times U_n$ is defined as the fuzzy set

$$Q = \{((u_1, u_2, ..., u_n), \mu_Q(u_1, u_2, ..., u_n)) | (u_1, u_2, ..., u_n) \in U_1 \times U_2 \times \cdots \times U_n\} \quad (4.8)$$

where $\mu_Q : U_1 \times U_2 \times \cdots \times U_n \to [0, 1]$.

As a special case, a *binary fuzzy relation* is a fuzzy set defined in the Cartesian product of two crisp sets. A binary relation on a finite Cartesian product is usually represented by a *fuzzy relational matrix*, that is, a matrix whose elements are the membership values of the corresponding pairs belonging to the fuzzy relation. For example, (4.7) is a fuzzy relational matrix representing the fuzzy relation named "very far" between the two groups of cities.

Example 4.3. Let U and V be the set of real numbers, that is, $U = V = R$. A fuzzy relation "x is approximately equal to y," denoted by AE, may be defined by the membership function

$$\mu_{AE}(x, y) = e^{-(x-y)^2} \quad (4.9)$$

Similarly, a fuzzy relation "x is much larger than y," denoted by ML, may be defined by the membership function

$$\mu_{ML}(x, y) = \frac{1}{1 + e^{-(x-y)}} \quad (4.10)$$

Of course, other membership functions may be used to represent these fuzzy relations. □

4.1.2 Projections and Cylindric Extensions

Because a crisp relation is defined in the product space of two or more sets, the concepts of projection and cylindric extension were proposed. For example, consider the set $A = \{(x, y) \in R^2 | (x-1)^2 + (y-1)^2 \leq 1\}$ which is a relation in $U \times V = R^2$. Then the projection of A on U is $A_1 = [0, 1] \subset U$, and the projection of A on V is $A_2 = [0, 1] \subset V$; see Fig. 4.1. The cylindric extension of A_1 to $U \times V = R^2$ is $A_{1E} = [0, 1] \times (-\infty, \infty) \subset R^2$. These concepts can be extended to fuzzy relations.

Definition 4.2. Let Q be a fuzzy relation in $U_1 \times \cdots \times U_n$ and $\{i_1, ..., i_k\}$ be a subsequence of $\{1, 2, ..., n\}$, then the *projection* of Q on $U_{i_1} \times \cdots \times U_{i_k}$ is a fuzzy relation Q_P in $U_{i_1} \times \cdots \times U_{i_k}$ defined by the membership function

$$\mu_{Q_P}(u_{i_1}, ..., u_{i_k}) = \max_{u_{j1} \in U_{j1}, \cdots, u_{j(n-k)} \in U_{j(n-k)}} \mu_Q(u_1, ..., u_n) \quad (4.11)$$

where $\{u_{j1}, ..., u_{j(n-k)}\}$ is the complement of $\{u_{i_1}, ..., u_{i_k}\}$ with respect to $\{u_1, ..., u_n\}$.

As a special case, if Q is a binary fuzzy relation in $U \times V$, then the projection of Q on U, denoted by Q_1, is a fuzzy set in U defined by

$$\mu_{Q_1}(x) = \max_{y \in V} \mu_Q(x, y) \quad (4.12)$$

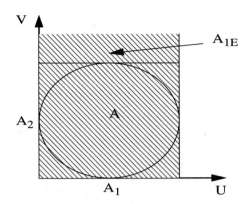

Figure 4.1. Projections and cylindric extensions of a relation.

Note that (4.12) is still valid if Q is a crisp relation. For example, if Q is the crisp relation A in Fig. 4.1, then its projection Q_1 defined by (4.12) is equal to the A_1 in Fig. 4.1. Hence, the projection of fuzzy relation defined by (4.11) is a natural extension of the projection of crisp relation.

Example 4.4. According to (4.12), the projection of fuzzy relation (4.7) on U and V are the fuzzy sets

$$Q_1 = 0.9/SF + 1/HK + 0.95/Tokyo \tag{4.13}$$

and

$$Q_2 = 1/Boston + 0.9/HK \tag{4.14}$$

respectively. Similarly, the projections of AE defined by (4.9) on U and V are the fuzzy sets

$$AE_1 = \int_U \max_{y \in V} e^{-(x-y)^2}/x = \int_U 1/x \tag{4.15}$$

and

$$AE_2 = \int_V \max_{x \in U} e^{-(x-y)^2}/y = \int_V 1/y \tag{4.16}$$

respectively. Note that AE_1 equals the crisp set U and AE_2 equals the crisp set V.
□

The projection constrains a fuzzy relation to a subspace; conversely, the cylindric extension extends a fuzzy relation (or fuzzy set) from a subspace to the whole space. Formally, we have the following definition.

Definition 4.3. Let Q_P be a fuzzy relation in $U_{i_1} \times \cdots \times U_{i_k}$ and $\{i_1, ..., i_k\}$ is a subsequence of $\{1, 2, ..., n\}$, then the *cylindric extension* of Q_P to $U_1 \times \cdots \times U_n$ is a fuzzy relation Q_{PE} in $U_1 \times \cdots \times U_n$ defined by

$$\mu_{Q_{PE}}(u_1, ..., u_n) = \mu_{Q_P}(u_{i_1}, ..., u_{i_k}) \tag{4.17}$$

As a special case, if Q_1 is a fuzzy set in U, then the cylindric extension of Q_1 to $U \times V$ is a fuzzy relation Q_{1E} in $U \times V$ defined by

$$\mu_{Q_{1E}}(x, y) = \mu_{Q_1}(x) \tag{4.18}$$

The definition (4.17) is also valid for crisp relations; check Fig. 4.1 for an example.

Example 4.5. Consider the projections Q_1 and Q_2 in Example 4.4 ((4.13) and (4.14)). According to (4.18), their cylindric extensions to $U \times V$ are

$$\begin{aligned} Q_{1E} = &\ 0.9/(SF, Boston) + 0.9/(SF, HK) + 1/(HK, Boston) \\ &+ 1/(HK, HK) + 0.95/(Tokyo, Boston) \\ &+ 0.95/(Tokyo, HK) \end{aligned} \tag{4.19}$$

and

$$\begin{aligned} Q_{2E} = &\ 1/(SF, Boston) + 1/(HK, Boston) + 1/(Tokyo, Boston) \\ &+ 0.9/(SF, HK) + 0.9/(HK, HK) + 0.9/(Tokyo, HK) \end{aligned} \tag{4.20}$$

Similarly, the cylindric extensions of AE_1 and AE_2 in (4.15) and (4.16) to $U \times V$ are

$$AE_{1E} = \int_{U \times V} 1/(x, y) = U \times V \tag{4.21}$$

and

$$AE_{2E} = \int_{U \times V} 1/(x, y) = U \times V \tag{4.22}$$

□

From Examples 4.4 and 4.5 we see that when we take the projection of a fuzzy relation and then cylindrically extend it, we obtain a fuzzy relation that is larger than the original one. To characterize this property formally, we first introduce the concept of Cartesian product of fuzzy sets. Let $A_1, ..., A_n$ be fuzzy sets in $U_1, ..., U_n$, respectively. The *Cartesian product* of $A_1, ..., A_n$, denoted by $A_1 \times \cdots \times A_n$, is a fuzzy relation in $U_1 \times \cdots \times U_n$ whose membership function is defined as

$$\mu_{A_1 \times \cdots \times A_n}(u_1, ..., u_n) = \mu_{A_1}(u_1) \star \cdots \star \mu_{A_n}(u_n) \tag{4.23}$$

where $\star$ represents any t-norm operator.

Lemma 4.1. If Q is a fuzzy relation in $U_1 \times \cdots \times U_n$ and $Q_1, ..., Q_n$ are its projections on $U_1, ..., U_n$, respectively, then (see Fig. 4.2 for illustration)

$$Q \subset Q_1 \times \cdots \times Q_n \tag{4.24}$$

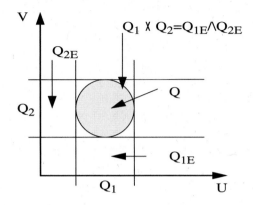

Figure 4.2. Relation between the Cartesian product and intersection of cylindric sets.

where we use *min* for the t-norm in the definition (4.23) of $Q_1 \times \cdots \times Q_n$.

    Proof: Substituting (4.11) into (4.17), we have

$$\mu_{Q_{PE}}(u_1, ..., u_n) = \max_{u_{j1}\in U_{j1},...,u_{j(n-k)}\in U_{j(n-k)}} \mu_Q(u_1, ..., u_n) \qquad (4.25)$$

Hence,

$$Q \subset Q_{iE} \qquad (4.26)$$

for all $i = 1, 2, ..., n$, where Q_{iE} is the cylindric extension of Q_i to $U_1 \times \cdots \times U_n$. Therefore, if we use *min* for intersection, we have

$$Q \subset Q_{1E} \cap \cdots \cap Q_{nE} = Q_1 \times \cdots \times Q_n \qquad (4.27)$$

$\square$

4.2   Compositions of Fuzzy Relations

Let $P(U, V)$ and $Q(V, W)$ be two crisp binary relations that share a common set V. The *composition* of P and Q, denoted by $P \circ Q$, is defined as a relation in $U \times W$ such that $(x, z) \in P \circ Q$ if and only if there exists at least one $y \in V$ such that $(x, y) \in P$ and $(y, z) \in Q$. Using the membership function representation of relations (see (4.5)), we have an equivalent definition for composition that is given in the following lemma.

Lemma 4.2. $P \circ Q$ is the composition of $P(U, V)$ and $Q(V, W)$ if and only if

$$\mu_{P \circ Q}(x, z) = \max_{y \in V} t[\mu_P(x, y), \mu_Q(y, z)] \qquad (4.28)$$

for any $(x, z) \in U \times W$, where t is any t-norm.

Proof: We first show that if $P \circ Q$ is the composition according to the definition, then (4.28) is true. If $P \circ Q$ is the composition, then $(x, z) \in P \circ Q$ implies that there exists $y \in V$ such that $\mu_P(x, y) = 1$ and $\mu_Q(y, z) = 1$. Hence, $\mu_{P \circ Q}(x, z) = 1 = \max_{y \in V} t[\mu_P(x, y), \mu_Q(y, z)]$, that is, (4.28) is true. If $(x, z) \notin P \circ Q$, then for any $y \in V$ either $\mu_P(x, y) = 0$ or $\mu_Q(y, z) = 0$. Hence, $\mu_{P \circ Q}(x, z) = 0 = \max_{y \in V} t[\mu_P(x, y), \mu_Q(y, z)]$. Therefore, (4.28) is true for any $(x, z) \in U \times W$.

Conversely, if (4.28) is true, then $(x, z) \in P \circ Q$ implies $\max_{y \in V} t[\mu_P(x, y), \mu_Q(y, z)] = 1$, which means that there exists at least one $y \in V$ such that $\mu_P(x, y) = \mu_Q(y, z) = 1$ (see Axiom t1 in Section 3.3); this is the definition. For $(x, z) \notin P \circ Q$, we have from (4.28) that $\max_{y \in V} t[\mu_P(x, y), \mu_Q(y, z)] = 0$, which means that there is no $y \in V$ such that $\mu_P(x, y) = \mu_Q(y, z) = 1$. Therefore, (4.28) implies that $P \circ Q$ is the composition according to the definition. $\square$

Now we generalize the concept of composition to fuzzy relations. From Lemma 4.2 we see that if we use (4.28) to define composition of fuzzy relations (suppose P and Q are fuzzy relatioins), then the definition is valid for the special case where P and Q are crisp relations. Therefore, we give the following definition.

Definition 4.4. The *composition of fuzzy relations* $P(U, V)$ and $Q(V, W)$, denoted by $P \circ Q$, is defined as a fuzzy relation in $U \times W$ whose membership function is given by (4.28).

Because the t-norm in (4.28) can take a variety of formulas, for each t-norm we obtain a particular composition. The two most commonly used compositions in the literature are the so-called *max-min* composition and *max-product* composition, which are defined as follows:

- The *max-min composition* of fuzzy relations $P(U, V)$ and $Q(V, W)$ is a fuzzy relation $P \circ Q$ in $U \times W$ defined by the membership function

$$\mu_{P \circ Q}(x, z) = \max_{y \in V} min[\mu_P(x, y), \mu_Q(y, z)] \qquad (4.29)$$

where $(x, z) \in U \times W$.

- The *max-product composition* of fuzzy relations $P(U, V)$ and $Q(V, W)$ is a fuzzy relation $P \circ Q$ in $U \times W$ defined by the membership function

$$\mu_{P \circ Q}(x, z) = \max_{y \in V} [\mu_P(x, y) \mu_Q(y, z)] \qquad (4.30)$$

where $(x, z) \in U \times W$.

We see that the max-min and max-product compositions use minimum and algebraic product for the t-norm in the definition (4.28), respectively. We now consider two examples for how to compute the compositions.

Example 4.6. Let U and V be defined as in Example 4.2 and $W = \{New\ York\ City,\ Beijing\}$. Let $P(U,V)$ denote the fuzzy relation "very far" defined by (4.7). Define the fuzzy relation "very near" in $V \times W$, denoted by $Q(V,W)$, by the relational matrix

$$
\begin{array}{ccc}
 & W & \\
 & NYC \quad Beijing & \\
V \quad Boston & 0.95 \qquad 0.1 & \\
HK & 0.1 \qquad 0.9 &
\end{array}
\qquad (4.31)
$$

Using the notation (2.7), we can write P and Q as

$$
\begin{aligned}
P = {} & 0.3/(SF, Boston) + 0.9/(SF, HK) + 1/(HK, Boston) \\
& + 0/(HK, HK) + 0.95/(Tokyo, Boston) + 0.1/(Tokyo, HK) \qquad (4.32) \\
Q = {} & 0.95/(Boston, NYC) + 0.1/(Boston, Beijing) + 0.1/(HK, NYC) \\
& + 0.9/(HK, Beijing) \qquad (4.33)
\end{aligned}
$$

We now compute the max-min and max-product compositions of P and Q. First, we note that $U \times W$ contains six elements: (SF,NYC), (SF,Beijing), (HK,NYC), (HK,Beijing), (Tokyo,NYC) and (Tokyo,Beijing). Thus, our task is to determine the membership values of $\mu_{P \circ Q}$ at these six elements. Using the max-min composition (4.29), we have

$$
\begin{aligned}
\mu_{P \circ Q}(SF, NYC) &= max\{min[\mu_P(SF, Boston), \mu_Q(Boston, NYC)], \\
& \qquad min[\mu_P(SF, HK), \mu_Q(HK, NYC)]\} \\
&= max[min(0.3, 0.95), min(0.9, 0.1)] \\
&= 0.3 \qquad\qquad (4.34)
\end{aligned}
$$

Similarly, we have

$$
\begin{aligned}
\mu_{P \circ Q}(SF, Beijing) &= max\{min[\mu_P(SF, Boston), \mu_Q(Boston, Beijing)], \\
& \qquad min[\mu_P(SF, HK), \mu_Q(HK, Beijing)]\} \\
&= max[min(0.3, 0.1), min(0.9, 0.9)] \\
&= 0.9 \qquad\qquad (4.35)
\end{aligned}
$$

The final $P \circ Q$ is

$$
\begin{aligned}
P \circ Q = {} & 0.3/(SF, NYC) + 0.9/(SF, Beijing) + 0.95/(HK, NYC) \\
& + 0.1/(HK, Beijing) + 0.95/(Tokyo, NYC) \\
& + 0.1/(Tokyo, Beijing) \qquad\qquad (4.36)
\end{aligned}
$$

If we use the max-product composition (4.30), then following the same procedure as above (replacing *min* by *product*), we obtain

$$P \circ Q = 0.285/(SF, NYC) + 0.81/(SF, Beijing) + 0.95/(HK, NYC)$$
$$+0.1/(HK, Beijing) + 0.9025/(Tokyo, NYC)$$
$$+0.095/(Tokyo, Beijing) \tag{4.37}$$

□

From (4.36), (4.37) and the relational matrices (4.7) and (4.31), we see that a simpler way to compute $P \circ Q$ is to use relational matrices and matrix product. Specifically, let P and Q be the relational matrices for the fuzzy relations $P(U, V)$ and $Q(V, W)$, respectively. Then, the relational matrix for the fuzzy composition $P \circ Q$ can be computed according to the following method:

- For max-min composition, write out each element in the matrix product PQ, but treat each multiplication as a *min* operation and each addition as a *max* operation.

- For max-product composition, write out each element in the matrix product PQ, but treat each addition as a *max* operation.

We now check that (4.36) and (4.37) can be obtained by this method. Specifically, we have

$$\begin{pmatrix} 0.3 & 0.9 \\ 1 & 0 \\ 0.95 & 0.1 \end{pmatrix} \circ \begin{pmatrix} 0.95 & 0.1 \\ 0.1 & 0.9 \end{pmatrix} = \begin{pmatrix} 0.3 & 0.9 \\ 0.95 & 0.1 \\ 0.95 & 0.1 \end{pmatrix} \tag{4.38}$$

for max-min composition, and

$$\begin{pmatrix} 0.3 & 0.9 \\ 1 & 0 \\ 0.95 & 0.1 \end{pmatrix} \circ \begin{pmatrix} 0.95 & 0.1 \\ 0.1 & 0.9 \end{pmatrix} = \begin{pmatrix} 0.285 & 0.81 \\ 0.95 & 0.1 \\ 0.9025 & 0.095 \end{pmatrix} \tag{4.39}$$

for max-product composition.

In Example 4.6, the universal sets U, V and W contain a finite number of elements. In most engineering applications, however, the U, V and W are real-valued spaces that contain an infinite number of elements. We now consider an example for computing the composition of fuzzy relations in continuous domains.

Example 4.7: Let $U = V = W = R$. Consider the fuzzy relation AE (approximately equal) and ML (much larger) defined by (4.9) and (4.10) in Example 4.3. We now want to determine the composition $AE \circ ML$. Using the max-product composition, we have

$$\mu_{AE \circ ML}(x, z) = \max_{y \in R} [\frac{e^{-(x-y)^2}}{1 + e^{-(y-z)}}] \tag{4.40}$$

To compute the right hand side of (4.40), we must determine the $y \in R$ at which $\frac{e^{-(x-y)^2}}{1+e^{-(y-z)}}$ achieves its maximum value, where x and z are considered to be fixed values in R. The necessary condition for such y is

$$\frac{\partial}{\partial y}\left[\frac{e^{-(x-y)^2}}{1+e^{-(y-z)}}\right] = 0 \tag{4.41}$$

Because it is impossible to obtain a closed form solution for (4.41), we cannot further simplify (4.40). In practice, for given values of x and z we can first determine the numerical solution of (4.41) and then substitute it into (4.40). Comparing this example with Example 4.6, we see that fuzzy compositions in continuous domains are much more difficult to compute than those in discrete domains. $\square$

4.3 The Extension Principle

The extension principle is a basic identity that allows the domain of a function to be extended from crisp points in U to fuzzy sets in U. More specifically, let $f : U \to V$ be a function from crisp set U to crisp set V. Suppose that a fuzzy set A in U is given and we want to determine a fuzzy set $B = f(A)$ in V that is induced by f. If f is an one-to-one mapping, then we can define

$$\mu_B(y) = \mu_A[f^{-1}(y)], \ y \in V \tag{4.42}$$

where $f^{-1}(y)$ is the inverse of f, that is, $f[f^{-1}(y)] = y$. If f is not one-to-one, then an ambiguity arises when two or more distinct points in U with different membership values in A are mapped into the same point in V. For example, we may have $f(x_1) = f(x_2) = y$ but $x_1 \neq x_2$ and $\mu_A(x_1) \neq \mu_A(x_2)$, so the right hand side of (4.42) may take two different values $\mu_A(x_1 = f^{-1}(y))$ or $\mu_A(x_2 = f^{-1}(y))$. To resolve this ambiguity, we assign the larger one of the two membership values to $\mu_B(y)$. More generally, the membership function for B is defined as

$$\mu_B(y) = \max_{x \in f^{-1}(y)} \mu_A(x), \ y \in V \tag{4.43}$$

where $f^{-1}(y)$ denotes the set of all points $x \in U$ such that $f(x) = y$. The identity (4.43) is called the *extension principle*.

   Example 4.8. Let $U = \{1, 2, ..., 10\}$ and $f(x) = x^2$. Let *small* be a fuzzy set in U defined by

$$small = 1/1 + 1/2 + 0.8/3 + 0.6/4 + 0.4/5 \tag{4.44}$$

Then, in consequence of (4.43), we have

$$small^2 = 1/1 + 1/4 + 0.8/9 + 0.6/16 + 0.4/25 \tag{4.45}$$

$\square$

4.4 Summary and Further Readings

In this chapter we have demonstrated the following:

- The concepts of fuzzy relations, projections, and cylindric extensions.
- The max-min and max-product compositions of fuzzy relations.
- The extension principle and its applications.

The basic ideas of fuzzy relations, projections, cylindric extensions, compositions of fuzzy relations, and the extension principle were proposed in Zadeh [1971b] and Zadeh [1975]. These original papers of Zadeh were very clearly written and are still the best sources to learn these fundamental concepts.

4.5 Exercises

Exercise 4.1. Given an n-ary relation, how many different projections of the relation can be taken?

Exercise 4.2. Consider the fuzzy relation Q defined in $U_1 \times \cdots \times U_4$ where $U_1 = \{a, b, c\}, U_2 = \{s, t\}, U_3 = \{x, y\}$ and $U_4 = \{i, j\}$:

$$Q = 0.4/(b,t,y,i) + 0.6/(a,s,x,i) + 0.9/(b,s,y,i) + 1/(b,s,y,j)$$
$$+0.6/(a,t,y,j) + 0.2/(c,s,y,i)$$

(a) Compute the projections of Q on $U_1 \times U_2 \times U_4$, $U_1 \times U_3$ and U_4.

(b) Compute the cylindric extensions of the projections in (a) to $U_1 \times U_2 \times U_3 \times U_4$.

Exercise 4.3. Consider the three binary fuzzy relations defined by the relational matrices:

$$Q_1 = \begin{pmatrix} 1 & 0 & 0.7 \\ 0.3 & 0.2 & 0 \\ 0 & 0.5 & 1 \end{pmatrix}, \quad Q_2 = \begin{pmatrix} 0.6 & 0.6 & 0 \\ 0 & 0.6 & 0.1 \\ 0 & 0.1 & 0 \end{pmatrix}, \quad Q_3 = \begin{pmatrix} 1 & 0 & 0.7 \\ 0 & 1 & 0 \\ 0.7 & 0 & 1 \end{pmatrix}$$

$$\tag{4.46}$$

Compute the max-min and max-product compositions $Q_1 \circ Q_2, Q_1 \circ Q_3$ and $Q_1 \circ Q_2 \circ Q_3$.

Exercise 4.4. Consider fuzzy set $A = 0.5/-1 + 0.8/0 + 1/1 + 0.4/2$ and function $f(x) = x^2$. Determine the fuzzy set $f(A)$ using the extension principle.

Exercise 4.5. Compute the $\mu_{AE \circ ML}(x, z)$ in Example 4.7 for $(x, z) = (0, 0), (0, 1), (1, 0), (1, 1)$.

Chapter 5

Linguistic Variables and Fuzzy IF-THEN Rules

5.1 From Numerical Variables to Linguistic Variables

In our daily life, words are often used to describe variables. For example, when we say "today is hot," or equivalently, "today's temperature is high," we use the word "high" to describe the variable "today's temperature." That is, the variable "today's temperature" takes the word "high" as its value. Clearly, the variable "today's temperature" also can take numbers like $25^o c, 19^o c$, etc., as its values. When a variable takes numbers as its values, we have a well-established mathematical framework to formulate it. But when a variable takes words as its values, we do not have a formal framework to formulate it in classical mathematical theory. In order to provide such a formal framework, the concept of linguistic variables was introduced. Roughly speaking, if a variable can take words in natural languages as its values, it is called a *linguistic variable*. Now the question is how to formulate the words in mathematical terms? Here we use fuzzy sets to characterize the words. Thus, we have the following definition.

Definition 5.1. If a variable can take words in natural languages as its values, it is called a *linguistic variable*, where the words are characterized by fuzzy sets defined in the universe of discourse in which the variable is defined.

Example 5.1. The speed of a car is a variable x that takes values in the interval $[0, V_{max}]$, where V_{max} is the maximum speed of the car. We now define three fuzzy sets "slow," "medium," and "fast" in $[0, V_{max}]$ as shown in Fig. 5.1. If we view x as a linguistic variable, then it can take "slow," "medium" and "fast" as its values. That is, we can say "x is slow," "x is medium," and "x is fast." Of course, x also can take numbers in the interval $[0, V_{max}]$ as its values, for example, $x = 50$mph, 35mph, etc. □

Definition 5.1 gives a simple and intuitive definition for linguistic variables. In the fuzzy theory literature, a more formal definition of linguistic variables was usu-

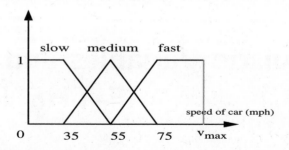

Figure 5.1. The speed of a car as a linguistic variable that can take fuzzy sets "slow," "medium" and "fast" as its values.

ally employed (Zadeh [1973] and [1975]). This definition is given below.

Definition 5.2. A *linguistic variable* is characterized by (X, T, U, M), where:

- X is the name of the linguistic variable; in Example 5.1, X is the speed of the car.

- T is the set of linguistic values that X can take; in Example 5.1, $T = \{$slow, medium, fast$\}$.

- U is the actual physical domain in which the linguistic variable X takes its quantitative (crisp) values; in Example 5.1, $U = [0, V_{max}]$.

- M is a semantic rule that relates each linguistic value in T with a fuzzy set in U; in Example 5.1, M relates "slow," "medium," and "fast" with the membership functions shown in Fig. 5.1.

Comparing Definitions 5.1 with 5.2, we see that they are essentially equivalent. Definition 5.1 is more intuitive, whereas Definition 5.2 looks more formal. From these definitions we see that linguistic variables are extensions of numerical variables in the sense that they are allowed to take fuzzy sets as their values; see Fig. 5.2.

Why is the concept of linguistic variable important? Because linguistic variables are the most fundamental elements in human knowledge representation. When we use sensors to measure a variable, they give us numbers; when we ask human experts to evaluate a variable, they give us words. For example, when we use a radar gun to measure the speed of a car, it gives us numbers like $39mph, 42mph$, etc.; when

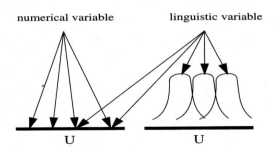

Figure 5.2. From numerical variable to linguistic variable.

we ask a human to tell us about the speed of the car, he/she often tells us in words like "it's slow," "it's fast," etc. Hence, by introducing the concept of linguistic variables, we are able to formulate vague descriptions in natural languages in precise mathematical terms. This is the first step to incorporate human knowledge into engineering systems in a systematic and efficient manner.

5.2   Linguistic Hedges

With the concept of linguistic variables, we are able to take words as values of (linguistic) variables. In our daily life, we often use more than one word to describe a variable. For example, if we view the speed of a car as a linguistic variable, then its values might be "not slow," "very slow," "slightly fast," "more or less medium," etc. In general, the value of a linguistic variable is a composite term $x = x_1 x_2 \cdots x_n$ that is a concatenation of atomic terms $x_1, x_2, ..., x_n$. These *atomic terms* may be classified into three groups:

- *Primary terms*, which are labels of fuzzy sets; in Example 5.1, they are "slow," "medium," and "fast."

- Complement "not" and connections "and" and "or."

- *Hedges*, such as "very," "slightly," "more or less," etc.

The terms "not," "and," and "or" were studied in Chapters 2 and 3. Our task now is to characterize hedges.

Although in its everyday use the hedge *very* does not have a well-defined meaning, in essence it acts as an intensifier. In this spirit, we have the following definition for the two most commonly used hedges: *very* and *more or less*.

Definition 5.3. Let A be a fuzzy set in U, then *very* A is defined as a fuzzy set in U with the membership function

$$\mu_{very\ A}(x) = [\mu_A(x)]^2 \tag{5.1}$$

and *more or less* A is a fuzzy set in U with the membership function

$$\mu_{more\ or\ less\ A}(x) = [\mu_A(x)]^{1/2} \tag{5.2}$$

Example 5.2. Let $U = \{1, 2, ..., 5\}$ and the fuzzy set *small* be defined as

$$small = 1/1 + 0.8/2 + 0.6/3 + 0.4/4 + 0.2/5 \tag{5.3}$$

Then, according to (5.1) and (5.2), we have

$$very\ small = 1/1 + 0.64/2 + 0.36/3 + 0.16/4 + 0.04/5 \tag{5.4}$$

$$\begin{aligned} very\ very\ small &= very\ (very\ small) \\ &= 1/1 + 0.4096/2 + 0.1296/3 + 0.0256/4 \\ &\quad +0.0016/5 \end{aligned} \tag{5.5}$$

$$\begin{aligned} more\ or\ less\ small &= 1/1 + 0.8944/2 + 0.7746/3 + 0.6325/4 \\ &\quad +0.4472/5 \end{aligned} \tag{5.6}$$

□

5.3   Fuzzy IF-THEN Rules

In Chapter 1 we mentioned that in fuzzy systems and control, human knowledge is represented in terms of fuzzy IF-THEN rules. A *fuzzy IF-THEN rule* is a conditional statement expressed as

$$IF\ <fuzzy\ proposition>,\ THEN\ <fuzzy\ proposition> \tag{5.7}$$

Therefore, in order to understand fuzzy IF-THEN rules, we first must know what are fuzzy propositions.

5.3.1   Fuzzy Propositions

There are two types of fuzzy propositions: atomic fuzzy propositions, and compound fuzzy propositions. An *atomic fuzzy proposition* is a single statement

$$x\ is\ A \tag{5.8}$$

where x is a linguistic variable, and A is a linguistic value of x (that is, A is a fuzzy set defined in the physical domain of x). A *compound fuzzy proposition* is a composition of atomic fuzzy propositions using the connectives "and," "or," and "not" which represent fuzzy intersection, fuzzy union, and fuzzy complement, respectively. For example, if x represents the speed of the car in Example 5.1, then the following are fuzzy propositions (the first three are atomic fuzzy propositions and the last three are compound fuzzy propositions):

$$x \; is \; S \tag{5.9}$$

$$x \; is \; M \tag{5.10}$$

$$x \; is \; F \tag{5.11}$$

$$x \; is \; S \; or \; x \; is \; not \; M \tag{5.12}$$

$$x \; is \; not \; S \; and \; x \; is \; not \; F \tag{5.13}$$

$$(x \; is \; S \; and \; x \; is \; not \; F) \; or \; x \; is \; M \tag{5.14}$$

where S, M and F denote the fuzzy sets "slow," "medium," and "fast," respectively.

Note that in a compound fuzzy proposition, the atomic fuzzy propositions are independent, that is, the x's in the same proposition of (5.12)-(5.14) can be different variables. Actually, the linguistic variables in a compound fuzzy proposition are in general not the same. For example, let x be the speed of a car and $y = \dot{x}$ be the acceleration of the car, then if we define fuzzy set $large(L)$ for the acceleration, the following is a compound fuzzy proposition

$$x \; is \; F \; and \; y \; is \; L$$

Therefore, *compound fuzzy propositions should be understood as fuzzy relations.* How to determine the membership functions of these fuzzy relations?

- *For connective "and" use fuzzy intersections.* Specifically, let x and y be linguistic variables in the physical domains U and V, and A and B be fuzzy sets in U and V, respectively, then the compound fuzzy proposition

$$x \; is \; A \; and \; y \; is \; B \tag{5.15}$$

is interpreted as the fuzzy relation $A \cap B$ [1] in $U \times V$ with membership function

$$\mu_{A \cap B}(x,y) = t[\mu_A(x), \mu_B(y)] \tag{5.16}$$

where $t : [0,1] \times [0,1] \to [0,1]$ is any t-norm.

- *For connective "or" use fuzzy unions.* Specifically, the compound fuzzy proposition

$$x \; is \; A \; or \; y \; is \; B \tag{5.17}$$

---

[1] Note that in Chapters 2 and 3, A and B are fuzzy sets defined in the same universal set U and $A \cup B$ and $A \cap B$ are fuzzy sets in U; here, $A \cup B$ and $A \cap B$ are fuzzy relations in $U \times V$, where U may or may not equal V.

is interpreted as the fuzzy relation $A \cup B$ in $U \times V$ with membership function

$$\mu_{A \cup B}(x, y) = s[\mu_A(x), \mu_B(y)] \tag{5.18}$$

where $s : [0, 1] \times [0, 1] \to [0, 1]$ is any s-norm.

- *For connective "not" use fuzzy complements.* That is, replace *not A* by $\bar{A}$, which is defined according to the complement operators in Chapter 3.

Example 5.3. The fuzzy proposition (5.14), that is,

$$FP = (x \ is \ S \ and \ x \ is \ not \ F) \ or \ x \ is \ M \tag{5.19}$$

is a fuzzy relation in the product space $[0, V_{max}]^3$ with the membership function

$$\mu_{FP}(x_1, x_2, x_3) = s\{t[\mu_S(x_1), c(\mu_F(x_2))], \mu_M(x_3)\} \tag{5.20}$$

where s, t and c are s-norm, t-norm and fuzzy complement operators, respectively, the fuzzy sets $S = small, M = medium$, and $F = fast$ are defined in Fig. 5.1, and $x_1 = x_2 = x_3 = x$. $\square$

We are now ready to interpret the fuzzy IF-THEN rules in the form of (5.7).

5.3.2 Interpretations of Fuzzy IF-THEN Rules

Because the fuzzy propositions are interpreted as fuzzy relations, the key question remaining is how to interpret the IF-THEN operation. In classical propositional calculus, the expression *IF p THEN q* is written as $p \to q$ with the implication $\to$ regarded as a connective defined by Table 5.1, where p and q are propositional variables whose values are either truth (T) or false (F). From Table 5.1 we see that if both p and q are true or false, then $p \to q$ is true; if p is true and q is false, then $p \to q$ is false; and, if p is false and q is true, then $p \to q$ is true. Hence, $p \to q$ is equivalent to

$$\bar{p} \vee q \tag{5.21}$$

and

$$(p \wedge q) \vee \bar{p} \tag{5.22}$$

in the sense that they share the same truth table (Table 5.1) as $p \to q$, where $\bar{\ }, \vee$ and $\wedge$ represent (classical) logic operations "not," "or," and "and," respectively.

Because fuzzy IF-THEN rules can be viewed as replacing the p and q with fuzzy propositions, we can interpret the fuzzy IF-THEN rules by replacing the $\bar{\ }, \vee$ and $\wedge$ operators in (5.21) and (5.22) with fuzzy complement, fuzzy union, and fuzzy intersection, respectively. Since there are a wide variety of fuzzy complement, fuzzy union, and fuzzy intersection operators, a number of different interpretations of fuzzy IF-THEN rules were proposed in the literature. We list some of them below.

Table 5.1. Truth table for $p \rightarrow q$

p	q	$p \rightarrow q$
T	T	T
T	F	F
F	T	T
F	F	T

In the following, we rewrite (5.7) as $IF \ < FP_1 > \ THEN \ < FP_2 >$ and replace the p and q in (5.21) and (5.22) by FP_1 and FP_2, respectively, where FP_1 and FP_2 are fuzzy propositions. We assume that FP_1 is a fuzzy relation defined in $U = U_1 \times \cdots \times U_n$, FP_2 is a fuzzy relation defined in $V = V_1 \times \cdots \times V_m$, and x and y are linguistic variables (vectors) in U and V, respectively.

- **Dienes-Rescher Implication**: If we replace the logic operators ⁻ and $\vee$ in (5.21) by the basic fuzzy complement (3.1) and the basic fuzzy union (3.2), respectively, then we obtain the so-called *Dienes-Rescher implication*. Specifically, the fuzzy IF-THEN rule $IF \ < FP_1 > \ THEN \ < FP_2 >$ is interpreted as a fuzzy relation Q_D in $U \times V$ with the membership function

$$\mu_{Q_D}(x,y) = max[1 - \mu_{FP_1}(x), \mu_{FP_2}(y)] \tag{5.23}$$

- **Lukasiewicz Implication**: If we use the Yager s-norm (3.10) with $w = 1$ for the $\vee$ and basic fuzzy complement (3.1) for the ⁻ in (5.21), we obtain the *Lukasiewicz implication*. Specifically, the fuzzy IF-THEN rule $IF \ < FP_1 > \ THEN \ < FP_2 >$ is interpreted as a fuzzy relation Q_L in $U \times V$ with the membership function

$$\mu_{Q_L}(x,y) = min[1, 1 - \mu_{FP_1}(x) + \mu_{FP_2}(y)] \tag{5.24}$$

- **Zadeh Implication**: Here the fuzzy IF-THEN rule $IF \ < FP_1 > \ THEN \ < FP_2 >$ is interpreted as a fuzzy relation Q_Z in $U \times V$ with the membership function

$$\mu_{Q_Z}(x,y) = max[min(\mu_{FP_1}(x), \mu_{FP_2}(y)), 1 - \mu_{FP_1}(x)] \tag{5.25}$$

Clearly, (5.25) is obtained from (5.22) by using basic fuzzy complement (3.1), basic fuzzy union (3.2), and basic fuzzy intersection (3.3) for ⁻$\vee$ and $\wedge$, respectively.

- **Gödel Implication**: The Gödel implication is a well-known implication formula in classical logic. By generalizing it to fuzzy propositions, we obtain the following: the fuzzy IF-THEN rule $IF \ < FP_1 > \ THEN \ < FP_2 >$ is interpreted as a fuzzy relation Q_G in $U \times V$ with the membership function

$$\mu_{Q_G}(x,y) = \begin{cases} 1 & if \ \mu_{FP_1}(x) \leq \mu_{FP_2}(y) \\ \mu_{FP_2}(y) & otherwise \end{cases} \tag{5.26}$$

It is interesting to explore the relationship among these implications. The following lemma shows that the Zadeh implication is smaller than the Dienes-Rescher implication, which is smaller than the Lukasiewicz implication.

Lemma 5.1. For all $(x, y) \in U \times V$, the following is true

$$\mu_{Q_Z}(x,y) \leq \mu_{Q_D}(x,y) \leq \mu_{Q_L}(x,y) \tag{5.27}$$

Proof: Since $0 \leq 1 - \mu_{FP_1}(x) \leq 1$ and $0 \leq \mu_{FP_2}(y) \leq 1$, we have $max[1 - \mu_{FP_1}(x), \mu_{FP_2}(y)] \leq 1 - \mu_{FP_1}(x) + \mu_{FP_2}(y)$ and $max[1 - \mu_{FP_1}(x), \mu_{FP_2}(y)] \leq 1$. Hence, $\mu_{Q_D}(x,y) = max[1 - \mu_{FP_1}(x), \mu_{FP_2}(y)] \leq min[1, 1 - \mu_{FP_1}(x) + \mu_{FP_2}(y)] = \mu_{Q_L}(x,y)$. Since $min[\mu_{FP_1}(x), \mu_{FP_2}(y)] \leq \mu_{FP_2}(y)$, we have $max[min(\mu_{FP_1}(x), \mu_{FP_2}(y)), 1 - \mu_{FP_1}(x)] \leq max[\mu_{FP_2}(y), 1 - \mu_{FP_1}(x)]$, which is $\mu_{Q_Z}(x,y) \leq \mu_{Q_D}(x,y)$. $\square$

Conceptually, we can replace the $\bar{\ }, \lor$ and $\land$ in (5.21) and (5.22) by any fuzzy complement, s-norm and t-norm, respectively, to obtain a particular interpretation. So a question arises: Based on what criteria do we choose the combination of fuzzy complements, s-norms, and t-norms? This is an important question and we will discuss it in Chapters 7-10. Another question is: Are (5.21) and (5.22) still "equivalent" to $p \rightarrow q$ when p and q are fuzzy propositions and what does this "equivalent" mean? We now try to answer this question. When p and q are crisp propositions (that is, p and q are either true or false), $p \rightarrow q$ is a *global* implication in the sense that Table 5.1 covers all the possible cases. However, when p and q are fuzzy propositions, $p \rightarrow q$ may only be a *local* implication in the sense that $p \rightarrow q$ has large truth value only when both p and q have large truth values. For example, when we say "IF speed is high, THEN resistance is high," we are concerned only with a local situation in the sense that this rule tells us nothing about the situations when "speed is slow," "speed is medium," etc. Therefore, the fuzzy IF-THEN rule

$$IF \ < FP_1 > \ THEN \ < FP_2 > \tag{5.28}$$

should be interpreted as

$$IF \ < FP_1 > \ THEN \ < FP_2 > \ ELSE \ < NOTHING > \tag{5.29}$$

where $NOTHING$ means that this rule does not exist. In logic terms, it becomes

$$p \rightarrow q = p \land q \tag{5.30}$$

Using *min* or *algebraic product* for the $\land$ in (5.30), we obtain the *Mamdani implications*.

- **Mamdani Implications**: The fuzzy IF-THEN rule (5.28) is interpreted as a fuzzy relation Q_{MM} or Q_{MP} in $U \times V$ with the membership function

$$\mu_{Q_{MM}}(x,y) = min[\mu_{FP_1}(x), \mu_{FP_2}(y)] \qquad (5.31)$$

or

$$\mu_{Q_{MP}}(x,y) = \mu_{FP_1}(x)\mu_{FP_2}(y) \qquad (5.32)$$

Mamdani implications are the most widely used implications in fuzzy systems and fuzzy control. They are supported by the argument that fuzzy IF-THEN rules are local. However, one may not agree with this argument. For example, one may argue that when we say "IF speed is high, THEN resistance is high," we *implicitly* indicate that "IF speed is slow, THEN resistance is low." In this sense, fuzzy IF-THEN rules are nonlocal. This kind of debate indicates that when we represent human knowledge in terms of fuzzy IF-THEN rules, different people have different interpretations. Consequently, different implications are needed to cope with the diversity of interpretations. For example, if the human experts think that their rules are local, then the Mamdani implications should be used; otherwise, the global implications (5.23)-(5.26) should be considered.

We now consider some examples for the computation of Q_D, Q_L, Q_Z, Q_{MM} and Q_{MP}.

Example 5.4. Let x_1 be the speed of a car, x_2 be the acceleration, and y be the force applied to the accelerator. Consider the following fuzzy IF-THEN rule:

$$IF\ x_1\ is\ slow\ and\ x_2\ is\ small,\ THEN\ y\ is\ large \qquad (5.33)$$

where "slow" is the fuzzy set defined in Fig. 5.1, that is,

$$\mu_{slow}(x_1) = \begin{cases} 1 & if\ x_1 \leq 35 \\ \frac{55-x_1}{20} & if\ 35 < x_1 \leq 55 \\ 0 & if\ x_1 > 55 \end{cases} \qquad (5.34)$$

"small" is a fuzzy set in the domain of acceleration with the membership function

$$\mu_{small}(x_2) = \begin{cases} \frac{10-x_2}{10} & if\ 0 \leq x_2 \leq 10 \\ 0 & if\ x_2 > 10 \end{cases} \qquad (5.35)$$

and "large" is a fuzzy set in the domain of force applied to the accelerator with the membership function

$$\mu_{large}(y) = \begin{cases} 0 & if\ y \leq 1 \\ y-1 & if\ 1 \leq y \leq 2 \\ 1 & if\ y > 2 \end{cases} \qquad (5.36)$$

Let the domains of x_1, x_2 and y be $U_1 = [0, 100], U_2 = [0, 30]$, and $V = [0, 3]$, respectively. If we use algebraic product for the t-norm in (5.16), then the fuzzy proposition

$$FP_1 = x_1\ is\ slow\ and\ x_2\ is\ small \qquad (5.37)$$

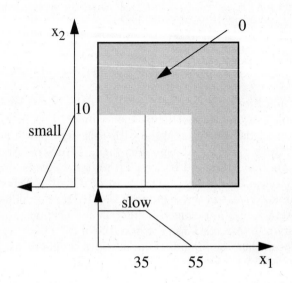

Figure 5.3.   Illustration for how to compute $\mu_{slow}(x_1)\mu_{small}(x_2)$ in Example 5.4.

is a fuzzy relation in $U_1 \times U_2$ with the membership function

$$\mu_{FP_1}(x_1, x_2) = \mu_{slow}(x_1)\mu_{small}(x_2)$$

$$= \begin{cases} 0 & if \ x_1 \geq 55 \ or \ x_2 > 10 \\ \frac{10-x_2}{10} & if \ x_1 \leq 35 \ and \ x_2 \leq 10 \\ \frac{(55-x_1)(10-x_2)}{200} & if \ 35 < x_1 \leq 55 \ and \ x_2 \leq 10 \end{cases} \quad (5.38)$$

Fig. 5.3 illustrates how to compute $\mu_{FP_1}(x_1, x_2)$.

If we use the Dienes-Rescher implication (5.23), then the fuzzy IF-THEN rule (5.33) is interpreted as a fuzzy relation $Q_D(x_1, x_2, y)$ in $U_1 \times U_2 \times V$ with the membership function

$$\mu_{Q_D}(x_1, x_2, y) = max[1 - \mu_{FP_1}(x_1, x_2), \mu_{large}(y)] \quad (5.39)$$

From (5.38) we have

$$1 - \mu_{FP_1}(x_1, x_2) = \begin{cases} 1 & if \quad x_1 \geq 55 \ or \ x_2 > 10 \\ x_2/10 & if \quad x_1 \leq 35 \ and \ x_2 \leq 10 \\ 1 - \frac{(55-x_1)(10-x_2)}{200} & if \quad 35 < x_1 \leq 55 \ and \ x_2 \leq 10 \end{cases} \quad (5.40)$$

To help us combining $1 - \mu_{FP_1}(x_1, x_2)$ of (5.40) with $\mu_{large}(y)$ of (5.36) using the *max* operator, we illustrate in Fig. 5.4 the division of the domains of $1 - \mu_{FP_1}(x_1, x_2)$

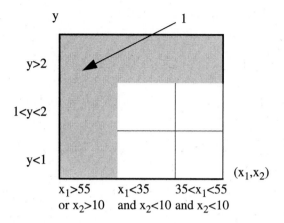

Figure 5.4. Division of the domains of $1 - \mu_{FP_1}(x_1, x_2)$ and $\mu_{large}(y)$ and their combinations for Example 5.4.

and $\mu_{large}(y)$ and their combinations. From Fig. 5.4, we obtain

$$\mu_{Q_D}(x_1, x_2, y) =$$

$$\begin{cases} 1 & if\ x_1 \geq 55\ or\ x_2 > 10\ or\ y > 2 \\ x_2/10 & if\ x_1 \leq 35\ and\ x_2 \leq 10\ and\ y \leq 1 \\ 1 - \frac{(55-x_1)(10-x_2)}{200} & if\ 35 < x_1 \leq 55\ and\ x_2 \leq 10 \\ & and\ y \leq 1 \\ max[y-1, x_2/10] & if\ x_1 \leq 35\ and\ x_2 \leq 10 \\ & and\ 1 < y \leq 2 \\ max[y-1, 1 - \frac{(55-x_1)(10-x_2)}{200}] & if\ 35 < x_1 \leq 55\ and\ x_2 \leq 10 \\ & and\ 1 < y \leq 2 \end{cases} \qquad (5.41)$$

For Lukasiewicz, Zadeh and Mamdani implications, we can use the same procedure to determine the membership functions. □

From Example 5.4 we see that if the membership functions in the atomic fuzzy propositions are not smooth functions (for example, (5.34)-(5.36)), the computation of the final membership functions μ_{Q_D}, μ_{Q_Z}, etc., is cumbersome, although it is straightforward. A way to resolve this complexity is to use a single smooth function to approximate the nonsmooth functions; see the following example.

Example 5.4 (Cont'd). Suppose we use

$$\mu_{slow}(x_1) = \frac{1}{1 + e^{\frac{x_1 - 45}{5}}} \tag{5.42}$$

to approximate the $\mu_{slow}(x_1)$ of (5.34),

$$\mu_{small}(x_2) = \frac{1}{1 + e^{\frac{x_2 - 5}{2}}} \tag{5.43}$$

to approximate the $\mu_{small}(x_2)$ of (5.35), and

$$\mu_{large}(y) = \frac{1}{1 + e^{2(-y+1.25)}} \tag{5.44}$$

to approximate the $\mu_{large}(y)$ of (5.36). Now if we use Mamdani product implication (5.32) and algebraic product for the t-norm in (5.16), then the membership function $\mu_{Q_{MP}}(x_1, x_2, y)$ can be easly computed as

$$\mu_{Q_{MP}}(x_1, x_2, y) = \mu_{slow}(x_1)\mu_{small}(x_2)\mu_{large}(y)$$
$$= \frac{1}{(1 + e^{\frac{x_1 - 45}{5}})(1 + e^{\frac{x_2 - 5}{2}})(1 + e^{2(-y+1.5)})} \tag{5.45}$$

□

Example 5.5. Let $U = \{1, 2, 3, 4\}$ and $V = \{1, 2, 3\}$. Suppose we know that $x \in U$ is somewhat inversely propositional to $y \in V$. To formulate this knowledge, we may use the following fuzzy IF-THEN rule:

$$IF\ x\ is\ large,\ THEN\ y\ is\ small \tag{5.46}$$

where the fuzzy sets "large" and "small" are defined as

$$large = 0/1 + 0.1/2 + 0.5/3 + 1/4 \tag{5.47}$$
$$small = 1/1 + 0.5/2 + 0.1/3 \tag{5.48}$$

If we use the Dienes-Rescher implication (5.23), then the fuzzy IF-THEN rule (5.46) is interpreted as the following fuzzy relation Q_D in $U \times V$:

$$Q_D = 1/(1,1) + 1/(1,2) + 1/(1,3) + 1/(2,1) + 0.9/(2,2) + 0.9/(2,3)$$
$$+ 1/(3,1) + 0.5/(3,2) + 0.5/(3,3) + 1/(4,1) + 0.5/(4,2)$$
$$+ 0.1/(4,3) \tag{5.49}$$

If we use the Lukasiewicz implication (5.24), the rule (5.46) becomes

$$Q_L = 1/(1,1) + 1/(1,2) + 1/(1,3) + 1/(2,1) + 1/(2,2) + 1/(2,3) + 1/(3,1)$$
$$+ 1/(3,2) + 0.6/(3,3) + 1/(4,1) + 0.5/(4,2) + 0.1/(4,3) \tag{5.50}$$

For the Zadeh implication (5.25) and the Gödel implication (5.26), we have

$$Q_Z = 1/(1,1) + 1/(1,2) + 1/(1,3) + 0.9/(2,1) + 0.9/(2,2) + 0.9/(2,3)$$
$$+0.5/(3,1) + 0.5/(3,2) + 0.5/(3,3) + 1/(4,1)$$
$$+0.5/(4,2) + 0.1/(4,3) \tag{5.51}$$

and

$$Q_G = 1/(1,1) + 1/(1,2) + 1/(1,3) + 1/(2,1) + 1/(2,2) + 1/(2,3) + 1/(3,1)$$
$$+1/(3,2) + 0.1/(3,3) + 1/(4,1) + 0.5/(4,2) + 0.1/(4,3) \tag{5.52}$$

Finally, if we use the Mamdani implication (5.31) and (5.32), then the fuzzy IF-THEN rule (5.46) becomes

$$Q_{MM} = 0/(1,1) + 0/(1,2) + 0/(1,3) + 0.1/(2,1) + 0.1/(2,2)$$
$$+0.1/(2,3) + 0.5/(3,1) + 0.5/(3,2) + 0.1/(3,3)$$
$$+1/(4,1) + 0.5/(4,2) + 0.1/(4,3) \tag{5.53}$$

and

$$Q_{MP} = 0/(1,1) + 0/(1,2) + 0/(1,3) + 0.1/(2,1) + 0.05/(2,2)$$
$$+0.01/(2,3) + 0.5/(3,1) + 0.25/(3,2) + 0.05/(3,3)$$
$$+1/(4,1) + 0.5/(4,2) + 0.1/(4,3) \tag{5.54}$$

From (5.49)-(5.52) we see that for the combinations not covered by the rule (5.46), that is, the pairs $(1,1),(1,2)$ and $(1,3)$ (because $\mu_{large}(1) = 0$), Q_D, Q_L, Q_Z and Q_G give full membership value to them, but Q_{MM} and Q_{MP} give zero membership value. This is consistent with our earlier discussion that Dienes-Rescher, Lukasiewicz, Zadeh and Gödel implications are global, whereas Mamdani implications are local. $\square$

5.4   Summary and Further Readings

In this chapter we have demonstrated the following:

- The concept of linguistic variables and the characterization of hedges.

- The concept of fuzzy propositions and fuzzy IF-THEN rules.

- Different interpretations of fuzzy IF-THEN rules, including Dienes-Rescher, Lukasiewicz, Zadeh, Gödel and Mamdani implications.

- Properties and computation of these implications.

Linguistic variables were introduced in Zadeh's seminal paper Zadeh [1973]. This paper is another piece of art and the reader is highly recommended to study it. The comprehensive three-part paper Zadeh [1975] summarized many concepts and principles associated with linguistic variables.

5.5 Exercises

Exercise 5.1. Give three examples of linguistic variables. Combine these linguistic variables into a compound fuzzy proposition and determine its membership function.

Exercise 5.2. Consider some other linguistic hedges than those in Section 5.2 and propose reasonable operations that represent them.

Exercise 5.3. Let Q_L, Q_G, Q_{MM} and Q_{MP} be the fuzzy relations defined in (5.24), (5.26), (5.31), and (5.32), respectively. Show that

$$Q_{MP} \subset Q_{MM} \subset Q_G \subset Q_L \tag{5.55}$$

Exercise 5.4. Use basic fuzzy operators (3.1)-(3.3) for "not," "or," and "and," respectively, and determine the membership functions for the fuzzy propositions (5.12) and (5.13). Plot these membership functions.

Exercise 5.5. Consider the fuzzy IF-THEN rule (5.33) with the fuzzy sets "slow," "small" and "large" defined by (5.42), (5.43) and (5.44), respectively. Use min for the t-norm in (5.16) and compute the fuzzy relations $Q_D, Q_L, Q_Z, Q_G, Q_{MM}$ and Q_{MP}.

Exercise 5.6. Let Q be a fuzzy relation in $U \times U$. Q is called *reflexive* if $\mu_Q(u,u) = 1$ for all $u \in U$. Show that if Q is reflexive, then: (a) $Q \circ Q$ is also reflexive, and (b) $Q \subseteq Q \circ Q$, where $\circ$ denotes $max - min$ composition.

Chapter 6

Fuzzy Logic and Approximate Reasoning

6.1 From Classical Logic to Fuzzy Logic

Logic is the study of methods and principles of reasoning, where *reasoning* means obtaining new propositions from existing propositions. In classical logic, the propositions are required to be either true or false, that is, the truth value of a proposition is either 0 or 1. Fuzzy logic generalizes classical two-value logic by allowing the truth values of a proposition to be any number in the interval $[0,1]$. This generalization allows us to perform *approximate reasoning*, that is, deducing imprecise conclusions (fuzzy propositions) from a collection of imprecise premises (fuzzy propositions). In this chapter, we first review some basic concepts and principles in classical logic and then study their generalization to fuzzy logic.

6.1.1 Short Primer on Classical Logic

In classical logic, the relationships between propositions are usually represented by a truth table. The fundamental truth table for conjunction $\vee$, disjunction $\wedge$, implication $\rightarrow$, equivalence $\leftrightarrow$ and negation $^-$ are collected together in Table 6.1, where the symbols T and F denote truth and false, respectively.

Given n basic propositions $p_1, ..., p_n$, a new proposition can be defined by a function that assigns a particular truth value to the new proposition for each combination of truth values of the given propositions. The new proposition is usually called a *logic function*. Since n propositions can assume 2^n possible combinations of truth values, there are 2^{2^n} possible logic functions defining n propositions. Because 2^{2^n} is a huge number for large n, a key issue in classical logic is to express all the logic functions with only a few basic logic operations; such basic logic operations are called *a complete set of primitives*. The most commonly used complete set of primitives is negation $^-$, conjunction $\vee$, and disjunction $\wedge$. By combining $^-$, $\vee$ and $\wedge$ in appropriate algebraic expressions, referred to as *logic formulas*, we can form any

Table 6.1. Truth table for five operations that are frequently applied to propositions.

p	q	$p \wedge q$	$p \vee q$	$p \to q$	$p \leftrightarrow q$	$\bar{p}$
T	T	T	T	T	T	F
T	F	F	T	F	F	F
F	T	F	T	T	F	T
F	F	F	F	T	T	T

other logic function. Logic formulas are defined recursively as follows:

- The truth values 0 and 1 are logic formulas.

- If p is a proposition, then p and $\bar{p}$ are logic formulas.

- If p and q are logic formulas, then $p \vee q$ and $p \wedge q$ are also logic formulas.

- The only logic formulas are those defined by (a)-(c).

When the proposition represented by a logic formula is always true regardless of the truth values of the basic propositions participating in the formula, it is called a *tautology*; when it is always false, it is called a *contradiction*.

Example 6.1. The following logic formulas are tautologies:

$$(p \to q) \leftrightarrow (\bar{p} \vee q) \tag{6.1}$$

$$(p \to q) \leftrightarrow ((p \wedge q) \vee \bar{p}) \tag{6.2}$$

To prove (6.1) and (6.2), we use the truth table method, that is, we list all the possible values of (6.1) and (6.2) and see whether they are all true. Table 6.2 shows the results, which indicates that (6.1) and (6.2) are tautologies. □

Table 6.2. Proof of $(p \to q) \leftrightarrow (\bar{p} \vee q)$ and $(p \to q) \leftrightarrow ((p \wedge q) \vee \bar{p})$.

p	q	$p \to q$	$\bar{p} \vee q$	$(p \wedge q) \vee \bar{p}$	$(p \to q) \leftrightarrow (\bar{p} \vee q)$	$(p \to q) \leftrightarrow ((p \wedge q) \vee \bar{p})$
T	T	T	T	T	T	T
T	F	F	F	F	T	T
F	T	T	T	T	T	T
F	F	T	T	T	T	T

Various forms of tautologies can be used for making deductive inferences. They are referred to as *inference rules*. The three most commonly used inference rules are:

- **Modus Ponens**: This inference rule states that given two propositions p and $p \to q$ (called the *premises*), the truth of the proposition q (called the *conclusion*) should be inferred. Symbolically, it is represented as

$$(p \wedge (p \to q)) \to q \qquad (6.3)$$

A more intuitive representation of modus ponens is

$$\begin{array}{ll} Premise\ 1: & x\ is\ A \\ Premise\ 2: & IF\ x\ is\ A\ THEN\ y\ is\ B \\ Conclusion: & y\ is\ B \end{array}$$

- **Modus Tollens**: This inference rule states that given two propositions $\bar{q}$ and $p \to q$, the truth of the proposition $\bar{p}$ should be inferred. Symbolically, it becomes

$$(\bar{q} \wedge (p \to q)) \to \bar{p} \qquad (6.4)$$

A more intuitive representation of modus tollens is

$$\begin{array}{ll} Premise\ 1: & y\ is\ not\ B \\ Premise\ 2: & IF\ x\ is\ A\ THEN\ y\ is\ B \\ Conclusion: & x\ is\ not\ A \end{array}$$

- **Hypothetical Syllogism**: This inference rule states that given two propositions $p \to q$ and $q \to r$, the truth of the proposition $p \to r$ should be inferred. Symbolically, we have

$$((p \to q) \wedge (q \to r)) \to (p \to r) \qquad (6.5)$$

A more intuitive representation of it is

$$\begin{array}{ll} Premise\ 1: & IF\ x\ is\ A\ THEN\ y\ is\ B \\ Premise\ 2: & IF\ y\ is\ B\ THEN\ z\ is\ C \\ Conclusion: & IF\ x\ is\ A\ THEN\ z\ is\ C \end{array}$$

6.1.2 Basic Principles in Fuzzy Logic

In fuzzy logic, the propositions are fuzzy propositions that, as defined in Chapter 5, are represented by fuzzy sets. The ultimate goal of fuzzy logic is to provide foundations for approximate reasoning with imprecise propositions using fuzzy set theory as the principal tool. To achieve this goal, the so-called *generalized modus ponens, generalized modus tollens,* and *generalized hypothetical syllogism* were proposed. They are the fundamental principles in fuzzy logic.

- **Generalized Modus Ponens**: This inference rule states that given two fuzzy propositions x *is* A' and *IF* x *is* A *THEN* y *is* B, we should infer a new fuzzy proposition y *is* B' such that the closer the A' to A, the closer the B' to B, where A, A', B and B' are fuzzy sets; that is,

$$Premise\ 1: \quad x\ is\ A'$$
$$Premise\ 2: \quad IF\ x\ is\ A\ THEN\ y\ is\ B$$
$$Conclusion: \quad y\ is\ B'$$

Table 6.3 shows the intuitive criteria relating Premise 1 and the Conclusion in generalized modus ponens. We note that if a causal relation between "x is A" and "y is B" is not strong in Premise 2, the satisfaction of criterion p3 and criterion p5 is allowed. Criterion p7 is interpreted as: "IF x is A THEN y is B, ELSE y is not B." Although this relation is not valid in classical logic, we often make such an interpretation in everyday reasoning.

Table 6.3. Intuitive criteria relating Premise 1 and the Conclusion for given Premise 2 in generalized modus ponens.

	x is A' (Premise 1)	y is B' (Conclusion)
criterion p1	x is A	y is B
criterion p2	x is very A	y is very B
criterion p3	x is very A	y is B
criterion p4	x is more or less A	y is more or less B
criterion p5	x is more or less A	y is B
criterion p6	x is not A	y is unknown
criterion p7	x is not A	y is not B

- **Generalized Modus Tollens**: This inference rule states that given two fuzzy propositions y *is* B' and *IF* x *is* A *THEN* y *is* B, we should infer a new fuzzy proposition x *is* A' such that the more difference between B' and B, the more difference between A' and A, where A', A, B' and B are fuzzy sets; that is,

$$Premise\ 1: \quad y\ is\ B'$$
$$Premise\ 2: \quad IF\ x\ is\ A\ THEN\ y\ is\ B$$
$$Conclusion: \quad x\ is\ A'$$

Table 6.4 shows some intuitive criteria relating Premise 1 and the Conclusion in generalized modus tollens. Similar to the criteria in Table 6.3, some criteria in Table 6.4 are not true in classical logic, but we use them approximately in our daily life.

Table 6.4. Intuitive criteria relating Premise 1 and the Conclusion for given Premise 2 in generalized modus tollens.

	y is B' (Premise 1)	x is A' (Conclusion)
criterion t1	y is not B	x is not A
criterion t2	y is not very B	x is not very A
criterion t3	y is not more or less B	x is not more or less A
criterion t4	y is B	x is unknown
criterion t5	y is B	x is A

- **Generalized Hypothetical Syllogism**: This inference rule states that given two fuzzy propositions *IF x is A THEN y is B* and *IF y is B' THEN z is C*, we could infer a new fuzzy proposition *IF x is A THEN z is C'* such that the closer the B to B', the closer the C' to C, where A, B, B', C and C' are fuzzy sets; that is,

$$Premise\ 1: \quad IF\ x\ is\ A\ THEN\ y\ is\ B$$
$$Premise\ 2: \quad IF\ y\ is\ B'\ THEN\ z\ is\ C$$
$$Conclusion: \quad IF\ x\ is\ A\ THEN\ z\ is\ C'$$

Table 6.5 shows some intuitive criteria relating *y is B'* with *z is C'* in the generalized hypothetical syllogism. Criteria s2 is obtained from the following intuition: To match the B in Premise 1 with the $B' = very\ B$ in Premise 2, we may change Premise 1 to *IF x is very A THEN y is very B*, so we have *IF x is very A THEN z is C*. By applying the hedge *more or less* to cancel the *very*, we have *IF x is A THEN z is more or less C*, which is criterion s2. Other criteria can be justified in a similar manner.

Table 6.5. Intuitive criteria relating *y is B'* in Premise 2 and *z is C'* in the Conclusion in generalized hypothetical syllogism.

	y is B' (Premise 2)	z is C' (Conclusion)
criterion s1	y is B	z is C
criterion s2	y is very B	z is more or less C
criterion s3	y is very B	z is C
criterion s4	y is more or less B	z is very C
criterion s5	y is more or less B	z is C
criterion s6	y is not B	z is unknown
criterion s7	y is not B	z is not C

We call the criteria in Tables 6.3-6.5 *intuitive* criteria because they are not necessarily true for a particular choice of fuzzy sets; this is what *approximate reasoning* means. Although these criteria are not absolutely correct, they do make some sense. They should be viewed as guidelines (or soft constraints) in designing specific inferences.

We have now shown the basic ideas of three fundamental principles in fuzzy logic: generalized modus ponens, generalized modus tollens, and generalized hypothetical syllogism. The next question is how to determine the membership functions of the fuzzy propositions in the conclusions given those in the premises. The *compositional rule of inference* was proposed to answer this question.

6.2   The Compositional Rule of Inference

The compositional rule of inference is a generalization of the following procedure (referring to Fig. 6.1): suppose we have a curve $y = f(x)$ from $x \in U$ to $y \in V$ and are given $x = a$, then from $x = a$ and $y = f(x)$ we can infer $y = b = f(a)$.

Let us generalize the above procedure by assuming that a is an interval and $f(x)$ is an interval-valued function as shown in Fig. 6.2. To find the interval b which is inferred from a and $f(x)$, we first construct a cylindrical set a_E with base a and find its intersection I with the interval-valued curve. Then we project I on V yielding the interval b.

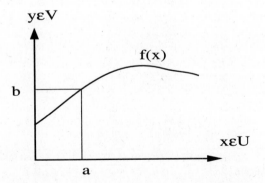

Figure 6.1. Inferring $y = b$ from $x = a$ and $y = f(x)$.

Going one step further in our chain of generalization, assume the A' is a fuzzy set in U and Q is a fuzzy relation in $U \times V$. Again, forming a cylindrical extension

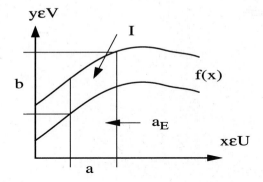

Figure 6.2. Inferring interval b from interval a and interval-valued function $f(x)$.

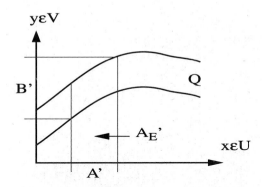

Figure 6.3. Inferring fuzzy set B' from fuzzy set A' and fuzzy relation Q.

A'_E of A' and intersecting it with the fuzzy relation Q (see Fig. 6.3), we obtain a fuzzy set $A'_E \cap Q$ which is analog of the intersection I in Fig. 6.2. Then, projecting $A'_E \cap Q$ on the y-axis, we obtain the fuzzy set B'.

More specifically, given $\mu_{A'}(x)$ and $\mu_Q(x, y)$, we have

$$\mu_{A'_E}(x, y) = \mu_{A'}(x) \tag{6.6}$$

(see (4.18)) and, consequently,

$$\mu_{A'_E \cap Q}(x,y) = t[\mu_{A'_E}(x,y), \mu_Q(x,y)]$$
$$= t[\mu_{A'}(x), \mu_Q(x,y)] \tag{6.7}$$

Finally, from (4.12) we obtain B', the projection of $A'_E \cap Q$ on V, as

$$\mu_{B'}(y) = \sup_{x \in U} t[\mu_{A'}(x), \mu_Q(x,y)] \tag{6.8}$$

(6.8) is called the *compositional rule of inference*. In the literature, the symbol "$\star$" is often used for the t-norm operator, so (6.8) is also written as

$$\mu_{B'}(y) = \sup_{x \in U}[\mu_{A'}(x) \star \mu_Q(x,y)] \tag{6.9}$$

The compositional rule of inference is also called the *sup-star composition*.

In Chapter 5, we learned that a fuzzy IF-THEN rule, for example, IF x is A THEN y is B, is interpreted as a fuzzy relation in the Cartesian product of the domains of x and y. Different implication principles give different fuzzy relations; see (5.23)-(5.26), (5.31), and (5.32). Therefore, the Premise 2s in the generalized modus ponens and generalized modus tollens can be viewed as the fuzzy relation Q in (6.9). For generalized hypothetical syllogism, we see that it is simply the composition of two fuzzy relations, so we can use the composition (4.28) to determine the conclusion. In summary, we obtain the detailed formulas for computing the conclusions in generalized modus ponens, generalized modus tollens, and generalized hypothetical syllogism, as follows:

- **Generalized Modus Ponens**: Given fuzzy set A' (which represents the premise *x is A'*) and fuzzy relation $A \to B$ in $U \times V$ (which represents the premise *IF x is A THEN y is B*), a fuzzy set B' in V (which represents the conclusion *y is B'*) is inferred as

$$\mu_{B'}(y) = \sup_{x \in U} t[\mu_{A'}(x), \mu_{A \to B}(x,y)] \tag{6.10}$$

- **Generalized Modus Tollens**: Given fuzzy set B' (which represents the premise *y is B'*) and fuzzy relation $A \to B$ in $U \times V$ (which represents the premise *IF x is A THEN y is B*), a fuzzy set A' in U (which represents the conclusion *x is A'*) is inferred as

$$\mu_{A'}(x) = \sup_{y \in V} t[\mu_{B'}(y), \mu_{A \to B}(x,y)] \tag{6.11}$$

- **Generalized Hypothetical Syllogism**: Given fuzzy relation $A \to B$ in $U \times V$ (which represents the premise *IF x is A THEN y is B*) and fuzzy relation $B' \to C$ in $V \times W$ (which represents the premise *IF y is B' THEN z* *is C*), a fuzzy relation $A \to C'$ in $U \times W$ (which represents the conclusion *IF x is A THEN z is C'*) is inferred as

$$\mu_{A \to C'}(x, z) = \sup_{y \in V} t[\mu_{A \to B}(x, y), \mu_{B' \to C}(y, z)] \tag{6.12}$$

Using different t-norms in (6.10)-(6.12) and different implication rules (5.23)-(5.26), (5.31) and (5.32), we obtain a diversity of results. These results show the properties of the implication rules. We now study some of these properties.

6.3   Properties of the Implication Rules

In this section, we apply specific implication rules and t-norms to (6.10)-(6.12) and see what the $\mu_{B'}(y), \mu_{A'}(x)$ and $\mu_{A \to C'}(x, z)$ look like for some typical cases of A' and B'. We consider the generalized modus ponens, generalized modus tollens, and generalized hypothetical syllogism in sequal.

6.3.1   Generalized Modus Ponens

Example 6.2. Suppose we use *min* for the t-norm and Mamdani's product implication (5.32) for the $\mu_{A \to B}(x, y)$ in the generalized modus ponens (6.10). Consider four cases of A': (a) $A' = A$, (b) $A' = very\ A$, (c) $A' = more\ or\ less\ A$, and (d) $A = \bar{A}$. Our task is to determine the corresponding B'. We assume that $\sup_{x \in U}[\mu_A(x)] = 1$ (the fuzzy set A is normal). If $A' = A$, we have

$$\mu_{B'}(y) = \sup_{x \in U}\{min[\mu_A(x), \mu_A(x)\mu_B(y)]\}$$
$$= \sup_{x \in U}[\mu_A(x)\mu_B(y)]$$
$$= \mu_B(y) \tag{6.13}$$

If $A' = very\ A$, we have

$$\mu_{B'}(y) = \sup_{x \in U}\{min[\mu_A^2(x), \mu_A(x)\mu_B(y)]\} \tag{6.14}$$

Since $\sup_{x \in U}[\mu_A(x)] = 1$ and x can take any values in U, for any $y \in V$ there exists $x \in U$ such that $\mu_A(x) \geq \mu_B(y)$. Thus (6.14) can be simplified to

$$\mu_{B'}(y) = \sup_{x \in U}[\mu_A(x)\mu_B(y)]$$
$$= \mu_B(y) \tag{6.15}$$

If $A' = more\ or\ less\ A$, then from $\mu_A^{1/2}(x) \geq \mu_A(x) \geq \mu_A(x)\mu_B(x)$, we have

$$\mu_{B'}(y) = \sup_{x \in U}\{min[\mu_A^{1/2}(x), \mu_A(x)\mu_B(y)]\}$$
$$= \mu_B(y) \tag{6.16}$$

Finally, if $A' = \bar{A}$, we obtain

$$\mu_{B'}(y) = \sup_{x \in U}\{min[1 - \mu_A(x), \mu_A(x)\mu_B(y)]\} \qquad (6.17)$$

Since for fixed $y \in V$, $\mu_A(x)\mu_B(y)$ is an increasing function with $\mu_A(x)$ while $1 - \mu_A(x)$ is a decreasing function with $\mu_A(x)$, the $\sup_{x \in U} min$ in (6.17) is achieved when $1 - \mu_A(x) = \mu_A(x)\mu_B(y)$, that is, when $\mu_A(x) = \frac{1}{1+\mu_B(y)}$. Hence,

$$\mu_{B'}(y) = \frac{\mu_B(y)}{1 + \mu_B(y)} \qquad (6.18)$$

From (6.13), (6.15), (6.16), (6.18), and Table 6.3 we see that the particular generalized modus ponens considered in this example satisfies critera p1, p3 and p5, but does not satisfy criteria p2, p4, p6, and p7. □

Example 6.3. In this example, we still use min for the t-norm but use Zadeh implication (5.25) for the $\mu_{A \to B}(x, y)$ in the generalized modus ponens (6.10). Again, we consider the four typical cases of A' in Example 6.2 and assume that $\sup_{x \in U}[\mu_A(x)] = 1$.

(a) For $A' = A$, we have

$$\mu_{B'}(y) = \sup_{x \in U} min\{\mu_A(x), max[min(\mu_A(x), \mu_B(y)), 1 - \mu_A(x)]\} \qquad (6.19)$$

Since $\sup_{x \in U} \mu_A(x) = 1$, the $\sup_{x \in U} min$ in (6.19) is achieved at the particular $x_0 \in U$ when

$$\mu_A(x_0) = max[min(\mu_A(x_0), \mu_B(y)), 1 - \mu_A(x_0)] \qquad (6.20)$$

If $\mu_A(x_0) < \mu_B(y)$, then (6.20) becomes

$$\mu_A(x_0) = max[\mu_A(x_0), 1 - \mu_A(x_0)] \qquad (6.21)$$

which is true when $\mu_A(x_0) \geq 0.5$; thus from (6.19) and (6.20) we have $\mu_{B'}(y) = \mu_A(x_0)$. Since $\sup_{x \in U}[\mu_A(x)] = 1$, it must be true that $\mu_A(x_0) = 1$, but this leads to $\mu_B(y) > \mu_A(x_0) = 1$, which is impossible. Thus, we cannot have $\mu_A(x_0) < \mu_B(y)$. Now consider the only possible case $\mu_A(x_0) \geq \mu_B(y)$. In this case, (6.20) becomes

$$\mu_A(x_0) = max[\mu_B(y), 1 - \mu_A(x_0)] \qquad (6.22)$$

If $\mu_B(y) < 1 - \mu_A(x_0)$, then $\mu_A(x_0) = 1 - \mu_A(x_0)$, which is true when $\mu_A(x_0) = 0.5$. If $\mu_B(y) \geq 1 - \mu_A(x_0)$, then from (6.22) we have $\mu_A(x_0) = \mu_B(y) \geq 0.5$. Hence, $\mu_A(x_0) = max[0.5, \mu_B(y)]$ and we obtain

$$\mu_{B'}(y) = \mu_A(x_0) = max[0.5, \mu_B(y)] \qquad (6.23)$$

(b) For $A' = very\ A$, we have

$$\mu_{B'}(y) = \sup_{x \in U} min\{\mu_A^2(x), max[min(\mu_A(x), \mu_B(y)), 1 - \mu_A(x)]\} \qquad (6.24)$$

Similar to the $A' = A$ case, the $\sup_{x \in U} min$ is achieved at $x_0 \in U$ when

$$\mu_A^2(x_0) = max[min(\mu_A(x_0), \mu_B(y)), 1 - \mu_A(x_0)] \qquad (6.25)$$

If $\mu_A(x_0) < \mu_B(y)$, then

$$\mu_A^2(x_0) = max[\mu_A(x_0), 1 - \mu_A(x_0)] \qquad (6.26)$$

which is true only when $\mu_A(x_0) = 1$, but this leads to the contradiction $\mu_B(y) > 1$. Thus $\mu_A(x_0) \geq \mu_B(y)$ is the only possible case. If $\mu_A(x_0) \geq \mu_B(y)$, then (6.25) becomes

$$\mu_A^2(x_0) = max[\mu_B(y), 1 - \mu_A(x_0)] \qquad (6.27)$$

If $\mu_B(y) < 1 - \mu_A(x_0)$, then $\mu_A^2(x_0) = 1 - \mu_A(x_0)$, which is true when $\mu_A(x_0) = \frac{\sqrt{5}-1}{2}$. Hence, if $\mu_B(y) < 1 - \mu_A(x_0) = \frac{3-\sqrt{5}}{2}$, we have $\mu_{B'}(y) = \mu_A^2(x_0) = \frac{3-\sqrt{5}}{2}$. If $\mu_B(y) \geq 1 - \mu_A(x_0)$, we have $\mu_B(y) = \mu_A^2(x_0) = \mu_B(y) \geq \frac{3-\sqrt{5}}{2}$. In summary, we obtain

$$\mu_{B'}(y) = \mu_A^2(x_0) = max[\frac{3 - \sqrt{5}}{2}, \mu_B(y)] \qquad (6.28)$$

(c) If $A' = $ *more or less* A, we have

$$\mu_{B'}(y) = \sup_{x \in U} min\{\mu_A^{1/2}(x), max[min(\mu_A(x), \mu_B(y)), 1 - \mu_A(x)]\} \qquad (6.29)$$

where the $\sup_{x \in U} min$ is achieved at $x_0 \in U$ when

$$\mu_A^{1/2}(x_0) = max[min(\mu_A(x_0), \mu_B(y)), 1 - \mu_A(x_0)] \qquad (6.30)$$

Similar to the $A' = $ *very* A case, we can show that $\mu_A(x_0) < \mu_B(y)$ is impossible. For $\mu_A(x_0) \geq \mu_B(y)$, we have

$$\mu_A^{1/2}(x_0) = max[\mu_B(y), 1 - \mu_A(x_0)] \qquad (6.31)$$

If $\mu_B(y) < 1 - \mu_A(x_0)$, then $\mu_A^{1/2}(x_0) = 1 - \mu_A(x_0)$, which is true when $\mu_A(x_0) = \frac{3-\sqrt{5}}{2}$. Thus, if $\mu_B(y) < 1 - \mu_A(x_0) = \frac{\sqrt{5}-1}{2}$, we have $\mu_{B'}(y) = \mu_A^{1/2}(x_0) = \frac{\sqrt{5}-1}{2}$. If $\mu_B(y) \geq 1 - \mu_A(x_0)$, we have $\mu_{B'}(y) = \mu_A^{1/2}(x_0) = \mu_B(y) \geq \frac{\sqrt{5}-1}{2}$. To summarize, we obtain

$$\mu_{B'}(y) = \mu_A^{1/2}(x_0) = max[\frac{\sqrt{5} - 1}{2}, \mu_B(y)] \qquad (6.32)$$

(d) Finally, when $A' = \bar{A}$, we have

$$\mu_{B'}(y) = \sup_{x \in U} min\{1 - \mu_A(x), max[min(\mu_A(x), \mu_B(y)), 1 - \mu_A(x)]\} \qquad (6.33)$$

By inspecting (6.33) we see that if we choose $x_0 \in U$ such that $\mu_A(x_0) = 0$, then $1 - \mu_A(x_0) = 1$ and $max[min(\mu_A(x), \mu_B(y)), 1 - \mu_A(x)] = 1$, thus the $\sup_{x \in U} min$ is achieved at $x = x_0$. Hence, in this case we have

$$\mu_{B'}(y) = 1 \tag{6.34}$$

From (6.23), (6.28), (6.32), and (6.34), we see that for all the criteria in Table 6.3, only criterion p6 is satisfied. (This approximate reasoning is truely *approximate*!) □

6.3.2 Generalized Modus Tollens

Example 6.4. Similar to Example 6.2, we use *min* for the t-norm and Mamdani's product implication (5.32) for the $\mu_{A \to B}(x, y)$ in the generalized modus tollens (6.11). Consider four cases of B': (a) $B' = \bar{B}$, (b) $B' = not\ very\ B$, (c) $B' = not\ more\ or\ less\ B$, and (d) $B' = B$. We assume that $\sup_{y \in V}[\mu_B(y)] = 1$. If $B' = \bar{B}$, we have from (6.11) that

$$\mu_{A'}(x) = \sup_{y \in V} min[1 - \mu_B(y), \mu_A(x)\mu_B(y)] \tag{6.35}$$

The $\sup_{y \in V} min$ is achieved at $y_0 \in V$ when $1 - \mu_B(y_0) = \mu_A(x)\mu_B(y_0)$, which implies $\mu_B(y_0) = \frac{1}{1 + \mu_A(x)}$, hence,

$$\mu_{A'}(x) = 1 - \mu_B(y_0) = \frac{\mu_A(x)}{1 + \mu_A(x)} \tag{6.36}$$

If $B' = not\ very\ B$, then

$$\mu_{A'}(x) = \sup_{y \in V} min[1 - \mu_B^2(y), \mu_A(x)\mu_B(y)] \tag{6.37}$$

where the $\sup_{y \in V} min$ is achieved at $y_0 \in V$ when $1 - \mu_B^2(y_0) = \mu_A(x)\mu_B(y_0)$, which gives $\mu_B(y_0) = \frac{\sqrt{\mu_A^2(x) + 4} - \mu_A(x)}{2}$. Hence,

$$\mu_{A'}(x) = \mu_A(x)\mu_B(y_0) = \frac{\mu_A(x)\sqrt{\mu_A^2(x) + 4} - \mu_A^2(x)}{2} \tag{6.38}$$

If $B' = not\ more\ or\ less\ B$, we have

$$\mu_{A'}(x) = \sup_{y \in V} min[1 - \mu_B^{1/2}(y), \mu_A(x)\mu_B(y)] \tag{6.39}$$

Again, the $\sup_{y \in V} min$ is achieved at $y_0 \in V$ when $1 - \mu_B^{1/2}(y_0) = \mu_A(x)\mu_B(y_0)$, which gives $\mu_B(y_0) = \frac{1 + 2\mu_A(x) - \sqrt{\mu_A^2(x) + 1}}{2\mu_A^2(x)}$. Hence,

$$\mu_{A'}(x) = \mu_A(x)\mu_B(y_0) = \frac{1 + 2\mu_A(x) - \sqrt{\mu_A^2(x) + 1}}{2\mu_A(x)} \tag{6.40}$$

Finally, when $B' = B$, we have

$$\mu_{A'}(x) = \sup_{y \in V} min[\mu_B(y), \mu_A(x)\mu_B(y)]$$

$$= \sup_{y \in V} \mu_A(x)\mu_B(y)$$

$$= \mu_A(x) \tag{6.41}$$

From (6.36), (6.38), (6.40) and (6.41) we see that among the seven intuitive criteria in Table 6.4 only criterion t5 is satisfied. □

6.3.3 Generalized Hypothetical Syllogism

Example 6.5. Similar to Examples 6.2 and 6.4, we use min for the t-norm and Mamdani product implication for the $\mu_{A \to B}(x, y)$ and $\mu_{B' \to C}(y, z)$ in the generalized hypothetical syllogism (6.12). We assume $\sup_{y \in V}[\mu_B(y)] = 1$ and consider four typical cases of B': (a) $B' = B$, (b) $B' = very\ B$, (c) $B' = more\ or\ less\ B$, and (d) $B' = \bar{B}$. If $B' = B$, we have from (6.12) that

$$\mu_{A \to C'}(x, z) = \sup_{y \in V} min[\mu_A(x)\mu_B(y), \mu_B(y)\mu_C(z)]$$

$$= (\sup_{y \in V} \mu_B(y))min[\mu_A(x), \mu_C(z)]$$

$$= min[\mu_A(x), \mu_C(z)] \tag{6.42}$$

If $B' = very\ B$, we have

$$\mu_{A \to C'}(x, z) = \sup_{y \in V} min[\mu_A(x)\mu_B(y), \mu_B^2(y)\mu_C(z)] \tag{6.43}$$

If $\mu_A(x) > \mu_C(z)$, then it is always true that $\mu_A(x)\mu_B(y) > \mu_B^2(y)\mu_C(z)$, thus, $\mu_{A \to C'}(x, z) = \sup_{y \in V} \mu_B^2(y)\mu_C(z) = \mu_C(z)$. If $\mu_A(x) \leq \mu_C(z)$, then the $\sup_{y \in V} min$ is achieved at $y_0 \in V$, when $\mu_A(x)\mu_B(y_0) = \mu_B^2(y_0)\mu_C(z)$, which gives $\mu_B(y_0) = \frac{\mu_A(x)}{\mu_C(z)}$; hence, in this case $\mu_{A \to C'}(x, z) = \mu_A(x)\mu_B(y_0) = \frac{\mu_A^2(x)}{\mu_C(z)}$. In summary, we obtain

$$\mu_{A \to C'}(x, z) = \begin{cases} \mu_C(z) & if \quad \mu_C(z) < \mu_A(x) \\ \frac{\mu_A^2(x)}{\mu_C(z)} & if \quad \mu_C(z) \geq \mu_A(x) \end{cases} \tag{6.44}$$

If $B' = more\ or\ less\ B$, then using the same method as for the $B' = very\ B$ case, we have

$$\mu_{A \to C'}(x, z) = \begin{cases} \mu_A(x) & if \quad \mu_A(x) < \mu_C(z) \\ \frac{\mu_C^2(z)}{\mu_A(x)} & if \quad \mu_A(x) \geq \mu_C(z) \end{cases} \tag{6.45}$$

Finally, when $B' = \bar{B}$, we have

$$\mu_{A \to C'}(x, z) = \sup_{y \in V} min[\mu_A(x)\mu_B(y), (1 - \mu_B(y))\mu_C(z)] \tag{6.46}$$

where the $\sup_{y \in V} min$ is achieved at $y_0 \in V$ when $\mu_A(x)\mu_B(y_0) = (1-\mu_B(y_0))\mu_C(z)$, that is, when $\mu_B(y_0) = \frac{\mu_C(z)}{\mu_A(x)+\mu_C(z)}$. Hence,

$$\mu_{A \to C'}(x, z) = \frac{\mu_A(x)\mu_C(z)}{\mu_A(x) + \mu_C(z)} \tag{6.47}$$

$\square$

6.4 Summary and Further Readings

In this chapter we have demonstrated the following:

- Using truth tables to prove the equivalence of propositions.

- Basic inference rules (Modus Ponens, Modus Tollens, and Hypothetical Syllogism) and their generalizations to fuzzy propositions (Generalized Modus Ponens, Generalized Modus Tollens, and Generalized Hypothetical Syllogism).

- The idea and applications of the compositional rule of inference.

- Determining the resulting membership functions from different implication rules and typical cases of premises.

A comprehensive treatment of many-valued logic was prepared by Rescher [1969]. The generalizations of classical logic principles to fuzzy logic were proposed in Zadeh [1973], Zadeh [1975] and other papers of Zadeh in the 1970s. The compositional rule of inference also can be found in these papers of Zadeh.

6.5 Exercises

Exercise 6.1. Use the truth table method to prove that the following are tautologies: (a) modus ponens (6.3), (b) modus tollens (6.4), and (c) hypothetical syllogism (6.5).

Exercise 6.2. Let $U = \{x_1, x_2, x_3\}$ and $V = \{y_1, y_2\}$, and assume that a fuzzy IF-THEN rule "IF x is A, THEN y is B" is given, where $A = .5/x_1 + 1/x_2 + .6/x_3$ and $B = 1/y_1 + .4/y_2$. Then, given a fact "x is A'," where $A' = .6/x_1 + .9/x_2 + .7/x_3$, use the generalized modus ponens (6.10) to derive a conclusion in the form "y is B'," where the fuzzy relation $A \to B$ is interpreted using:

(a) Dienes-Rescher implication (5.23),

(b) Lukasiewicz implication (5.24),

(c) Zadeh implication (5.25), and

(d) Mamdani Product implication (5.32).

Exercise 6.3. Repeat Exercise 6.2 with $A = .6/x_1 + 1/x_2 + .9/x_3$, $B = .6/y_1 + 1/y_2$, and $A' = .5/x_1 + .9/x_2 + 1/x_3$.

Exercise 6.4. Let U, V, A, and B be the same as in Exercise 6.2. Now given a fact "y is B'," where $B' = .9/y_1 + .7/y_2$, use the generalized modus tollens (6.11) to derive a conclusion "x is A'," where the fuzzy relation $A \rightarrow B$ is interpreted using:

(a) Lukasiewicz implication (5.24), and

(b) Mamdani Product implication (5.32).

Exercise 6.5. Use *min* for the t-norm and Lukasiewicz implication (5.24) for the $\mu_{A \rightarrow B}(x, y)$ in the generalized modus ponens (6.10), and determine the membership function $\mu_{B'}(y)$ in terms of $\mu_B(y)$ for: (a) $A' = A$, (b) $A' = very\ A$, (c) $A' = more\ or\ less\ A$, and (d) $A' = \bar{A}$.

Exercise 6.6. Use *min* for the t-norm and Dienes-Rescher implication (5.23) for the $\mu_{A \rightarrow B}(x, y)$ in the generalized modus ponens (6.10), and determine the membership function $\mu_{B'}(y)$ in terms of $\mu_B(y)$ for: (a) $A' = A$, (b) $A' = very\ A$, (c) $A' = more\ or\ less\ A$, and (d) $A' = \bar{A}$.

Exercise 6.7. With *min* as the t-norm and Mamdani minimum implication (5.31) for the $\mu_{A \rightarrow B}(x, y)$ in the generalized modus tollens (6.11), determine the membership function $\mu_{A'}(x)$ in terms of $\mu_A(x)$ for: (a) $B' = \bar{B}$, (b) $B' = not\ very\ B$, (c) $B' = not\ more\ or\ less\ B$, and (d) $B' = B$.

Exercise 6.8. Consider a fuzzy logic based on the standard operation ($min, max, 1 - a$). For any two arbitrary propositions, A and B, in the logic, assume that we require that the equality

$$\overline{A \wedge \bar{B}} = B \vee (\bar{A} \wedge \bar{B}) \tag{6.48}$$

holds. Imposing such requirement means that pairs of truth values of A and B become restricted to a subset of $[0, 1]^2$. Show exactly how they are restricted.

Part II

Fuzzy Systems and Their Properties

We learned in Chapter 1 that a fuzzy system consists of four components: fuzzy rule base, fuzzy inference engine, fuzzifier and defuzzifier, as shown in Fig. 1.5. In this part (Chapters 7-11), we will study each of the four components in detail. We will see how the fuzzy mathematical and logic principles we learned in Part I are used in the fuzzy systems. We will derive the compact mathematical formulas for different types of fuzzy systems and study their approximation properties.

In Chapter 7, we will analyze the structure of fuzzy rule base and propose a number of specific fuzzy inference engines. In Chapter 8, a number of fuzzifiers and defuzzifiers will be proposed and analyzed in detail. In Chapter 9, the fuzzy inference engines, fuzzifiers, and defuzzifiers proposed in Chapters 7 and 8 will be combined to obtain some specific fuzzy systems that will be proven to have the universal approximation property. In Chapters 10 and 11, the approximation accuracy of fuzzy systems will be studied in detail and we will show how to design fuzzy systems to achieve any specified accuracy.

Chapter 7

Fuzzy Rule Base and Fuzzy Inference Engine

Consider the fuzzy system shown in Fig. 1.5, where $U = U_1 \times U_2 \times \cdots \times U_n \subset R^n$ and $V \subset R$. We consider only the multi-input-single-output case, because a multi-output system can always be decomposed into a collection of single-output systems. For example, if we are asked to design a 4-input-3-output fuzzy system, we can first design three 4-input-1-output fuzzy systems separately and then put them together as in Fig. 7.1.

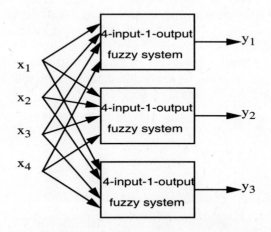

Figure 7.1. A multi-input-multi-output fuzzy system can be decomposed into a collection of multi-input-single-output fuzzy systems.

In this chapter, we will study the details inside the fuzzy rule base and fuzzy inference engine; fuzzifiers and defuzzifiers will be studied in the next chapter.

7.1 Fuzzy Rule Base

7.1.1 Structure of Fuzzy Rule Base

A *fuzzy rule base* consists of a set of fuzzy IF-THEN rules. It is the heart of the fuzzy
system in the sense that all other components are used to implement these rules in
a reasonable and efficient manner. Specifically, the fuzzy rule base comprises the
following fuzzy IF-THEN rules:

$$Ru^{(l)} : IF \ x_1 \ is \ A_1^l \ and \ ... \ and \ x_n \ is \ A_n^l, \ THEN \ y \ is \ B^l \qquad (7.1)$$

where A_i^l and B^l are fuzzy sets in $U_i \subset R$ and $V \subset R$, respectively, and $\mathbf{x} =
(x_1, x_2, ..., x_n)^T \in U$ and $y \in V$ are the input and output (linguistic) variables of
the fuzzy system, respectively. Let M be the number of rules in the fuzzy rule base;
that is, $l = 1, 2, ..., M$ in (7.1). We call the rules in the form of (7.1) *canonical fuzzy
IF-THEN rules* because they include many other types of fuzzy rules and fuzzy
propositions as special cases, as shown in the following lemma.

Lemma 7.1. The canonical fuzzy IF-THEN rules in the form of (7.1) include
the following as special cases:

(a) "Partial rules":

$$IF \ x_1 \ is \ A_1^l \ and \ ... \ and \ x_m \ is \ A_m^l, \ THEN \ y \ is \ B^l \qquad (7.2)$$

where $m < n$.

(b) "Or rules":

$$IF \ x_1 \ is \ A_1^l \ and \ ... \ and \ x_m \ is \ A_m^l \ or \ x_{m+1} \ is \ A_{m+1}^l \ and \ ... \ and \ x_n \ is \ A_n^l,$$
$$THEN \ y \ is \ B^l \qquad (7.3)$$

(c) Single fuzzy statement:
$$y \ is \ B^l \qquad (7.4)$$

(d) "Gradual rules," for example:

$$The \ smaller \ the \ x, \ the \ bigger \ the \ y \qquad (7.5)$$

(e) Non-fuzzy rules (that is, conventional production rules).

Proof: The partial rule (7.2) is equivalent to

$$IF x_1 \ is \ A_1^l \ and \ ... \ and \ x_m \ is \ A_m^l \ and \ x_{m+1} \ is \ I \ and \ ... \ and \ x_n \ is \ I,$$
$$THEN \ y \ is \ B^l \qquad (7.6)$$

where I is a fuzzy set in R with $\mu_I(x) \equiv 1$ for all $x \in R$. The preceding rule is in the form of (7.1); this proves (a). Based on intuitive meaning of the logic operator "or," the "Or rule" (7.3) is equivalent to the following two rules:

$$IF x_1 \ is \ A_1^l \ and \ ... \ and \ x_m \ is \ A_m^l, \ THEN \ y \ is \ B^l \qquad (7.7)$$

$$IF x_{m+1} \ is \ A_{m+1}^l \ and \ ... \ and \ x_n \ is \ A_n^l, \ THEN \ y \ is \ B^l \qquad (7.8)$$

From (a) we have that the two rules (7.7) and (7.8) are special cases of (7.1); this proves (b). The fuzzy statement (7.4) is equivalent to

$$IF \ x_1 \ is \ I \ and \ ... \ and \ x_n \ is \ I, \ THEN \ y \ is \ B^l \qquad (7.9)$$

which is in the form of (7.1); this proves (c). For (d), let S be a fuzzy set representing "smaller," for example, $\mu_S(x) = 1/(1 + exp(5(x + 2)))$, and B be a fuzzy set representing "bigger," for example, $\mu_B(y) = 1/(1 + exp(-5(y - 2)))$, then the "Gradual rule" (7.5) is equivalent to

$$IF \ x \ is \ S, \ THEN \ y \ is \ B \qquad (7.10)$$

which is a special case of (7.1); this proves (d). Finally, if the membership functions of A_i^l and B^l can only take values 1 or 0, then the rules (7.1) become non-fuzzy rules. $\square$

In our fuzzy system framework, human knowledge has to be represented in the form of the fuzzy IF-THEN rules (7.1). That is, we can only utilize human knowledge that can be formulated in terms of the fuzzy IF-THEN rules. Fortunately, Lemma 7.1 ensures that these rules provide a quite general knowledge representation scheme.

7.1.2   Properties of Set of Rules

Because the fuzzy rule base consists of a set of rules, the relationship among these rules and the rules as a whole impose interesting questions. For example, do the rules cover all the possible situations that the fuzzy system may face? Are there any conflicts among these rules? To answer these sorts of questions, we introduce the following concepts.

Definition 7.1. A set of fuzzy IF-THEN rules is *complete* if for any $\mathbf{x} \in U$, there exists at least one rule in the fuzzy rule base, say rule $Ru^{(l)}$ (in the form of (7.1)), such that

$$\mu_{A_i^l}(x_i) \neq 0 \qquad (7.11)$$

for all $i = 1, 2, ..., n$.

Intuitively, the completeness of a set of rules means that at any point in the input space there is at least one rule that "fires"; that is, the membership value of the IF part of the rule at this point is non-zero.

Example 7.1. Consider a 2-input-1-output fuzzy system with $U = U_1 \times U_2 = [0,1] \times [0,1]$ and $V = [0,1]$. Define three fuzzy sets S_1, M_1 and L_1 in U_1, and two fuzzy sets S_2 and L_2 in U_2, as shown in Fig. 7.2. In order for a fuzzy rule base to be complete, it must contain the following six rules whose IF parts constitute all the possible combinations of S_1, M_1, L_1 with S_2, L_2:

$$IF\ x_1\ is\ S_1\ and\ x_2\ is\ S_2,\ THEN\ y\ is\ B^1$$
$$IF\ x_1\ is\ S_1\ and\ x_2\ is\ L_2,\ THEN\ y\ is\ B^2$$
$$IF\ x_1\ is\ M_1\ and\ x_2\ is\ S_2,\ THEN\ y\ is\ B^3 \tag{7.12}$$
$$IF\ x_1\ is\ M_1\ and\ x_2\ is\ L_2,\ THEN\ y\ is\ B^4$$
$$IF\ x_1\ is\ L_1\ and\ x_2\ is\ S_2,\ THEN\ y\ is\ B^5$$
$$IF\ x_1\ is\ L_1\ and\ x_2\ is\ L_2,\ THEN\ y\ is\ B^6$$

where B^l $(l = 1, 2, ..., 6)$ are fuzzy sets in V. If any rule in this group is missing, then we can find point $x^* \in U$, at which the IF part propositions of all the remaining rules have zero membership value. For example, if the second rule in (7.12) is missing, then this $x^* = (0, 1)$ (Why?). $\square$

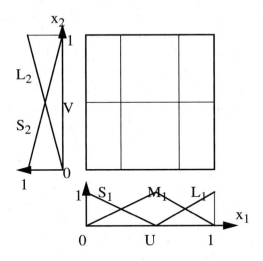

Figure 7.2. An example of membership functions for a two-input fuzzy system.

From Example 7.1 we see that if we use the triangular membership functions as in Fig. 7.2, the number of rules in a *complete* fuzzy rule base increases exponentially with the dimension of the input space U. This problem is called the *curse of dimensionality* and will be further discussed in Chapter 22.

Definition 7.2. A set of fuzzy IF-THEN rules is *consistent* if there are no rules with the same IF parts but different THEN parts.

For nonfuzzy production rules, consistence is an important requirement because it is difficult to continue the search if there are conflicting rules. For fuzzy rules, however, consistence is not critical because we will see later that if there are conflicting rules, the fuzzy inference engine and the defuzzifier will automatically average them out to produce a compromised result. Of course, it is better to have a consistent fuzzy rule base in the first place.

Definition 7.3. A set of fuzzy IF-THEN rules is *continuous* if there do not exist such neighboring rules whose THEN part fuzzy sets have empty intersection.

Intuitively, continuity means that the input-output behavior of the fuzzy system should be smooth. It is difficult to explain this concept in more detail at this point, because we have not yet derived the complete formulas of the fuzzy systems, but it will become clear as we move into Chapter 9.

7.2 Fuzzy Inference Engine

In a *fuzzy inference engine*, fuzzy logic principles are used to combine the fuzzy IF-THEN rules in the fuzzy rule base into a mapping from a fuzzy set A' in U to a fuzzy set B' in V. In Chapter 6 we learned that a fuzzy IF-THEN rule is interpreted as a fuzzy relation in the input-output product space $U \times V$, and we proposed a number of implications that specify the fuzzy relation. If the fuzzy rule base consists of only a single rule, then the generalized modus ponens (6.10) specifies the mapping from fuzzy set A' in U to fuzzy set B' in V. Because any practical fuzzy rule base constitutes more than one rule, the key question here is how to infer with a set of rules. There are two ways to infer with a set of rules: composition based inference and individual-rule based inference, which we will discuss next.

7.2.1 Composition Based Inference

In *composition based inference*, all rules in the fuzzy rule base are combined into a single fuzzy relation in $U \times V$, which is then viewed as a single fuzzy IF-THEN rule. So the key question is how to perform this combination. We should first understand what a set of rules mean intuitively, and then we can use appropriate logic operators to combine them.

There are two opposite arguments for what a set of rules should mean. The first one views the rules as independent conditional statements. If we accept this point of view, then a reasonable operator for combining the rules is *union*. The second one views the rules as strongly coupled conditional statements such that the conditions of all the rules must be satisfied in order for the whole set of rules to have an impact. If we adapt this view, then we should use the operator *intersection* to combine the rules. The second view may look strange, but for some implications, for example, the Gödel implication (5.26), it makes sense, as we will see very soon in this section. We now show the details of these two schemes.

Let $Ru^{(l)}$ be a fuzzy relation in $U \times V$, which represents the fuzzy IF-THEN rule (7.1); that is, $Ru^{(l)} = A_1^l \times \cdots \times A_n^l \to B^l$. From Chapter 5 we know that $A_1^l \times \cdots \times A_n^l$ is a fuzzy relation in $U = U_1 \times \cdots \times U_n$ defined by

$$\mu_{A_1^l \times \cdots \times A_n^l}(x_1, ..., x_n) = \mu_{A_1^l}(x_1) \star \cdots \star \mu_{A_n^l}(x_n) \qquad (7.13)$$

where $\star$ represents any t-norm operator. The implication $\to$ in $Ru^{(l)}$ is defined according to various implications (5.23)-(5.26), (5.31), and (5.32). If we accept the first view of a set of rules, then the M rules in the form of (7.1) are interpreted as a single fuzzy relation Q_M in $U \times V$ defined by

$$Q_M = \bigcup_{l=1}^{M} Ru^{(l)} \qquad (7.14)$$

This combination is called the *Mamdani combination*. If we use the symbol $\dot{+}$ to represent the s-norm, then (7.14) can be rewritten as

$$\mu_{Q_M}(\mathbf{x}, y) = \mu_{Ru^{(1)}}(\mathbf{x}, y) \dot{+} \cdots \dot{+} \mu_{Ru^{(M)}}(\mathbf{x}, y) \qquad (7.15)$$

For the second view of a set of rules, the M fuzzy IF-THEN rules of (7.1) are interpreted as a fuzzy relation Q_G in $U \times V$, which is defined as

$$Q_G = \bigcap_{l=1}^{M} Ru^{(l)} \qquad (7.16)$$

or equivalently,

$$\mu_{Q_G}(\mathbf{x}, y) = \mu_{Ru^{(1)}}(\mathbf{x}, y) \star \cdots \star \mu_{Ru^{(M)}}(\mathbf{x}, y) \qquad (7.17)$$

where $\star$ denotes t-norm. This combination is called the *Gödel combination*.

Let A' be an arbitrary fuzzy set in U and be the input to the fuzzy inference engine. Then, by viewing Q_M or Q_G as a single fuzzy IF-THEN rule and using the generalized modus ponens (6.10), we obtain the output of the fuzzy inference engine as

$$\mu_{B'}(y) = \sup_{\mathbf{x} \in U} t[\mu_{A'}(\mathbf{x}), \mu_{Q_M}(\mathbf{x}, y)] \qquad (7.18)$$

if we use the Mamdani combination, or as

$$\mu_{B'}(y) = \sup_{\mathbf{x} \in U} t[\mu_{A'}(\mathbf{x}), \mu_{Q_G}(\mathbf{x}, y)] \qquad (7.19)$$

if we use the Gödel combination.

In summary, the computational procedure of the composition based inference is given as follows:

- **Step 1**: For the M fuzzy IF-THEN rules in the form of (7.1), determine the membership functions $\mu_{A_1^l \times \cdots \times A_n^l}(x_1, ..., x_n)$ for $l = 1, 2, ..., M$ according to (7.13).

- **Step 2**: View $A_1^l \times \cdots \times A_n^l$ as the FP_1 and B^l as the FP_2 in the implications (5.23)-(5.26), (5.31) and (5.32), and determine $\mu_{Ru^{(l)}}(x_1, ..., x_n, y) = \mu_{A_1^l \times \cdots \times A_n^l \to B^l}(x_1, ..., x_n, y)$ for $l = 1, 2, ..., M$ according to any one of these implications.

- **Step 3**: Determine $\mu_{Q_M}(\mathbf{x}, y)$ or $\mu_{Q_G}(\mathbf{x}, y)$ according to (7.15) or (7.17).

- **Step 4**: For given input A', the fuzzy inference engine gives output B' according to (7.18) or (7.19).

7.2.2   Individual-Rule Based Inference

In *individual-rule based inference*, each rule in the fuzzy rule base determines an output fuzzy set and the output of the whole fuzzy inference engine is the combination of the M individual fuzzy sets. The combination can be taken either by union or by intersection.

The computational procedure of the individual-rule based inference is summarized as follows:

- **Steps 1 and 2**: Same as the Steps 1 and 2 for the composition based inference.

- **Step 3**: For given input fuzzy set A' in U, compute the output fuzzy set B_l' in V for each individual rule $Ru^{(l)}$ according to the generalized modus ponens (6.10), that is,

$$\mu_{B_l'}(y) = \sup_{\mathbf{x} \in U} t[\mu_{A'}(\mathbf{x}), \mu_{Ru^{(l)}}(\mathbf{x}, y)] \qquad (7.20)$$

for $l = 1, 2, ..., M$.

- **Step 4**: The output of the fuzzy inference engine is the combination of the M fuzzy sets $\{B_1', ..., B_M'\}$ either by union, that is,

$$\mu_{B'}(y) = \mu_{B_1'}(y) \dotplus \cdots \dotplus \mu_{B_M'}(y) \qquad (7.21)$$

or by intersection, that is,

$$\mu_{B'}(y) = \mu_{B_1'}(y) \star \cdots \star \mu_{B_M'}(y) \qquad (7.22)$$

where $\dotplus$ and $\star$ denote s-norm and t-norm operators, respectively.

7.2.3 The Details of Some Inference Engines

From the previous two subsections we see that there are a variety of choices in the fuzzy inference engine. Specifically, we have the following alternatives: (i) composition based inference or individual-rule based inference, and within the composition based inference, Mamdani inference or Gödel inference, (ii) Dienes-Rescher implication (5.23), Lukasiewicz implication (5.24), Zadeh implication (5.25), Gödel implication (5.26), or Mamdani implications (5.31)-(5.32), and (iii) different operations for the t-norms and s-norms in the various formulas. So a natural question is: how do we select from these alternatives?

In general, the following three criteria should be considered:

- *Intuitive appeal*: The choice should make sense from an intuitive point of view. For example, if a set of rules are given by a human expert who believes that these rules are independent of each other, then they should be combined by union.

- *Computational efficiency*: The choice should result in a formula relating B' with A', which is simple to compute.

- *Special properties*: Some choice may result in an inference engine that has special properties. If these properties are desirable, then we should make this choice.

We now show the detailed formulas of a number of fuzzy inference engines that are commonly used in fuzzy systems and fuzzy control.

- **Product Inference Engine**: In *product inference engine*, we use: (i) individual-rule based inference with union combination (7.21), (ii) Mamdani's product implication (5.32), and (iii) algebraic product for all the t-norm operators and *max* for all the s-norm operators. Specifically, from (7.20), (7.21), (5.32), and (7.13), we obtain the product inference engine as

$$\mu_{B'}(y) = \max_{l=1}^{M}[\sup_{\mathbf{x} \in U}(\mu_{A'}(\mathbf{x}) \prod_{i=1}^{n} \mu_{A_i^l}(x_i)\mu_{B^l}(y))] \tag{7.23}$$

  That is, given fuzzy set A' in U, the product inference engine gives the fuzzy set B' in V according to (7.23).

- **Minimum Inference Engine**: In *minimum inference engine*, we use: (i) individual-rule based inference with union combination (7.21), (ii) Mamdani's minimum implication (5.31), and (iii) *min* for all the t-norm operators and *max* for all the s-norm operators. Specifically, from (7.20), (7.21), (5.31), and (7.13) we have

$$\mu_{B'}(y) = \max_{l=1}^{M}[\sup_{\mathbf{x} \in U} min(\mu_{A'}(\mathbf{x}), \mu_{A_1^l}(x_1), ..., \mu_{A_n^l}(x_n), \mu_{B^l}(y))] \tag{7.24}$$

That is, given fuzzy set A' in U, the minimum inference engine gives the fuzzy set B' in V according to (7.24).

The product inference engine and the minimum inference engine are the most commonly used fuzzy inference engines in fuzzy systems and fuzzy control. The main advantage of them is their computational simplicity; this is especially true for the product inference engine (7.23). Also, they are intuitively appealing for many practical problems, especially for fuzzy control. We now show some properties of the product and minimum inference engines.

Lemma 7.2. The product inference engine is unchanged if we replace "individual-rule based inference with union combination (7.21)" by "composition based inference with Mamdani combination (7.15)."

Proof: From (7.15) and (7.18) we have

$$\mu_{B'}(y) = \sup_{\mathbf{x} \in U} [\mu_{A'}(\mathbf{x}) \, \max_{l=1}^{M} (\mu_{R u^{(l)}}(\mathbf{x}, y))] \tag{7.25}$$

Using (5.32) and (7.13), we can rewrite (7.25) as

$$\mu_{B'}(y) = \sup_{\mathbf{x} \in U} \max_{l=1}^{M} [\mu_{A'}(\mathbf{x}) \prod_{i=1}^{n} \mu_{A_i^l}(x_i) \mu_{B^l}(y)] \tag{7.26}$$

Because the $\max_{l=1}^{M}$ and $\sup_{\mathbf{x} \in U}$ are interchangeable, (7.26) is equivalent to (7.23). $\square$

Lemma 7.3. If the fuzzy set A' is a fuzzy singleton, that is, if

$$\mu_{A'}(\mathbf{x}) = \begin{cases} 1 \; if \; \mathbf{x} = \mathbf{x}^* \\ 0 \; otherwise \end{cases} \tag{7.27}$$

where $\mathbf{x}^*$ is some point in U, then the product inference engine is simplified to

$$\mu_{B'}(y) = \max_{l=1}^{M} [\prod_{i=1}^{n} \mu_{A_i^l}(x_i^*) \mu_{B^l}(y)] \tag{7.28}$$

and the minimum inference engine is simplified to

$$\mu_{B'}(y) = \max_{l=1}^{M} [min(\mu_{A_1^l}(x_1^*), ..., \mu_{A_n^l}(x_n^*), \mu_{B^l}(y))] \tag{7.29}$$

Proof: Substituting (7.27) into (7.23) and (7.24), we see that the $\sup_{\mathbf{x} \in U}$ is achieved at $\mathbf{x} = \mathbf{x}^*$. Hence, (7.23) reduces (7.28) and (7.24) reduces (7.29). $\square$

Lemma 7.2 shows that although the individual-rule based and composition based inferences are conceptually different, they produce the same fuzzy inference engine in certain important cases. Lemma 7.3 indicates that the computation within the fuzzy inference engine can be greatly simplified if the input is a fuzzy singleton (the most difficult computation in (7.23) and (7.24) is the $\sup_{\mathbf{x} \in U}$, which disappears in (7.28) and (7.29)).

A disadvantage of the product and minimum inference engines is that if at some $\mathbf{x} \in U$ the $\mu_{A_i^l}(x_i)$'s are very small, then the $\mu_{B'}(y)$ obtained from (7.23) or (7.24) will be very small. This may cause problems in implementation. The following three fuzzy inference engines overcome this disadvantage.

- **Lukasiewicz Inference Engine**: In *Lukasiewicz inference engine*, we use: (i) individual-rule based inference with intersection combination (7.22), (ii) Lukasiewicz implication (5.24), and (iii) *min* for all the t-norm operators. Specifically, from (7.22), (7.20), (5.24) and (7.13) we obtain

$$\mu_{B'}(y) = \min_{l=1}^{M}[\sup_{\mathbf{x} \in U} min(\mu_{A'}(\mathbf{x}), \mu_{Ru^{(l)}}(\mathbf{x}, y))]$$

$$= \min_{l=1}^{M}\{\sup_{\mathbf{x} \in U} min[\mu_{A'}(\mathbf{x}), min(1, 1 - \min_{i=1}^{n}(\mu_{A_i^l}(x_i)) + \mu_{B^l}(y))]\}$$

$$= \min_{l=1}^{M}\{\sup_{\mathbf{x} \in U} min[\mu_{A'}(\mathbf{x}), 1 - \min_{i=1}^{n}(\mu_{A_i^l}(x_i)) + \mu_{B^l}(y)]\} \qquad (7.30)$$

That is, for given fuzzy set A' in U, the Lukasiewicz inference engine gives the fuzzy set B' in V according to (7.30).

- **Zadeh Inference Engine**: In *Zadeh inference engine*, we use: (i) individual-rule based inference with intersection combination (7.22), (ii) Zadeh implication (5.25), and (iii) *min* for all the t-norm operators. Specifically, from (7.22), (7.20), (5.25), and (7.13) we obtain

$$\mu_{B'}(y) = \min_{l=1}^{M}\{\sup_{\mathbf{x} \in U} min[\mu_{A'}(\mathbf{x}), max(min(\mu_{A_1^l}(x_1), ..., \mu_{A_n^l}(x_n), \mu_{B^l}(y)),$$

$$1 - \min_{i=1}^{n}(\mu_{A_i^l}(x_i)))]\} \qquad (7.31)$$

- **Dienes-Rescher Inference Engine**: In *Dienes-Rescher inference engine*, we use the same operations as in the Zadeh inference engine, except that we replace the Zadeh implication (5.25) with the Dienes-Rescher implication (5.23). Specifically, we obtain from (7.22), (7.20), (5.23), and (7.13) that

$$\mu_{B'}(y) = \min_{l=1}^{M}\{\sup_{\mathbf{x} \in U} min[\mu_{A'}(\mathbf{x}), max(1 - \min_{i=1}^{n}(\mu_{A_i^l}(x_i)), \mu_{B^l}(y))]\} \qquad (7.32)$$

Similar to Lemma 7.3, we have the following results for the Lukasiewicz, Zadeh and Dienes-Rescher inference engines.

Lemma 7.4: If A' is a fuzzy singleton as defined by (7.27), then the Lukasiewicz, Zadeh and Dienes-Rescher inference engines are simplified to

$$\mu_{B'}(y) = \min_{l=1}^{M}[1, 1 - \min_{i=1}^{n}(\mu_{A_i^l}(x_i^*)) + \mu_{B^l}(y)] \tag{7.33}$$

$$\mu_{B'}(y) = \min_{l=1}^{M}\{max[min(\mu_{A_1^l}(x_1^*), ..., \mu_{A_n^l}(x_n^*), \mu_{B^l}(y)),$$
$$1 - \min_{i=1}^{n}(\mu_{A_i^l}(x_i^*))]\} \tag{7.34}$$

$$\mu_{B'}(y) = \min_{l=1}^{M}\{max[1 - \min_{i=1}^{n}(\mu_{A_i^l}(x_i^*)), \mu_{B^l}(y)]\} \tag{7.35}$$

respectively.

Proof: Using the same arguments as in the proof of Lemma 7.3, we can prove this lemma. □

We now have proposed five fuzzy inference engines. Next, we compare them through two examples.

Example 7.2: Suppose that a fuzzy rule base consists of only one rule

$$IF\ x_1\ is\ A_1\ and\ \cdots\ and\ x_n\ is\ A_n,\ THEN\ y\ is\ B \tag{7.36}$$

where

$$\mu_B(y) = \begin{cases} 1 - |y|\ if\ -1 \le y \le 1 \\ 0\ otherwise \end{cases} \tag{7.37}$$

Assume that A' is a fuzzy singleton defined by (7.27). We would like to plot the $\mu_{B'}(y)$ obtained from the five fuzzy inference engines. Let B'_P, B'_M, B'_L, B'_Z and B'_D be the fuzzy set B' using the product, minimum, Lukasiewicz, Zadeh and Dienes-Rescher inference engines, respectively, and let $min[\mu_{A_1}(x_1^*), ..., \mu_{A_n}(x_n^*)] = \mu_{A_p}(x_p^*)$ and $\prod_{i=1}^{n}\mu_{A_i}(x_i^*) = \mu_A(\mathbf{x}^*)$. Then from (7.28), (7.29), and (7.33)-(7.35) we have

$$\mu_{B'_P}(y) = \mu_A(\mathbf{x}^*)\mu_B(y) \tag{7.38}$$

$$\mu_{B'_M}(y) = min[\mu_{A_p}(x_p^*), \mu_B(y)] \tag{7.39}$$

$$\mu_{B'_L}(y) = min[1, 1 - \mu_{A_p}(x_p^*) + \mu_B(y)] \tag{7.40}$$

$$\mu_{B'_Z}(y) = max[min(\mu_{A_p}(x_p^*), \mu_B(y)), 1 - \mu_{A_p}(x_p^*)] \tag{7.41}$$

$$\mu_{B'_D}(y) = max[1 - \mu_{A_p}(x_p^*), \mu_B(y)] \tag{7.42}$$

For the case of $\mu_{A_p}(x_p^*) \ge 0.5$, $\mu_{B'_P}(y)$ and $\mu_{B'_M}(y)$ are plotted in Fig. 7.3, and $\mu_{B'_L}(y), \mu_{B'_Z}(y)$ and $\mu_{B'_D}(y)$ are plotted in Fig. 7.4. For the case of $\mu_{A_p}(x_p^*) < 0.5$, $\mu_{B'_P}(y)$ and $\mu_{B'_M}(y)$ are plotted in Fig. 7.5, and $\mu_{B'_L}(y), \mu_{B'_Z}(y)$ and $\mu_{B'_D}(y)$ are plotted in Fig. 7.6.

From Figs.7.3-7.6 we have the following observations: (i) if the membership value of the IF part at point $\mathbf{x}^*$ is small (say, $\mu_{A_p}(x_p^*) < 0.5$), then the product and minimum inference engines give very small membership values, whereas the Lukasiewicz, Zadeh and Dienes-Rescher inference engines give very large membership values; (ii) the product and minimum inference engines are similar, while the Lukasiewicz, Zadeh and Dienes-Rescher inference engines are similar, but there are big differences between these two groups; and (iii) the Lukasiewicz inference engine gives the largest output membership function, while the product inference engine gives the smallest output membership function in all the cases; the other three inference engines are in between. □

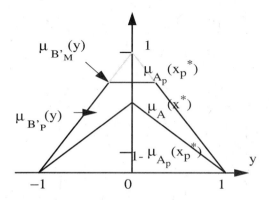

Figure 7.3. Output membership functions using the Lukasiewicz, Zadeh and Dienes-Rescher inference engines for the $\mu_{A_p^*}(x_p^*) \geq 0.5$ case.

Example 7.3: In this example, we consider that the fuzzy system contains two rules: one is the same as (7.36), and the other is

$$IF\ x_1\ is\ C_1\ and\ \cdots\ and\ x_n\ is\ C_n,\ THEN\ y\ is\ D \tag{7.43}$$

where

$$\mu_D(y) = \begin{cases} 1 - |y - 1| & if\ 0 \leq y \leq 2 \\ 0\ otherwise \end{cases} \tag{7.44}$$

Assume again that A' is the fuzzy singleton defined by (7.27). We would like to plot the $\mu_{B'}(y)$ using the product inference engine, that is, $\mu_{B'_P}(y)$. Fig. 7.7 shows the $\mu_{B'_P}(y)$, where $\mu_A(\mathbf{x}^*) = \prod_{i=1}^n \mu_{A_i}(x_i^*)$ and $\mu_C(\mathbf{x}^*) = \prod_{i=1}^n \mu_{C_i}(x_i^*)$. □

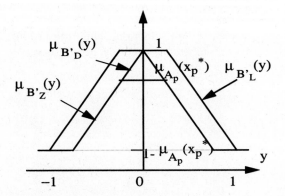

Figure 7.4. Output membership functions using the product and minimum inference engines for the $\mu_{A_p^*}(x_p^*) \geq 0.5$ case.

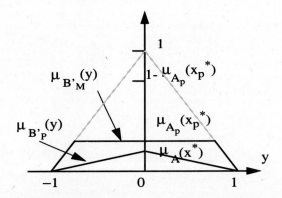

Figure 7.5. Output membership functions using the product and minimum inference engines for the $\mu_{A_p^*}(x_p^*) < 0.5$ case.

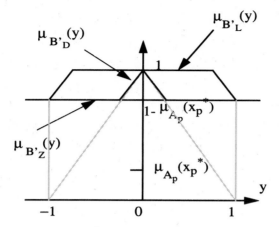

Figure 7.6. Output membership functions using the Lukasiewicz, Zadeh and Dienes-Rescher inference engines for the $\mu_{A_p^*}(x_p^*) < 0.5$ case.

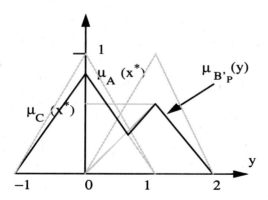

Figure 7.7. Output membership function using the product inference engine for the case of two rules.

7.3   Summary and Further Readings

In this chapter we have demonstrated the following:

- The structure of the canonical fuzzy IF-THEN rules and the criteria for evaluating a set of rules.

- The computational procedures for the composition based and individual-rule based inferences.

- The detailed formulas of the five specific fuzzy inference engines: product, minimum, Lukasiewicz, Zadeh and Dienes-Rescher inference engines, and their computations for particular examples.

Lee [1990] provided a very good survey on fuzzy rule bases and fuzzy inference engines. This paper gives intuitive analyses for various issues associated with fuzzy rule bases and fuzzy inference engines. A mathematical analysis of fuzzy inference engines, similar to the approach in this chapter, were prepared by Driankov, Hellendoorn and Reinfrank [1993].

7.4   Exercises

Exercise 7.1. If the third and sixth rules in (7.12) (Example 7.1) are missing, at what points do the IF part propositions of all the remaining rules have zero membership values?

Exercise 7.2. Give an example of fuzzy sets $B^1, ..., B^6$ such that the set of the six rules (7.12) is continuous.

Exercise 7.3. Suppose that a fuzzy rule base consists of only one rule (7.36) with

$$\mu_B(y) = exp(-y^2) \qquad (7.45)$$

Let the input A' to the fuzzy inference engine be the fuzzy singleton (7.27). Plot the output membership functions $\mu_{B'}(y)$ using: (a) product, (b) minimum, (c) Lukasiewicz, (d) Zadeh, and (e) Dienes-Rescher inference engines.

Exercise 7.4. Consider Example 7.3 and plot the $\mu_{B'}(y)$ using: (a) Lukasiewicz inference engine, and (b) Zadeh inference engine.

Exercise 7.5. Use the Gödel implication to propose a so-called Gödel inference engine.

Chapter 8

Fuzzifiers and Defuzzifiers

We learned from Chapter 7 that the fuzzy inference engine combines the rules in the fuzzy rule base into a mapping from fuzzy set A' in U to fuzzy set B' in V. Because in most applications the input and output of the fuzzy system are real-valued numbers, we must construct interfaces between the fuzzy inference engine and the environment. The interfaces are the fuzzifier and defuzzifier in Fig. 1.5.

8.1   Fuzzifiers

The *fuzzifier* is defined as a mapping from a real-valued point $\mathbf{x}^* \in U \subset R^n$ to a fuzzy set A' in U. What are the criteria in designing the fuzzifier? First, the fuzzifier should consider the fact that the input is at the crisp point $\mathbf{x}^*$, that is, the fuzzy set A' should have large membership value at $\mathbf{x}^*$. Second, if the input to the fuzzy system is corrupted by noise, then it is desireable that the fuzzifier should help to suppress the noise. Third, the fuzzifier should help to simplify the computations involved in the fuzzy inference engine. From (7.23), (7.24) and (7.30)-(7.32) we see that the most complicated computation in the fuzzy inference engine is the $\sup_{\mathbf{x} \in U}$, therefore our objective is to simplify the computations involving $\sup_{\mathbf{x} \in U}$.

We now propose three fuzzifiers:

- **Singleton fuzzifier**: The *singleton fuzzifier* maps a real-valued point $\mathbf{x}^* \in U$ into a fuzzy singleton A' in U, which has membership value 1 at $\mathbf{x}^*$ and 0 at all other points in U; that is,

$$\mu_{A'}(\mathbf{x}) = \begin{cases} 1 \ if \ \mathbf{x} = \mathbf{x}^* \\ 0 \ otherwise \end{cases} \tag{8.1}$$

- **Gaussian fuzzifier**: The *Gaussian fuzzifier* maps $\mathbf{x}^* \in U$ into fuzzy set A' in U, which has the following Gaussian membership function:

$$\mu_{A'}(\mathbf{x}) = e^{-(\frac{x_1 - x_1^*}{a_1})^2} \star \cdots \star e^{-(\frac{x_n - x_n^*}{a_n})^2} \tag{8.2}$$

where a_i are positive parameters and the t-norm $\star$ is usually chosen as algebraic product or *min*.

- **Triangular fuzzifier**: The *triangular fuzzifier* maps $\mathbf{x}^* \in U$ into fuzzy set A' in U, which has the following triangular membership function

$$\mu_{A'}(\mathbf{x}) = \begin{cases} (1 - \frac{|x_1 - x_1^*|}{b_1}) \star \cdots \star (1 - \frac{|x_n - x_n^*|}{b_n}) \; if \; |x_i - x_i^*| \leq b_i, \; i = 1, 2, ..., n \\ 0 \; otherwise \end{cases}$$

(8.3)

where b_i are positive parameters and the t-norm $\star$ is usually chosen as algebraic product or *min*.

From (8.1)-(8.3) we see that all three fuzzifiers satisfy $\mu_{A'}(\mathbf{x}^*) = 1$; that is, they satisfy the first criterion mentioned before. From Lemmas 7.3 and 7.4 we see that the singleton fuzzifier greatly simplifies the computations involved in the fuzzy inference engine. Next, we show that if the fuzzy sets A_i^l in the rules (7.1) have Gaussian or triangular membership functions, then the Gaussian or triangular fuzzifier also will simplify some fuzzy inference engines.

Lemma 8.1. Suppose that the fuzzy rule base consists of M rules in the form of (7.1) and that

$$\mu_{A_i^l}(x_i) = e^{-(\frac{x_i - \bar{x}_i^l}{\sigma_i^l})^2}$$

(8.4)

where $\bar{x}_i^l$ and σ_i^l are constant parameters, $i = 1, 2, ..., n$ and $l = 1, 2, ..., M$. If we use the Gaussian fuzzifier (8.2), then:

(a) If we choose algebraic product for the t-norm $\star$ in (8.2), the product inference engine (7.23) is simplified to

$$\mu_{B'}(y) = \max_{l=1}^{M} [\prod_{i=1}^{n} e^{-(\frac{x_{iP}^l - \bar{x}_i^l}{\sigma_i^l})^2} e^{-(\frac{x_{iP}^l - x_i^*}{a_i})^2} \mu_{B^l}(y)]$$

(8.5)

where

$$x_{iP}^l = \frac{a_i^2 \bar{x}_i^l + (\sigma_i^l)^2 x_i^*}{a_i^2 + (\sigma_i^l)^2}$$

(8.6)

(b) If we choose *min* for the t-norm $\star$ in (8.2), the minimum inference engine (7.24) is simplified to

$$\mu_{B'}(y) = \max_{l=1}^{M} [min(e^{-(\frac{x_{1M}^l - \bar{x}_1^l}{\sigma_1^l})^2}, ..., e^{-(\frac{x_{nM}^l - \bar{x}_n^l}{\sigma_n^l})^2}, \mu_{B^l}(y))]$$

(8.7)

where

$$x_{iM}^l = \frac{a_i \bar{x}_i^l + \sigma_i^l x_i^*}{a_i + \sigma_i^l}$$

(8.8)

Proof: (a) Substituting (8.2) and (8.4) into (7.23) and noticing that the $\star$ is an algebraic product, we obtain

$$\mu_{B'}(y) = \max_{l=1}^{M}[\sup_{\mathbf{x} \in U} \prod_{i=1}^{n} e^{-(\frac{x_i - x_i^*}{a_i})^2} e^{-(\frac{x_i - \bar{x}_i^l}{\sigma_i^l})^2} \mu_{B^l}(y)]$$

$$= \max_{l=1}^{M}[\prod_{i=1}^{n} \sup_{\mathbf{x} \in U} e^{-(\frac{x_i - x_i^*}{a_i})^2 - (\frac{x_i - \bar{x}_i^l}{\sigma_i^l})^2} \mu_{B^l}(y)] \qquad (8.9)$$

Since

$$-(\frac{x_i - x_i^*}{a_i})^2 - (\frac{x_i - \bar{x}_i^l}{\sigma_i^l})^2 = -k_1(x_i - \frac{a_i^2 \bar{x}_i^l + (\sigma_i^l)^2 x_i^*}{a_i^2 + (\sigma_i^l)^2})^2 + k_2 \qquad (8.10)$$

where k_1 and k_2 are not functions of x_i, the $\sup_{\mathbf{x} \in U}$ in (8.9) is achieved at $x_{iP} \in U$ $(i = 1, 2, ..., n)$, which is exactly (8.6).

(b) Substituting (8.2) and (8.4) into (7.24) and noticing that the $\star$ is min, we obtain

$$\mu_{B'}(y) = \max_{l=1}^{M}\{min[\sup_{\mathbf{x} \in U} min(e^{-(\frac{x_1 - x_1^*}{a_1})^2}, e^{-(\frac{x_1 - \bar{x}_1^l}{\sigma_1^l})^2}), \cdots,$$

$$\sup_{\mathbf{x} \in U} min(e^{-(\frac{x_n - x_n^*}{a_n})^2}, e^{-(\frac{x_n - \bar{x}_n^l}{\sigma_n^l})^2}), \mu_{B^l}(y)]\} \qquad (8.11)$$

Clearly, the $\sup_{\mathbf{x} \in U} min$ is achieved at $x_i = x_{iM}^l$ when

$$e^{-(\frac{x_i - x_i^*}{a_i})^2} = e^{-(\frac{x_i - \bar{x}_i^l}{\sigma_i^l})^2} \qquad (8.12)$$

which gives (8.8). Substituting $x_i = x_{iM}^l$ into (8.11) gives (8.7). $\square$

We can obtain similar results as in Lemma 8.1 when the triangular fuzzifier is used. If $a_i = 0$, then from (8.6) and (8.8) we have $x_{iP}^l = x_{iM}^l = x_i^*$; that is, in this case the Gaussian fuzzifier becomes the singleton fuzzifier. If a_i is much larger than σ_i^l, then from (8.6) and (8.8) we see that x_{iP}^l and x_{iM}^l will be very close to $\bar{x}_i^l$; that is, x_{iP}^l and x_{iM}^l will be insensitive to the changes in the input x_i^*. Therefore, by choosing large a_i, the Gaussian fuzzifier can suppress the noise in the input x_i^*. More specifically, suppose that the input x_i^* is corrupted by noise, that is,

$$x_i^* = x_{i0}^* + n_i^* \qquad (8.13)$$

where x_{io}^* is the useful signal and n_i^* is noise. Substituting (8.13) into (8.6), we have

$$x_{iP}^l = \frac{a_i^2 \bar{x}_i^l + (\sigma_i^l)^2 x_{io}^*}{a_i^2 + (\sigma_i^l)^2} + \frac{(\sigma_i^l)^2}{a_i^2 + (\sigma_i^l)^2} n_i^* \qquad (8.14)$$

From (8.14) we see that after passing through the Gaussian fuzzifier, the noise is suppressed by the factor $\frac{(\sigma_i^l)^2}{a_i^2+(\sigma_i^l)^2}$. When a_i is much larger than σ_i^l, the noise will be greatly suppressed. Similarly, we can show that the triangular fuzzifier also has this kind of noise suppressing capability.

In summary, we have the following remarks about the three fuzzifiers:

- The singleton fuzzifier greatly simplifies the computation involved in the fuzzy inference engine for any type of membership functions the fuzzy IF-THEN rules may take.

- The Gaussian or triangular fuzzifiers also simplify the computation in the fuzzy inference engine, if the membership functions in the fuzzy IF-THEN rules are Gaussian or triangular, respectively.

- The Gaussian and triangular fuzzifiers can suppress noise in the input, but the singleton fuzzifier cannot.

8.2   Defuzzifiers

The *defuzzifier* is defined as a mapping from fuzzy set B' in $V \subset R$ (which is the output of the fuzzy inference engine) to crisp point $y^* \in V$. Conceptually, the task of the defuzzifier is to specify a point in V that best represents the fuzzy set B'. This is similar to the mean value of a random variable. However, since the B' is constructed in some special ways (see Chapter 7), we have a number of choices in determining this representing point. The following three criteria should be considered in choosing a defuzzification scheme:

- *Plausibility*: The point y^* should represent B' from an intuitive point of view; for example, it may lie approximately in the middle of the support of B' or has a high degree of membership in B'.

- *Computational simplicity*: This criterion is particularly important for fuzzy control because fuzzy controllers operate in real-time.

- *Continuity*: A small change in B' should not result in a large change in y^*.

We now propose three types of defuzzifiers. For all the defuzzifiers, we assume that the fuzzy set B' is obtained from one of the five fuzzy inference engines in Chapter 7, that is, B' is given by (7.23), (7.24), (7.30), (7.31), or (7.32). From these equations we see that B' is the union or intersection of M individual fuzzy sets.

8.2.1   center of gravity Defuzzifier

The *center of gravity defuzzifier* specifies the y^* as the center of the area covered by the membership function of B', that is,

$$y^* = \frac{\int_V y \mu_{B'}(y) dy}{\int_V \mu_{B'}(y) dy} \tag{8.15}$$

where $\int_V$ is the conventional integral. Fig. 8.1 shows this operation graphically.

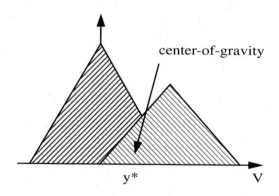

Figure 8.1. A graphical representation of the center of gravity defuzzifier.

If we view $\mu_{B'}(y)$ as the probability density function of a random variable, then the center of gravity defuzzifier gives the mean value of the random variable. Sometimes it is desirable to eliminate the $y \in V$, whose membership values in B' are too small; this results in the *indexed center of gravity defuzzifier*, which gives

$$y^* = \frac{\int_{V_\alpha} y \mu_{B'}(y) dy}{\int_{V_\alpha} \mu_{B'}(y) dy} \tag{8.16}$$

where V_α is defined as

$$V_\alpha = \{y \in V | \mu_{B'}(y) \geq \alpha\} \tag{8.17}$$

and α is a constant.

The advantage of the center of gravity defuzzifier lies in its intuitive plausibility. The disadvantage is that it is computationally intensive. In fact, the membership function $\mu_{B'}(y)$ is usually irregular and therefore the integrations in (8.15) and (8.16) are difficult to compute. The next defuzzifier tries to overcome this disadvantage by approximating (8.15) with a simpler formula.

8.2.2   Center Average Defuzzifier

Because the fuzzy set B' is the union or intersection of M fuzzy sets, a good approximation of (8.15) is the weighted average of the centers of the M fuzzy sets, with the weights equal the heights of the corresponding fuzzy sets. Specifically, let $\bar{y}^l$ be the center of the $l'th$ fuzzy set and w_l be its height, the *center average defuzzifier* determines y^* as

$$y^* = \frac{\sum_{l=1}^{M} \bar{y}^l w_l}{\sum_{l=1}^{M} w_l} \tag{8.18}$$

Fig. 8.2 illustrates this operation graphically for a simple example with $M = 2$.

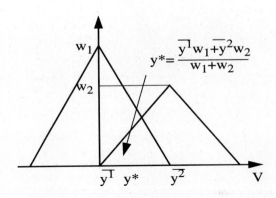

Figure 8.2. A graphical representation of the center average defuzzifier.

The center average defuzzifier is the most commonly used defuzzifier in fuzzy systems and fuzzy control. It is computationally simple and intuitively plausible. Also, small changes in $\bar{y}^l$ and w_l result in small changes in y^*. We now compare the center of gravity and center average defuzzifiers for a simple example.

Example 8.1. Suppose that the fuzzy set B' is the union of the two fuzzy sets shown in Fig. 8.2 with $\bar{y}^1 = 0$ and $\bar{y}^2 = 1$. Then the center average defuzzifier gives

$$y^* = \frac{w_2}{w_1 + w_2} \tag{8.19}$$

Table 8.1. Comparison of the center of gravity and center average defuzzifiers for Example 8.1.

w_1	w_2	y^* (center of gravity)	y^* (center average)	relative error
0.9	0.7	0.4258	0.4375	0.0275
0.9	0.5	0.5457	0.5385	0.0133
0.9	0.2	0.7313	0.7000	0.0428
0.6	0.7	0.3324	0.3571	0.0743
0.6	0.5	0.4460	0.4545	0.0192
0.6	0.2	0.6471	0.6250	0.0342
0.3	0.7	0.1477	0.1818	0.2308
0.3	0.5	0.2155	0.2500	0.1600
0.3	0.2	0.3818	0.4000	0.0476

We now compute the y^* resulting from the center of gravity defuzzifier. First, we notice that the two fuzzy sets intersect at $y = \frac{w_1}{w_1+w_2}$, hence,

$$\int_V \mu_{B'}(y)dy = \text{area of the first fuzzy set } + \text{ area of the second fuzzy set}$$

$$- \text{ their intersection}$$

$$= w_1 + w_2 - \frac{1}{2}\frac{w_1 w_2}{w_1 + w_2} \tag{8.20}$$

From Fig. 8.2 we have

$$\int_V y\mu_{B'}(y)dy = \int_{-1}^0 yw_1(1+y)dy + \int_0^{\frac{w_1}{w_1+w_2}} yw_1(1-y)dy + \int_{\frac{w_1}{w_1+w_2}}^1 yw_2 y\,dy$$

$$+ \int_1^2 yw_2(2-y)dy$$

$$= w_1(\frac{1}{2}y^2 + \frac{1}{3}y^3)|_{-1}^0 + w_1(\frac{1}{2}y^2 - \frac{1}{3}y^3)|_0^{\frac{w_1}{w_1+w_2}} + w_2\frac{1}{3}y^3|_{\frac{w_1}{w_1+w_2}}^1$$

$$+ w_2(y^2 - \frac{1}{3}y^3)|_1^2$$

$$= -\frac{1}{6}w_1 + w_2 + \frac{1}{6}\frac{w_1^3}{(w_1 + w_2)^2} \tag{8.21}$$

Dividing (8.21) by (8.20) we obtain the y^* of the center of gravity defuzzifier. Table 8.1 shows the values of y^* using these two defuzzifiers for certain values of w_1 and w_2. We see that the computation of the center of gravity defuzzifier is much more intensive than that of the center average defuzzifier. $\square$

8.2.3 Maximum Defuzzifier

Conceptually, the maximum defuzzifier chooses the y^* as the point in V at which $\mu_{B'}(y)$ achieves its maximum value. Define the set

$$hgt(B') = \{y \in V | \mu_{B'}(y) = \sup_{y \in V} \mu_{B'}(y)\} \tag{8.22}$$

that is, $hgt(B')$ is the set of all points in V at which $\mu_{B'}(y)$ achieves its maximum value. The *maximum defuzzifier* defines y^* as an arbitrary element in $hgt(B')$, that is,

$$y^* = any\ point\ in\ hgt(B') \tag{8.23}$$

If $hgt(B')$ contains a single point, then y^* is uniquely defined. If $hgt(B')$ contains more than one point, then we may still use (8.23) or use the smallest of maxima, largest of maxima, or mean of maxima defuzzifiers. Specifically, the *smallest of maxima defuzzifier* gives

$$y^* = inf\{y \in hgt(B')\} \tag{8.24}$$

the *largest of maxima defuzzifier* gives

$$y^* = sup\{y \in hgt(B')\} \tag{8.25}$$

and the *mean of maxima defuzzifier* is defined as

$$y^* = \frac{\int_{hgt(B')} y\,dy}{\int_{hgt(B')} dy} \tag{8.26}$$

where $\int_{hgt(B')}$is the usual integration for the continuous part of $hgt(B')$ and is summation for the discrete part of $hgt(B')$. We feel that the mean of maxima defuzzifier may give results which are contradictory to the intuition of maximum membership. For example, the y^* from the mean of maxima defuzzifier may have very small membership value in B'; see Fig. 8.3 for an example. This problem is due to the nonconvex nature of the membership function $\mu_{B'}(y)$.

The maximum defuzzifiers are intuitively plausible and computationally simple. But small changes in B' may result in large changes in y^*; see Fig.8.4 for an example. If the situation in Fig.8.4 is unlikely to happen, then the maximum defuzzifiers are good choices.

8.2.4 Comparison of the Defuzzifiers

Table 8.2 compares the three types of defuzzifiers according to the three criteria: plausibility, computational simplicity, and continuity. From Table 8.2 we see that the center average defuzzifier is the best.

Finally, we consider an example for the computation of the defuzzifiers with some particular membership functions.

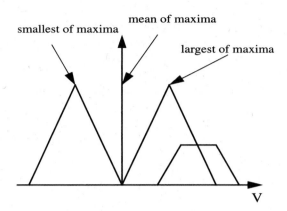

Figure 8.3. A graphical representation of the maximum defuzzifiers. In this example, the mean of maxima defuzzifier gives a result that is contradictory to the maximum-membership intuition.

Table 8.2. Comparison of the center of gravity, center average, and maximum defuzzifiers with respect to plausibility, computational simplicity, and continuity.

	center of gravity	center average	maximum
plausibility	yes	yes	yes
computational simplicity	no	yes	yes
continuity	yes	yes	no

Example 8.2. Consider a two-input-one-output fuzzy system that is constructed from the following two rules:

$$IF\ x_1\ is\ A_1\ and\ x_2\ is\ A_2,\ THEN\ y\ is\ A_1 \qquad (8.27)$$
$$IF\ x_1\ is\ A_2\ and\ x_2\ is\ A_1,\ THEN\ y\ is\ A_2 \qquad (8.28)$$

where A_1 and A_2 are fuzzy sets in R with membership functions

$$\mu_{A_1}(u) = \begin{cases} 1 - |u|\ if\ -1 \le u \le 1 \\ 0\ otherwise \end{cases} \qquad (8.29)$$

$$\mu_{A_2}(u) = \begin{cases} 1 - |u-1|\ if\ 0 \le u \le 2 \\ 0\ otherwise \end{cases} \qquad (8.30)$$

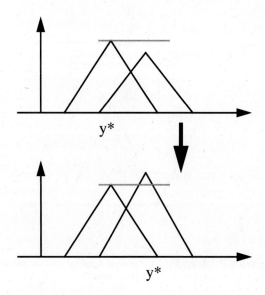

Figure 8.4. An example of maximum defuzzifier where small change in B' results in large change in y^*.

Suppose that the input to the fuzzy system is $(x_1^*, x_2^*) = (0.3, 0.6)$ and we use the singleton fuzzifier. Determine the output of the fuzzy system y^* in the following situations: (a) product inference engine (7.23) and center average defuzzifier (8.18); (b) product inference engine (7.23) and center of gravity defuzzifier (8.15); (c) Lakasiewicz inference engine (7.30) and mean of maxima defuzzifier (8.26); and (d) Lakasiewicz inference engine (7.30) and center average defuzzifier (8.18).

(a) Since we use singleton fuzzifier, from Lemma 7.3 ((7.28)) and (8.29)-(8.30) we have

$$\mu_{B'}(y) = max[\mu_{A_1}(0.3)\mu_{A_2}(0.6)\mu_{A_1}(y), \mu_{A_2}(0.3)\mu_{A_1}(0.6)\mu_{A_2}(y)]$$
$$= max[0.7 * 0.6 * \mu_{A_1}(y), 0.3 * 0.4 * \mu_{A_2}(y)]$$
$$= max[0.42\mu_{A_1}(y), 0.12\mu_{A_2}(y)] \qquad (8.31)$$

which is shown in Fig. 8.2 with $\bar{y}^1 = 0, \bar{y}^2 = 1, w_1 = 0.42$, and $w_2 = 0.12$. Hence, from (8.19) we obtain that the center average defuzzifier gives

$$y^* = \frac{0.12}{0.42 + 0.12} = 0.2222 \qquad (8.32)$$

(b) Following (a) and (8.20)-(8.21), we have

$$\int_V \mu_{B'}(y)dy = 0.42 + 0.12 - \frac{1}{2}\frac{0.42 * 0.12}{0.12 + 0.42} = 0.4933 \qquad (8.33)$$

$$\int_V y\mu_{B'}(y)dy = -\frac{1}{6}0.42 + 0.12 + \frac{1}{6}\frac{(0.42)^3}{(0.42 + 0.12)^2} = 0.0923 \qquad (8.34)$$

Hence, the y^* in this case is

$$y^* = 0.0923/0.4933 = 0.1871 \qquad (8.35)$$

(c) If we use the Lukasiewicz inference engine, then from Lemma 7.4 ((7.33)) we have

$$\begin{aligned}\mu_{B'}(y) &= min\{1, 1 - min[\mu_{A_1}(0.3), \mu_{A_2}(0.6)] + \mu_{A_1}(y),\\ &\quad 1 - min[\mu_{A_2}(0.3), \mu_{A_1}(0.6)] + \mu_{A_2}(y)\}\\ &= min[1, 0.4 + \mu_{A_1}(y), 0.7 + \mu_{A_2}(y)] \end{aligned} \qquad (8.36)$$

which is plotted in Fig. 8.5. From Fig. 8.5 we see that $sup_{y \in V} \mu_{B'}(y)$ is achieved in the interval $[0.3, 0.4]$, so the mean of maxima defuzzifier gives

$$y^* = 0.35 \qquad (8.37)$$

(d) From Fig. 8.5 we see that in this case $\bar{y}^1 = 0, \bar{y}^2 = 1, w_1 = 1$ and $w_2 = 1$. So the center average defuzzifier (8.18) gives

$$y^* = \frac{0 * 1 + 1 * 1}{1 + 1} = 0.5 \qquad (8.38)$$

□

8.3   Summary and Further Readings

In this chapter we have demonstrated the following:

- The definitions and intuitive meanings of the singleton, Gaussian and triangular fuzzifiers, and the center of gravity, center average and maximum defuzzifiers.

- Computing the outputs of the fuzzy systems for different combinations of the fuzzifiers, defuzzifiers, and fuzzy inference engines for specific examples.

Different defuzzification schemes were studied in detail in the books Driankov, Hellendoorn and Reifrank [1993] and Yager and Filev [1994].

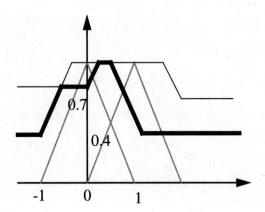

Figure 8.5. A graphical representation of the membership function $\mu_{B'}(y)$ of (8.36).

8.4 Exercises

Exercise 8.1. Suppose that a fuzzy rule base consists of the M rules (7.1) with

$$\mu_{A_i^l}(x_i) = \begin{cases} 1 - \frac{|x_i - \bar{x}_i^l|}{\sigma_i^l} & if \ |x_i - \bar{x}_i^l| \le \sigma_i^l \\ 0 \ otherwise \end{cases} \tag{8.39}$$

and that we use the triangular fuzzifier (8.3). Determine the output of the fuzzy inference engine $\mu_{B'}(y)$ for:

(a) product inference engine (7.23) with all $\star = algebraic \ product$, and

(b) minimum inference engine (7.24) with all $\star = min$.

Exercise 8.2. Consider Example 8.1 and determine the y^* using the indexed center of gravity defuzzifier with $\alpha = 0.1$. Compute the y^* for the specific values of w_1 and w_2 in Table 8.1.

Exercise 8.3. Consider Example 8.2 and determine the fuzzy system output y^* (with input $(x_1^*, x_2^*) = (0.3, 0.6)$) for:

(a) Zadeh inference engine (7.31) and maximum (or mean of maxima) defuzzifier, and

(b) Dienes-Rescher inference engine (7.32) with maximum (or mean of maxima) defuzzifier.

Exercise 8.4. Use a practical example, such as the mobile robot path planning problem, to show that the center of gravity and center average defuzzifiers may create problems when the fuzzy set B' is non-convex.

Exercsie 8.5. When the fuzzy set B' is non-convex, the so-called center of largest area defuzzifier is proposed. This method determines the convex part of B' that has the largest area and defines the crisp output y^* to be the center of gravity of this convex part. Create a specific non-convex B' and use the center of largest area defuzzifier to determine the defuzzified output y^*.

Chapter 9

Fuzzy Systems as Nonlinear Mappings

9.1 The Formulas of Some Classes of Fuzzy Systems

From Chapters 7 and 8 we see that there are a variety of choices in the fuzzy inference engine, fuzzifier, and defuzzifier modules. Specifically, we proposed five fuzzy inference engines (product, minimum, Lukasiewicz, Zadeh, and Dienes-Rescher), three fuzzifiers (singleton, Gaussian and triangular), and three types of defuzzifiers (center-of-gravity, center average, and maximum). Therefore, we have at least $5*3*3 = 45$ types of fuzzy systems by combining these different types of inference engines, fuzzifiers, and defuzzifiers. In this chapter, we will derive the detailed formulas of certain classes of fuzzy systems. We will see that some classes of fuzzy systems are very useful, while others do not make a lot of sense. That is, not every combination results in useful fuzzy systems. Because in Chapter 8 we showed that the center-of-gravity defuzzifier is computationally expensive and the center average defuzzifier is a good approximation of it, we will not consider the center-of-gravity defuzzifier in this chapter. We classify the fuzzy systems to be considered into two groups: fuzzy systems with center average defuzzifier, and fuzzy systems with maximum defuzzifier.

9.1.1 Fuzzy Systems with Center Average Defuzzifier

Lemma 9.1. Suppose that the fuzzy set B^l in (7.1) is normal with center $\bar{y}^l$. Then the fuzzy systems with fuzzy rule base (7.1), product inference engine (7.23), singleton fuzzifier (8.1), and center average defuzzifier (8.18) are of the following form:

$$f(x) = \frac{\sum_{l=1}^{M} \bar{y}^l (\prod_{i=1}^{n} \mu_{A_i^l}(x_i))}{\sum_{l=1}^{M} (\prod_{i=1}^{n} \mu_{A_i^l}(x_i))} \tag{9.1}$$

where $x \in U \subset R^n$ is the input to the fuzzy system, and $f(x) \in V \subset R$ is the output of the fuzzy system.

Proof: Substituting (8.1) into (7.23), we have

$$\mu_{B'}(y) = \max_{l=1}^{M}[\prod_{i=1}^{n} \mu_{A_i^l}(x_i^*)\mu_{B^l}(y)] \tag{9.2}$$

Since for a given input x_i^*, the center of the $l'th$ fuzzy set in (9.2) (that is, the fuzzy set with membership function $\mu_{A_i^l}(x_i^*)\mu_{B^l}(y)$) is the center of B^l, we see that the $\bar{y}^l$ in (8.18) is the same $\bar{y}^l$ in this lemma. Additionally, the height of the $l'th$ fuzzy set in (9.2), denoted by w_l in (8.18), is $\prod_{i=1}^{n} \mu_{A_i^l}(x_i^*)\mu_{B^l}(\bar{y}^l) = \prod_{i=1}^{n} \mu_{A_i^l}(x_i^*)$ (since B^l is normal). Hence, using the center average defuzzifier (8.18) for (9.2), we obtain

$$y^* = \frac{\sum_{l=1}^{M} \bar{y}^l (\prod_{i=1}^{n} \mu_{A_i^l}(x_i^*))}{\sum_{l=1}^{M} (\prod_{i=1}^{n} \mu_{A_i^l}(x_i^*))} \tag{9.3}$$

Using the notion of this lemma, we have $x^* = x$ and $y^* = f(x)$; thus, (9.3) becomes (9.1). $\square$

From Lemma 9.1 we see that the fuzzy system is a nonlinear mapping from $x \in U \subset R^n$ to $f(x) \in V \subset R$, and (9.1) gives the detailed formula of this mapping. The fuzzy systems in the form of (9.1) are the most commonly used fuzzy systems in the literature. They are computationally simple and intuitively appealing. From (9.1) we see that the output of the fuzzy system is a weighted average of the centers of the fuzzy sets in the THEN parts of the rules, where the weights equal the membership values of the fuzzy sets in the IF parts of the rules at the input point. Consequently, the more the input point agrees with the IF part of a rule, the larger weight this rule is given; this makes sense intuitively.

Lemma 9.1 also reveals an important contribution of fuzzy systems theory that is summarized as follows:

- **The dual role of fuzzy systems**: On one hand, fuzzy systems are rule-based systems that are constructed from a collection of linguistic rules; on the other hand, fuzzy systems are nonlinear mappings that in many cases can be represented by precise and compact formulas such as (9.1). *An important contribution of fuzzy systems theory is to provide a systematic procedure for transforming a set of linguistic rules into a nonlinear mapping.* Because non-linear mappings are easy to implement, fuzzy systems have found their way into a variety of engineering applications.

By choosing different forms of membership functions for $\mu_{A_i^l}$ and μ_{B^l}, we obtain different subclasses of fuzzy systems. One choice of $\mu_{A_i^l}$ and μ_{B^l} is Gaussian membership function. Specifically, if we choose the following *Gaussian membership function* for $\mu_{A_i^l}$ and μ_{B^l}:

$$\mu_{A_i^l}(x_i) = a_i^l exp[-(\frac{x_i - \bar{x}_i^l}{\sigma_i^l})^2] \tag{9.4}$$

$$\mu_{B^l}(y) = exp[-(y - \bar{y}^l)^2] \tag{9.5}$$

where $a_i^l \in (0,1], \sigma_i^l \in (0,\infty)$ and $\bar{x}_i^l, \bar{y}^l \in R$ are real-valued parameters, then the fuzzy systems in Lemma 9.1 become

$$f(x) = \frac{\sum_{l=1}^{M} \bar{y}^l[\prod_{i=1}^{n} a_i^l exp(-(\frac{x_i - \bar{x}_i^l}{\sigma_i})^2)]}{\sum_{l=1}^{M}[\prod_{i=1}^{n} a_i^l exp(-(\frac{x_i - \bar{x}_i^l}{\sigma_i})^2)]} \tag{9.6}$$

We call the fuzzy systems in the form of (9.6) *fuzzy systems with product inference engine, singleton fuzzifier, center average defuzzifier, and Gaussian membership functions*. Other popular choices of $\mu_{A_i^l}$ and μ_{B^l} are triangular and trapezoid membership functions. We will study the fuzzy systems with these types of membership functions in detail in Chapters 10 and 11.

Another class of commonly used fuzzy systems is obtained by replacing the product inference engine in Lemma 9.1 by the minimum inference engine. Using the same procedure as in the proof of Lemma 9.1, we obtain that *the fuzzy systems with fuzzy rule base (7.1), minimum inference engine (7.24), singleton fuzzifier (8.1), and center average defuzzifier (8.18)* are of the following form:

$$f(x) = \frac{\sum_{l=1}^{M} \bar{y}^l(\min_{i=1}^{n} \mu_{A_i^l}(x_i))}{\sum_{l=1}^{M}(\min_{i=1}^{n} \mu_{A_i^l}(x_i))} \tag{9.7}$$

where the variables have the same meaning as in (9.1).

We showed in Chapter 8 (Lemma 8.1) that if the membership functions for A_i^l are Gaussian, then the Gaussian fuzzifier also significantly simplifies the fuzzy inference engine. What do the fuzzy systems look like in this case?

Lemma 9.2. The fuzzy systems with fuzzy rule base (7.1), product inference engine (7.23), Gaussian fuzzifier (8.2) with ⋆=product, center average defuzzifier (8.18), and Gaussian membership functions (9.4) and (9.5) (with $a_i^l = 1$) are of the following form:

$$f(x) = \frac{\sum_{l=1}^{M} \bar{y}^l[\prod_{i=1}^{n} exp(-\frac{(x_i - \bar{x}_i^l)^2}{a_i^2 + (\sigma_i^l)^2})]}{\sum_{l=1}^{M}[\prod_{i=1}^{n} exp(-\frac{(x_i - \bar{x}_i^l)^2}{a_i^2 + (\sigma_i^l)^2})]} \tag{9.8}$$

If we replace the product inference engine (7.23) with the minimum inference engine

(7.24) and use $\star = min$ in (8.2), then the fuzzy systems become

$$f(x) = \frac{\sum_{l=1}^{M} \bar{y}^l [\min_{i=1}^n exp(-(\frac{x_i - \bar{x}_i^l}{a_i + \sigma_i^l})^2)]}{\sum_{l=1}^{M} [\min_{i=1}^n exp(-(\frac{x_i - \bar{x}_i^l}{a_i + \sigma_i^l})^2)]} \tag{9.9}$$

Proof: Substituting (8.6) into (8.5) and use x for x^*, we have

$$\mu_{B'}(y) = \max_{l=1}^{M} \{ \prod_{i=1}^n exp[-(\frac{\frac{a_i^2 \bar{x}_i^l + (\sigma_i^l)^2 x_i}{a_i^2 + (\sigma_i^l)^2} - \bar{x}_i^l}{\sigma_i^l})^2 - (\frac{\frac{a_i^2 \bar{x}_i^l + (\sigma_i^l)^2 x_i}{a_i^2 + (\sigma_i^l)^2} - x_i}{a_i})^2] \mu_{B^l}(y) \}$$

$$= \max_{l=1}^{M} \{ \prod_{i=1}^n exp[-\frac{(x_i - \bar{x}_i^l)^2}{a_i^2 + (\sigma_i^l)^2}] \mu_{B^l}(y) \} \tag{9.10}$$

Using the same arguments as in the proof of Lemma 9.1 and applying the center average defuzzifier (8.18) to (9.10), we obtain (9.8). Similarly, substituting (8.8) into (8.7), we have

$$\mu_{B'}(y) = \max_{l=1}^{M} \{ min[exp(-(\frac{\frac{a_1 \bar{x}_1^l + \sigma_1^l x_1}{a_1 + \sigma_1^l} - \bar{x}_1^l}{\sigma_1^l})^2), ..., exp(-(\frac{\frac{a_n \bar{x}_n^l + \sigma_n^l x_n}{a_n + \sigma_n^l} - \bar{x}_n^l}{\sigma_n^l})^2), \mu_{B^l}(y)] \}$$

$$= \max_{l=1}^{M} \{ min[exp(-(\frac{x_1 - \bar{x}_1^l}{a_1 + \sigma_1^l})^2), ..., exp(-(\frac{x_n - \bar{x}_n^l}{a_n + \sigma_n^l})^2), \mu_{B^l}(y)] \} \tag{9.11}$$

Applying the center average defuzzifier (8.18) to (9.11), we obtain (9.9). $\square$

In Chapter 7 we saw that the product and minimum inference engines are quite different from the Lukasiewicz, Zadeh and Dienes-Rescher inference engines. What do the fuzzy systems with these inference engines look like?

Lemma 9.3. If the fuzzy set B^l in (7.1) are normal with center $\bar{y}^l$, then the fuzzy systems with fuzzy rule base (7.1), Lukasiewicz inference engine (7.30) or Dienes-Rescher inference engine (7.32), singleton fuzzifier (8.1) or Gaussian fuzzifier (8.2) or triangular fuzzifier (8.3), and center average defuzzifier (8.18) are of the following form:

$$f(x) = \frac{1}{M} \sum_{l=1}^{M} \bar{y}^l \tag{9.12}$$

Proof: Since $\mu_{B^l}(\bar{y}^l) = 1$, we have $1 - \min_{i=1}^n (\mu_{A_i^l}(x_i)) + \mu_{B^l}(\bar{y}^l) \geq 1$; therefore, the height of the $l'th$ fuzzy set in (7.30) is

$$w_l = \sup_{x \in U} min[\mu_{A'}(x), 1 - \min_{i=1}^n (\mu_{A_i^l}(x_i)) + \mu_{B^l}(\bar{y}^l)]$$

$$= \sup_{x \in U} \mu_{A'}(x)$$

$$= 1 \tag{9.13}$$

where we use the fact that for all the three fuzzifiers (8.1)-(8.3) we have $\sup_{x \in U} \mu_{A'}(x)$ $= 1$. Similarly, the height of the $l'th$ fuzzy set in (7.32) is also equal to one. Hence, with the center average defuzzifier (8.18) we obtain (9.12). $\square$

The fuzzy systems in the form of (9.12) do not make a lot of sense because it gives a constant output no matter what the input is. Therefore, the combinations of fuzzy inference engine, fuzzifier, and defuzzifier in Lemma 9.3 do not result in useful fuzzy systems.

9.1.2   Fuzzy Systems with Maximum Defuzzifier

Lemma 9.4. Suppose the fuzzy set B^l in (7.1) is normal with center $\bar{y}^l$, then the fuzzy systems with fuzzy rule base (7.1), product inference engine (7.23), singleton fuzzifier (8.1), and maximum defuzzifier (8.23) are of the following form:

$$f(x) = \bar{y}^{l*} \tag{9.14}$$

where $l* \in \{1, 2, ..., M\}$ is such that

$$\prod_{i=1}^{n} \mu_{A_i^{l*}}(x_i) \geq \prod_{i=1}^{n} \mu_{A_i^{l}}(x_i) \tag{9.15}$$

for all $l = 1, 2, ..., M$.

Proof: From (7.28) (Lemma 7.3) and noticing that $x^* = x$ in this case, we have

$$\sup_{y \in V} \mu_{B'}(y) = \sup_{y \in V} \max_{l=1}^{M} [\prod_{i=1}^{n} \mu_{A_i^{l}}(x_i) \mu_{B^l}(y)] \tag{9.16}$$

Since $\sup_{y \in V}$ and $\max_{l=1}^{M}$ are interchangeable and B^l is normal, we have

$$\sup_{y \in V} \mu_{B'}(y) = \max_{l=1}^{M} [\sup_{y \in V} \prod_{i=1}^{n} \mu_{A_i^{l}}(x_i) \mu_{B^l}(y)]$$

$$= \max_{l=1}^{M} [\prod_{i=1}^{n} \mu_{A_i^{l}}(x_i)]$$

$$= \prod_{i=1}^{n} \mu_{A_i^{l*}}(x_i) \tag{9.17}$$

where $l*$ is defined according to (9.15). Since $\mu_{B^l}(\bar{y}^{l*}) \leq 1$ when $l \neq l*$ and $\mu_{B^{l*}}(\bar{y}^{l*}) = 1$, we have

$$\mu_{B'}(\bar{y}^{l*}) = \max_{l=1}^{M} [\prod_{i=1}^{n} \mu_{A_i^{l}}(x_i) \mu_{B^l}(\bar{y}^{l*})]$$

$$= \prod_{i=1}^{n} \mu_{A_i^{l*}}(x_i) \tag{9.18}$$

Hence, the $\sup_{y \in V}$ in (9.16) is achieved at $\bar{y}^{l*}$. Using the maximum defuzzifier (8.23) we obtain (9.14). $\square$

From Lemma 9.4 we see that the fuzzy systems in this case are simple functions, that is, they are piece-wise constant functions, and these constants are the centers of the membership functions in the THEN parts of the rules. From (9.15) we see that as long as the product of membership values of the IF-part fuzzy sets of the rule is greater than or equal to those of the other rules, the output of the fuzzy system remains unchanged. Therefore, these kinds of fuzzy systems are robust to small disturbances in the input and in the membership functions $\mu_{A_i^l}(x_i)$. However, these fuzzy systems are not continuous, that is, when $l*$ changes from one number to the other, $f(x)$ changes in a discrete fashion. If the fuzzy systems are used in decision making or other open-loop applications, this kind of abrupt change may be tolerated, but it is usually unacceptable in closed-loop control.

The next lemma shows that we can obtain a similar result if we use the minimum inference engine.

Lemma 9.5. If we change the product inference engine in Lemma 9.4 to the minimum inference engine (7.24) and keep the others unchanged, then the fuzzy systems are of the same form as (9.14) with $l*$ determined by

$$\min_{i=1}^{n}[\mu_{A_i^{l*}}(x_i)] \geq \min_{i=1}^{n}[\mu_{A_i^l}(x_i)] \tag{9.19}$$

where $l = 1, 2, ..., M$.

Proof: From (7.29) (Lemma 7.3) and using the facts that $\sup_{y \in V}$ and $\max_{l=1}^{M}$ are interchangeable and that B^l are normal, we have

$$\sup_{y \in V} \mu_{B'}(y) = \max_{l=1}^{M}[\sup_{y \in V} min(\mu_{A_1^l}(x_1), ..., \mu_{A_n^l}(x_n), \mu_{B^l}(y))]$$

$$= \max_{l=1}^{M}[\min_{i=1}^{n}(\mu_{A_i^l}(x_i))]$$

$$= \min_{i=1}^{n}(\mu_{A_i^{l*}}(x_i)) \tag{9.20}$$

Also from (7.29) we have that $\mu_{B'}(\bar{y}^{l*}) = \min_{i=1}^{n}(\mu_{A_i^{l*}}(x_i))$, thus the $\sup_{y \in V}$ in (9.20) is achieved at $\bar{y}^{l*}$. Hence, the maximum defuzzifier (8.23) gives (9.14). $\square$

Again, we obtain a class of fuzzy systems that are simple functions.

It is difficult to obtain closed-form formulas for fuzzy systems with maximum defuzzifier and Lukasiewicz, Zadeh, or Dienes-Rescher inference engines. The difficulty comes from the fact that the $\sup_{y \in V}$ and min operators are not interchangeable in general, therefore, from (7.30)-(7.32) we see that the maximum defuzzification becomes an optimization problem for a non-smooth function. In these cases, for a given input x, the output of the fuzzy system has to be computed in a step-by-step fashion, that is, computing the outputs of fuzzifier, fuzzy inference engine, and defuzzifier in sequel. Note that the output of the fuzzy inference engine is a function, not a single value, so the computation is very complex. We will not use this type of fuzzy systems (maximum defuzzifier with Lukasiewicz, Zadeh, or Dienes-rescher inference engine) in the rest of this book.

9.2   Fuzzy Systems As Universal Approximators

In the last section we showed that certain types of fuzzy systems can be written as compact nonlinear formulas. On one hand, these compact formulas simplify the computation of the fuzzy systems; on the other hand, they give us a chance to analyze the fuzzy systems in more details. We see that the fuzzy systems are particular types of nonlinear functions, so no matter whether the fuzzy systems are used as controllers or decision makers or signal processors or any others, it is interesting to know the capability of the fuzzy systems from a function approximation point of view. For example, what types of nonlinear functions can the fuzzy systems represent or approximate and to what degree of accuracy? If the fuzzy systems can approximate only certain types of nonlinear functions to a limited degree of accuracy, then the fuzzy systems would not be very useful in general applications. But if the fuzzy systems can approximate any nonlinear function to arbitrary accuracy, then they would be very useful in a wide variety of applications. In this section, we prove that certain classes of fuzzy systems that we studied in the last section have this universal approximation capability. Specifically, we have the following main theorem.

Theorem 9.1 (Universal Approximation Theorem). Suppose that the input universe of discourse U is a compact set in R^n. Then, for any given real continuous function $g(x)$ on U and arbitrary $\epsilon > 0$, there exists a fuzzy system $f(x)$ in the form of (9.6) such that

$$\sup_{x \in U} |f(x) - g(x)| < \epsilon \tag{9.21}$$

That is, the fuzzy systems with product inference engine, singleton fuzzifier, center average defuzzifier, and Gaussian membership functions are universal approximators.

One proof of this theorem is based on the following Stone-Weierstrass Theorem, which is well known in analysis.

Stone-Weierstrass Theorem (Rudin [1976]). Let Z be a set of real continuous functions on a compact set U. If (i) Z is an *algebra*, that is, the set Z is closed under addition, multiplication, and scalar multiplication; (ii) Z *separates points on U*, that is, for every $x, y \in U, x \neq y$, there exists $f \in Z$ such that $f(x) \neq f(y)$; and (iii) Z *vanishes at no point of U*, that is, for each $x \in U$ there exists $f \in Z$ such that $f(x) \neq 0$; then for any real continuous function $g(x)$ on U and arbitrary $\epsilon > 0$, there exists $f \in Z$ such that $\sup_{x \in U} |f(x) - g(x)| < \epsilon$.

Proof of Theorem 9.1: Let Y be the set of all fuzzy systems in the form of

(9.6). We now show that Y is an algebra, Y separates points on U, and Y vanishes at no point of U.

Let $f_1, f_2 \in Y$, so that we can write them as

$$f_1(x) = \frac{\sum_{l=1}^{M1} \bar{y}1^l [\prod_{i=1}^n a1_i^l exp(-(\frac{x_i - \bar{x}1_i^l}{\sigma1_i^l})^2)]}{\sum_{l=1}^{M1} [\prod_{i=1}^n a1_i^l exp(-(\frac{x_i - \bar{x}1_i^l}{\sigma1_i^l})^2)]} \tag{9.22}$$

$$f_2(x) = \frac{\sum_{l=1}^{M2} \bar{y}2^l [\prod_{i=1}^n a2_i^l exp(-(\frac{x_i - \bar{x}2_i^l}{\sigma2_i^l})^2)]}{\sum_{l=1}^{M2} [\prod_{i=1}^n a2_i^l exp(-(\frac{x_i - \bar{x}2_i^l}{\sigma2_i^l})^2)]} \tag{9.23}$$

Hence,

$$f_1(x) + f_2(x) = \frac{\sum_{l1=1}^{M1} \sum_{l2=1}^{M2} (\bar{y}1^{l1} + \bar{y}2^{l2}) [\prod_{i=1}^n a1_i^{l1} a2_i^{l2} exp(-(\frac{x_i - \bar{x}1_i^{l1}}{\sigma1_i^{l1}})^2 - (\frac{x_i - \bar{x}2_i^{l2}}{\sigma2_i^{l2}})^2)]}{\sum_{l1=1}^{M1} \sum_{l2=1}^{M2} [\prod_{i=1}^n a1_i^{l1} a2_i^{l2} exp(-(\frac{x_i - \bar{x}1_i^{l1}}{\sigma1_i^{l1}})^2 - (\frac{x_i - \bar{x}2_i^{l2}}{\sigma2_i^{l2}})^2)]} \tag{9.24}$$

Since $a1_i^{l1} a2_i^{l2} exp(-(\frac{x_i - \bar{x}1_i^{l1}}{\sigma1_i^{l1}})^2 - (\frac{x_i - \bar{x}2_i^{l2}}{\sigma2_i^{l2}})^2)$ can be represented in the form of (9.4) and $\bar{y}1^{l1} + \bar{y}2^{l2}$ can be viewed as the center of a fuzzy set in the form of (9.5), $f_1(x) + f_2(x)$ is in the form of (9.6); that is, $f_1 + f_2 \in Y$. Similarly,

$$f_1(x) f_2(x) = \frac{\sum_{l1=1}^{M1} \sum_{l2=1}^{M2} (\bar{y}1^{l1} \bar{y}2^{l2}) [\prod_{i=1}^n a1_i^{l1} a2_i^{l2} exp(-(\frac{x_i - \bar{x}1_i^{l1}}{\sigma1_i^{l1}})^2 - (\frac{x_i - \bar{x}2_i^{l2}}{\sigma2_i^{l2}})^2)]}{\sum_{l1=1}^{M1} \sum_{l2=1}^{M2} [\prod_{i=1}^n a1_i^{l1} a2_i^{l2} exp(-(\frac{x_i - \bar{x}1_i^{l1}}{\sigma1_i^{l1}})^2 - (\frac{x_i - \bar{x}2_i^{l2}}{\sigma2_i^{l2}})^2)]} \tag{9.25}$$

which also is in the form of (9.6), hence, $f_1 f_2 \in Y$. Finally, for arbitrary $c \in R$,

$$cf_1(x) = \frac{\sum_{l=1}^{M1} c\bar{y}1^l [\prod_{i=1}^n a1_i^l exp(-(\frac{x_i - \bar{x}1_i^l}{\sigma1_i^l})^2)]}{\sum_{l=1}^{M1} [\prod_{i=1}^n a1_i^l exp(-(\frac{x_i - \bar{x}1_i^l}{\sigma1_i^l})^2)]} \tag{9.26}$$

which is again in the form of (9.6), so $cf_1 \in Y$. Hence, Y is an algebra.

We show that Y separates points on U by constructing a required fuzzy system $f(x)$. Let $x^0, z^0 \in U$ be two arbitrary points and $x^0 \neq z^0$. We choose the parameters of the $f(x)$ in the form of (9.6) as follows: $M = 2, \bar{y}^1 = 0, \bar{y}^2 = 1, a_i^l = 1, \sigma_i^l = 1, \bar{x}_i^1 = x_i^0$ and $\bar{x}_i^2 = z_i^0$, where $i = 1, 2, ..., n$ and $l = 1, 2$. This specific fuzzy system is

$$f(x) = \frac{exp(-||x - z^0||_2^2)}{exp(-||x - x^0||_2^2) + exp(-||x - z^0||_2^2)} \tag{9.27}$$

from which we have

$$f(x^0) = \frac{exp(-||x^0 - z^0||_2^2)}{1 + exp(-||x^0 - z^0||_2^2)} \tag{9.28}$$

and

$$f(z^0) = \frac{1}{1 + exp(-||x^0 - z^0||_2^2)} \tag{9.29}$$

Since $x^0 \neq z^0$, we have $exp(-||x^0 - z^0||_2^2) \neq 1$ which, from (9.28) and (9.29), gives $f(x^0) \neq f(z^0)$. Hence, Y separates points on U.

To show that Y vanishes at no point of U, we simply observe that any fuzzy system $f(x)$ in the form of (9.6) with all $\bar{y}^l > 0$ has the property that $f(x) > 0, \forall x \in U$. Hence, Y vanishes at no point of U.

In summary of the above and the Stone-Weierstrass Theorem, we obtain the conclusion of this theorem. $\square$

Theorem 9.1 shows that fuzzy systems can approximate continuous functions to arbitrary accuracy; the following corollary extends the result to discrete functions.

Corollary 9.1. For any square-integrable function $g(x)$ on the compact set $U \subset R^n$, that is, for any $g \in L_2(U) = \{g : U \to R| \int_U |g(x)|^2 dx < \infty\}$, there exists fuzzy system $f(x)$ in the form of (9.6) such that

$$(\int_U |f(x) - g(x)|^2 dx)^{1/2} < \epsilon \tag{9.30}$$

Proof: Since U is compact, $\int_U dx = E < \infty$. Since continuous functions on U form a dense subset of $L_2(U)$ (Rudin [1976]), for any $g \in L_2(U)$ there exists a continuous function $\bar{g}$ on U such that $(\int_U |g(x) - \bar{g}(x)|^2 dx)^{1/2} < \epsilon/2$. By Theorem 9.1, there exists $f \in Y$ such that $\sup_{x \in U} |f(x) - \bar{g}(x)| < \epsilon/(2E^{1/2})$. Hence, we have

$$(\int_U |f(x) - g(x)|^2 dx)^{1/2} \leq (\int_U |f(x) - \bar{g}(x)|^2 dx)^{1/2} + (\int_U |\bar{g}(x) - g(x)|^2 dx)^{1/2}$$

$$< (\int_U (\sup_{x \in U} |f(x) - \bar{g}(x)|)^2 dx)^{1/2} + \epsilon/2$$

$$< (\frac{\epsilon^2}{2^2 E} E)^{1/2} + \epsilon/2 = \epsilon \tag{9.31}$$

$\square$

Theorem 9.1 and Corollary 9.1 provide a justification for using fuzzy systems in a variety of applications. Specifically, they show that for any kind of nonlinear operations the problem may require, it is always possible to design a fuzzy system that performs the required operation with any degree of accuracy. They also provide a theoretical explanation for the success of fuzzy systems in practical applications.

However, Theorem 9.1 and Corollary 9.1 give only existence result; that is, they show that there *exists* a fuzzy system in the form of (9.6) that can approximate any function to arbitrary accuracy. They do not show how to find such a fuzzy system. For engineering applications, knowing the existence of an ideal fuzzy system is not enough; we must develop approaches that can find good fuzzy systems for the particular applications. Depending upon the information provided, we may or may not find the ideal fuzzy system. In the next few chapters, we will develop a variety of approaches to designing the fuzzy systems.

9.3   Summary and Further Readings

In this chapter we have demonstrated the following:

- The compact formulas of some useful classes of fuzzy systems.

- How to derive compact formulas for any classes of fuzzy systems if such compact formulas exist.

- How to use the Stone-Weierstrass Theorem.

The derivations of the mathematical formulas of the fuzzy systems are new. A related reference is Wang [1994]. The Universal Approximation Theorem and its proof are taken from Wang [1992]. Other approaches to this problem can be found in Buckley [1992b] and Zeng and Singh [1994].

9.4   Exercises

Exercise 9.1. Derive the compact formula for the fuzzy systems with fuzzy rule base (7.1), Zadeh inference engine (7.31), singleton fuzzifier (8.1), and center average defuzzifier (8.18).

Exercise 9.2. Repeat Exercise 9.1 using Lukasiewicz inference engine rather than Zadeh inference engine.

Exercise 9.3. Show that the fuzzy systems in the form of (9.1) have the universal approximation property in Theorem 9.1.

Exercise 9.4. Can you use the Stone-Weierstrass Theorem to prove that fuzzy systems in the form of (9.7) or (9.6) with $a_i^l = 1$ are universal approximators? Explain your answer.

Exercise 9.5. Use the Stone-Weierstrass Theorem to prove that polynomials are universal appproximators.

Exercise 9.6. Plot the fuzzy systems $f_1(x)$ and $f_2(x)$ for $x \in U = [-1, 2] \times [-1, 2]$, where $f_1(x)$ is the fuzzy system with the two rules (8.27) and (8.28), product inference engine (7.23), singleton fuzzifier (8.1), and maximum defuzzifier (8.23), and $f_2(x)$ is the same as $f_1(x)$ except that the maximum defuzzifier is replaced by the center average defuzzifier (8.18).

Chapter 10

Approximation Properties of Fuzzy Systems I

In Chapter 9 we proved that fuzzy systems are universal approximators; that is, they can approximate any function on a compact set to arbitrary accuracy. However, this result showed only the existence of the optimal fuzzy system and did not provide methods to find it. In fact, finding the optimal fuzzy system is much more difficult than proving its existence. Depending upon the information provided, we may or may not find the optimal fuzzy system.

To answer the question of how to find the optimal fuzzy system, we must first see what information is available for the nonlinear function $g(x) : U \subset R^n \to R$, which we are asked to approximate. Generally speaking, we may encounter the following three situations:

- The analytic formula of $g(x)$ is known.

- The analytic formula of $g(x)$ is unknown, but for any $x \in U$ we can determine the corrspending $g(x)$. That is, $g(x)$ is a black box—we know the input-output behavior of $g(x)$ but do not know the details inside it.

- The analytic formula of $g(x)$ is unknown and we are provided only a limited number of input-output pairs $(x^j, g(x^j))$, where $x^j \in U$ cannot be arbitrarily chosen.

The first case is not very interesting because if the analytic formula of $g(x)$ is known, we can use it for whatever purpose the fuzzy system tries to achieve. In the rare case where we want to replace $g(x)$ by a fuzzy system, we can use the methods for the second case because the first case is a special case of the second one. Therefore, we will not consider the first case separately.

The second case is more realistic. We will study it in detail in this and the following chapters.

The third case is the most general one in practice. This is especially true for fuzzy control because stability requirements for control systems may prevent us from choosing the input values arbitrarily. We will study this case in detail in Chapters 12-15.

So, in this chapter we assume that the analytic formula of $g(x)$ is unknown but we can determine the input-output pairs $(x; g(x))$ for any $x \in U$. Our task is to design a fuzzy system that can approximate $g(x)$ in some optimal manner.

10.1 Preliminary Concepts

We first introduce some concepts.

Definition 10.1: *Pseudo-Trapezoid Membership Function.* Let $[a, d] \subset R$. The *pseudo-trapezoid membership function* of fuzzy set A is a continuous function in R given by

$$\mu_A(x; a, b, c, d, H) = \begin{cases} I(x), & x \in [a, b) \\ H, & x \in [b, c] \\ D(x), & x \in (c, d] \\ 0, & x \in R - (a, d) \end{cases} \tag{10.1}$$

where $a \leq b \leq c \leq d, a < d, 0 < H \leq 1, 0 \leq I(x) \leq 1$ is a nondecreasing function in $[a, b)$ and $0 \leq D(x) \leq 1$ is a nonincreasing function in $(c, d]$. When the fuzzy set A is normal (that is, $H = 1$), its membership function is simply written as $\mu_A(x; a, b, c, d)$.

Fig. 10.1 shows some examples of pseudo-trapezoid membership functions. If the universe of discourse is bounded, then a, b, c, d are finite numbers. Pseudo-trapezoid membership functions include many commonly used membership functions as special cases. For example, if we choose

$$I(x) = \frac{x - a}{b - a}, D(x) = \frac{x - d}{c - d} \tag{10.2}$$

then the pseudo-trapezoid membership functions become the *trapezoid membership functions*. If $b = c$ and $I(x)$ and $D(x)$ are as in (10.2), we obtain the *triangular membership functions*; a triangular membership function is denoted as $\mu_A(x; a, b, d)$. If we choose $a = \infty, b = c = \bar{x}, d = \infty$, and

$$I(x) = D(x) = exp(-(\frac{x - \bar{x}}{\sigma})^2) \tag{10.3}$$

then the pseudo-trapezoid membership functions become the *Gaussian membership functions*. Therefore, the pseudo-trapezoid membership functions constitute a very rich family of membership functions.

Definition 10.2: *Completeness of Fuzzy Sets.* Fuzzy sets $A^1, A^2, ..., A^N$ in $W \subset R$ are said to be *complete on* W if for any $x \in W$, there exists A^j such that $\mu_{A^j}(x) > 0$.

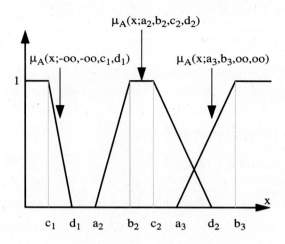

Figure 10.1. Examples of pseudo-trapezoid membership functions.

Definition 10.3: *Consistency of Fuzzy Sets.* Fuzzy sets $A^1, A^2, ..., A^N$ in $W \subset R$ are said to be *consistent on* W if $\mu_{A^j}(x) = 1$ for some $x \in W$ implies that $\mu_{A^i}(x) = 0$ for all $i \neq j$.

Definition 10.4: *High Set of Fuzzy Set.* The *high set* of a fuzzy set A in $W \subset R$ is a subset in W defined by

$$hgh(A) = \{x \in W | \mu_A(x) = \sup_{x' \in W} \mu_A(x')\} \qquad (10.4)$$

If A is a normal fuzzy set with pseudo-trapezoid membership function $\mu_A(x; a, b, c, d)$, then $hgh(A) = [b, c]$.

Definition 10.5: *Order Between Fuzzy Sets.* For two fuzzy sets A and B in $W \subset R$, we say $A > B$ if $hgh(A) > hgh(B)$ (that is, if $x \in hgh(A)$ and $x' \in hgh(B)$, then $x > x'$).

We now show some properties of fuzzy sets with pseudo-trapezoid membership functions.

Lemma 10.1. If $A^1, A^2, ..., A^N$ are consistent and normal fuzzy sets in $W \subset R$ with pseudo-trapezoid membership functions $\mu_{A^i}(x; a_i, b_i, c_i, d_i)$ $(i = 1, 2, ..., N)$, then there exists a rearrangement $\{i_1, i_2, ..., i_N\}$ of $\{1, 2, ..., N\}$ such that

$$A^{i_1} < A^{i_2} < \cdots < A^{i_N} \qquad (10.5)$$

Proof: For arbitrary $i, j \in \{1, 2, ..., N\}$, it must be true that $[b_i, c_i] \cap [b_j, c_j] = 0$, since otherwise the fuzzy sets $A^1, ..., A^N$ would not be consistent. Thus, there exists a rearrangement $\{i_1, i_2, ..., i_N\}$ of $\{1, 2, ..., N\}$ such that

$$[b_{i_1}, c_{i_1}] < [b_{i_2}, c_{i_2}] < \cdots < [b_{i_N}, c_{i_N}] \tag{10.6}$$

which implies (10.5). $\square$

Lemma 10.1 shows that we can always assume $A^1 < A^2 < \cdots < A^N$ without loss of generality.

Lemma 10.2. Let the fuzzy sets $A^1, A^2, ..., A^N$ in $W \subset R$ be normal, consistent and complete with pseudo-trapezoid membership functions $\mu_{A^i}(x; a_i, b_i, c_i, d_i)$. If $A^1 < A^2 < \cdots < A^N$, then

$$c_i \leq a_{i+1} < d_i \leq b_{i+1} \tag{10.7}$$

for $i = 1, 2, ..., N - 1$.

Fig. 10.2 illustrates the assertion of Lemma 10.2. The proof of Lemma 10.2 is straightforward and is left as an exercise.

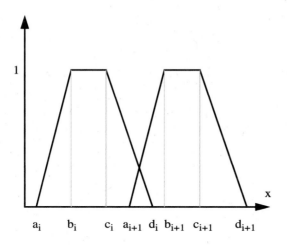

Figure 10.2. An example of Lemma 10.2: $c_i \leq a_{i+1} < d_i \leq b_{i+1}$.

10.2   Design of Fuzzy System

We are now ready to design a particular type of fuzzy systems that have some nice properties. For notational simplicity and ease of graphical explanation, we consider two-input fuzzy systems; however, the approach and results are all valid for general n-input fuzzy systems. That is, we can use exactly the same procedure to design n-input fuzzy systems. We first specify the problem.

The Problem: Let $g(x)$ be a function on the compact set $U = [\alpha_1, \beta_1] \times [\alpha_2, \beta_2] \subset R^2$ and the analytic formula of $g(x)$ be unknown. Suppose that for any $x \in U$, we can obtain $g(x)$. Our task is to design a fuzzy system that approximates $g(x)$.

We now design such a fuzzy system in a step-by-step manner.

Design of a Fuzzy System:

- **Step 1**. Define N_i $(i = 1, 2)$ fuzzy sets $A_i^1, A_i^2, ..., A_i^{N_i}$ in $[\alpha_i, \beta_i]$ which are normal, consistent, complete with pesudo-trapezoid membership functions $\mu_{A_i^1}(x_i; a_i^1, b_i^1, c_i^1, d_i^1), ..., \mu_{A_i^{N_i}}(x_i; a_i^{N_i}, b_i^{N_i}, c_i^{N_i}, d_i^{N_i})$, and $A_i^1 < A_i^2 < \cdots < A_i^{N_i}$ with $a_i^1 = b_i^1 = \alpha_i$ and $c_i^{N_i} = d_i^{N_i} = \beta_i$. Define $e_1^1 = \alpha_1, e_1^{N_1} = \beta_1$, and $e_1^j = \frac{1}{2}(b_1^j + c_1^j)$ for $j = 2, 3, ..., N_1 - 1$. Similarly, define $e_2^1 = \alpha_2, e_2^{N_2} = \beta_2$, and $e_2^j = \frac{1}{2}(b_2^j + c_2^j)$ for $j = 2, 3, ..., N_2 - 1$. Fig. 10.3 shows an example with $N_1 = 3, N_2 = 4, \alpha_1 = \alpha_2 = 0$ and $\beta_1 = \beta_2 = 1$.

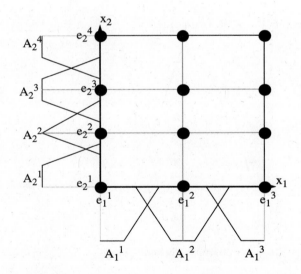

Figure 10.3. An example of the fuzzy sets defined in Step 1 of the design procedure.

- **Step 2**. Construct $M = N_1 \times N_2$ fuzzy IF-THEN rules in the following form:

$$Ru^{i_1 i_2} : \; IF \; x_1 \; is \; A_1^{i_1} \; and \; x_2 \; is \; A_2^{i_2}, \; THEN \; y \; is \; B^{i_1 i_2} \qquad (10.8)$$

where $i_1 = 1, 2, ..., N_1, i_2 = 1, 2, ..., N_2$, and the center of the fuzzy set $B^{i_1 i_2}$, denoted by $\bar{y}^{i_1 i_2}$, is chosen as

$$\bar{y}^{i_1 i_2} = g(e_1^{i_1}, e_2^{i_2}) \tag{10.9}$$

For the example in Fig. 10.3, we have $3 \times 4 = 12$ rules, and the centers of $B^{i_1 i_2}$ are equal to the $g(x)$ evaluated at the 12 dark points shown in the figure.

- **Step 3**. Construct the fuzzy system $f(x)$ from the $N_1 \times N_2$ rules of (10.8) using product inference engine (7.23), singleton fuzzifier (8.1), and center average defuzzifier (8.18) (see Lemma 9.1):

$$f(x) = \frac{\sum_{i_1=1}^{N_1} \sum_{i_2=1}^{N_2} \bar{y}^{i_1 i_2} (\mu_{A_1^{i_1}}(x_1) \mu_{A_2^{i_2}}(x_2))}{\sum_{i_1=1}^{N_1} \sum_{i_2=1}^{N_2} (\mu_{A_1^{i_1}}(x_1) \mu_{A_2^{i_2}}(x_2))} \tag{10.10}$$

Since the fuzzy sets $A_i^1, ... A_i^{N_i}$ are complete, at every $x \in U$ there exist i_1 and i_2 such that $\mu_{A_1^{i_1}}(x_1) \mu_{A_2^{i_2}}(x_2) \neq 0$. Hence, the fuzzy system (10.10) is well defined, that is, its denominator is always nonzero.

From Step 2 we see that the IF parts of the rules (10.8) constitute all the possible combinations of the fuzzy sets defined for each input variable. So, if we generalize the design procedure to n-input fuzzy systems and define N fuzzy sets for each input variable, then the total number of rules is N^n; that is, by using this design method, the number of rules increases exponentially with the dimension of the input space. This is called the *curse of dimensionality* and is a general problem for all high-dimensional approximation problems. We will address this issue again in Chapter 22.

The final observation of the design procedure is that we must know the values of $g(x)$ at $x = (e_1^{i_1}, e_2^{i_2})$ for $i_1 = 1, 2, ..., N_1$ and $i_2 = 1, 2, ..., N_2$. Since $(e_1^{i_1}, e_2^{i_2})$ can be arbitrary points in U, this is equivalent to say that we need to know the values of $g(x)$ at any $x \in U$.

Next, we study the approximation accuracy of the $f(x)$ designed above to the unknown function $g(x)$.

10.3   Approximation Accuracy of the Fuzzy System

Theorem 10.1. Let $f(x)$ be the fuzzy system in (10.10) and $g(x)$ be the unknown function in (10.9). If $g(x)$ is continuously differentiable on $U = [\alpha_1, \beta_1] \times [\alpha_2, \beta_2]$, then

$$||g - f||_\infty \leq ||\frac{\partial g}{\partial x_1}||_\infty h_1 + ||\frac{\partial g}{\partial x_2}||_\infty h_2 \tag{10.11}$$

where the infinite norm $|| * ||_\infty$ is defined as $||d(x)||_\infty = \sup_{x \in U} |d(x)|$, and $h_i = \max_{1 \leq j \leq N_i - 1} |e_i^{j+1} - e_i^j|$ $(i = 1, 2)$.

Proof: Let $U^{i_1 i_2} = [e_1^{i_1}, e_1^{i_1+1}] \times [e_2^{i_2}, e_2^{i_2+1}]$, where $i_1 = 1, 2, ..., N_1 - 1$ and $i_2 = 1, 2, ..., N_2 - 1$. Since $[\alpha_i, \beta_i] = [e_i^1, e_i^2] \cup [e_i^2, e_i^3] \cup \cdots \cup [e_i^{N_i-1}, e_i^{N_i}]$, $i = 1, 2$, we have

$$U = [\alpha_1, \beta_1] \times [\alpha_2, \beta_2] = \bigcup_{i_1=1}^{N_1-1} \bigcup_{i_2=1}^{N_2-1} U^{i_1 i_2} \tag{10.12}$$

which implies that for any $x \in U$, there exists $U^{i_1 i_2}$ such that $x \in U^{i_1 i_2}$. Now suppose $x \in U^{i_1 i_2}$, that is, $x_1 \in [e_1^{i_1}, e_1^{i_1+1}]$ and $x_2 \in [e_2^{i_2}, e_2^{i_2+1}]$ (since x is fixed, i_1 and i_2 are also fixed in the sequel). Since the fuzzy sets $A_1^1, ..., A_1^{N_1}$ are normal, consistent and complete, at least one and at most two $\mu_{A_1^{j_1}}(x_1)$ are nonzero for $j_1 = 1, 2, ..., N_1$. From the definition of e^{j_1} ($j_1 = 1, 2, ..., N_1 - 1$), these two possible nonzero $\mu_{A_1^{j_1}}(x_1)$'s are $\mu_{A_1^{i_1}}(x_1)$ and $\mu_{A_1^{i_1+1}}(x_1)$. Similarly, the two possible nonzero $\mu_{A_2^{j_2}}(x_2)$'s (for $j_2 = 1, 2, ..., N_2$) are $\mu_{A_2^{i_2}}(x_2)$ and $\mu_{A_2^{i_2+1}}(x_2)$. Hence, the fuzzy system $f(x)$ of (10.10) is simplified to

$$f(x) = \frac{\sum_{j_1=i_1}^{i_1+1} \sum_{j_2=i_2}^{i_2+1} \bar{y}^{j_1 j_2} (\mu_{A_1^{j_1}}(x_1) \mu_{A_2^{j_2}}(x_2))}{\sum_{j_1=i_1}^{i_1+1} \sum_{j_2=i_2}^{i_2+1} (\mu_{A_1^{j_1}}(x_1) \mu_{A_2^{j_2}}(x_2))}$$

$$= \sum_{j_1=i_1}^{i_1+1} \sum_{j_2=i_2}^{i_2+1} \left[\frac{\mu_{A_1^{j_1}}(x_1) \mu_{A_2^{j_2}}(x_2)}{\sum_{j_1=i_1}^{i_1+1} \sum_{j_2=i_2}^{i_2+1} (\mu_{A_1^{j_1}}(x_1) \mu_{A_2^{j_2}}(x_2))} \right] g(e_1^{j_1}, e_2^{j_2}) \tag{10.13}$$

where we use (10.9). Since

$$\sum_{j_1=i_1}^{i_1+1} \sum_{j_2=i_2}^{i_2+1} \left[\frac{\mu_{A_1^{j_1}}(x_1) \mu_{A_2^{j_2}}(x_2)}{\sum_{j_1=i_1}^{i_1+1} \sum_{j_2=i_2}^{i_2+1} (\mu_{A_1^{j_1}}(x_1) \mu_{A_2^{j_2}}(x_2))} \right] = 1 \tag{10.14}$$

we have

$$|g(x) - f(x)| \le \sum_{j_1=i_1}^{i_1+1} \sum_{j_2=i_2}^{i_2+1} \left[\frac{\mu_{A_1^{j_1}}(x_1) \mu_{A_2^{j_2}}(x_2)}{\sum_{j_1=i_1}^{i_1+1} \sum_{j_2=i_2}^{i_2+1} (\mu_{A_1^{j_1}}(x_1) \mu_{A_2^{j_2}}(x_2))} \right] |g(x) - g(e_1^{j_1}, e_2^{j_2})|$$

$$\le \max_{j_1=i_1,i_1+1; j_2=i_2,i_2+1} |g(x) - g(e_1^{j_1}, e_2^{j_2})| \tag{10.15}$$

Using the Mean Value Theorem, we can further write (10.15) as

$$|g(x) - f(x)| \le \max_{j_1=i_1,i_1+1; j_2=i_2,i_2+1} (\|\frac{\partial g}{\partial x_1}\|_\infty |x_1 - e_1^{j_1}| + \|\frac{\partial g}{\partial x_2}\|_\infty |x_2 - e_2^{j_2}|) \tag{10.16}$$

Since $x \in U^{i_1 i_2}$, which means that $x_1 \in [e_1^{i_1}, e_1^{i_1+1}]$ and $x_2 \in [e_2^{i_2}, e_2^{i_2+1}]$, we have that $|x_1 - e_1^{j_1}| \le |e_1^{i_1+1} - e_1^{i_1}|$ and $|x_2 - e_2^{j_2}| \le |e_2^{i_2+1} - e_2^{i_2}|$ for $j_1 = i_1, i_1 + 1$ and $j_2 = i_2, i_2 + 1$. Thus, (10.16) becomes

$$|g(x) - f(x)| \le \|\frac{\partial g}{\partial x_1}\|_\infty |e_1^{i_1+1} - e_1^{i_1}| + \|\frac{\partial g}{\partial x_2}\|_\infty |e_2^{i_2+1} - e_2^{i_2}| \tag{10.17}$$

from which we have

$$\|g - f\|_\infty = \sup_{x \in U} |g(x) - f(x)|$$

$$\leq \|\frac{\partial g}{\partial x_1}\|_\infty \max_{1 \leq i_1 \leq N_1 - 1} |e_1^{i_1+1} - e_1^{i_1}| + \|\frac{\partial g}{\partial x_2}\|_\infty \max_{1 \leq i_2 \leq N_2 - 1} |e_2^{i_2+1} - e_2^{i_2}|$$

$$= \|\frac{\partial g}{\partial x_1}\|_\infty h_1 + \|\frac{\partial g}{\partial x_2}\|_\infty h_2 \tag{10.18}$$

□

Theorem 10.1 is an important theorem. We can draw a number of conclusions from it, as follows:

- From (10.11) we can conclude that fuzzy systems in the form of (10.10) are universal approximators. Specifically, since $\|\frac{\partial g}{\partial x_1}\|_\infty$ and $\|\frac{\partial g}{\partial x_2}\|_\infty$ are finite numbers (a continuous function on a compact set is bounded by a finite number), for any given $\epsilon > 0$ we can choose h_1 and h_2 small enough such that $\|\frac{\partial g}{\partial x_1}\|_\infty h_1 + \|\frac{\partial g}{\partial x_2}\|_\infty h_2 < \epsilon$. Hence from (10.11) we have $\sup_{x \in U} |g(x) - f(x)| = \|g - f\|_\infty < \epsilon$.

- From (10.11) and the definition of h_1 and h_2 we see that more accurate approximation can be obtained by defining more fuzzy sets for each x_i. This confirms the intuition that more rules result in more powerful fuzzy systems.

- From (10.11) we see that in order to design a fuzzy system with a prespecified accuracy, we must know the bounds of the derivatives of $g(x)$ with respect to x_1 and x_2, that is, $\|\frac{\partial g}{\partial x_1}\|_\infty$ and $\|\frac{\partial g}{\partial x_2}\|_\infty$. In the design process, we need to know the value of $g(x)$ at $x = (e_1^{i_1}, e_2^{i_2})$ for $i_1 = 1, 2, ..., N_1$ and $i_2 = 1, 2, ..., N_2$. Therefore, this approach requires these two pieces of information in order for the designed fuzzy system to achieve any prespecified degree of accuracy.

- From the proof of Theorem 10.1 we see that if we change $\mu_{A_1^{i_1}}(x_1)\mu_{A_2^{i_2}}(x_2)$ to $min[\mu_{A_1^{i_1}}(x_1), \mu_{A_2^{i_2}}(x_2)]$, the proof is still vaild. Therefore, if we use minimum inference engine in the design procedure and keep the others unchanged, the designed fuzzy system still has the approximation property in Theorem 10.1. Consequently, the fuzzy systems with minimum inference engine, singleton fuzzifier, center average defuzzifier and pseudo-trapezoid membership functions are universal approximators.

Theorem 10.1 gives the accuracy of $f(x)$ as an approximator to $g(x)$. The next lemma shows at what points $f(x)$ and $g(x)$ are exactly equal.

Lemma 10.3. Let $f(x)$ be the fuzzy system (10.10) and $e_1^{i_1}$ and $e_2^{i_2}$ be the points defined in the design procedure for $f(x)$. Then,

$$f(e_1^{i_1}, e_2^{i_2}) = g(e_1^{i_1}, e_2^{i_2}) \tag{10.19}$$

for $i_1 = 1, 2, ..., N_1$ and $i_2 = 1, 2, ..., N_2$.

Proof: From the definition of $e_1^{i_1}$ and $e_2^{i_2}$ and the fact that $A_i^{j_i}$'s are normal, we have $\mu_{A_1^{i_1}}(e_1^{i_1}) = \mu_{A_2^{i_2}}(e_2^{i_2}) = 1$. Since the fuzzy sets $A_i^1, A_i^2, ..., A_i^{N_i}$ $(i = 1, 2)$ are consistent, we have that $\mu_{A_1^{j_1}}(e_1^{i_1}) = \mu_{A_2^{j_2}}(e_2^{i_2}) = 0$ for $j_1 \neq i_1$ and $j_2 \neq i_2$. Hence from (10.10) and (10.9) we have $f(e_1^{i_1}, e_2^{i_2}) = \bar{y}^{i_1 i_2} = g(e_1^{i_1}, e_2^{i_2})$. $\square$

Lemma 10.3 shows that the fuzzy system (10.10) can be viewed as an interpolation of function $g(x)$ at some regular points $(e_1^{i_1}, e_2^{i_2})$ $(i_1 = 1, 2, ..., N_1, i_2 = 1, 2, ..., N_2)$ in the universe of discourse U. This is intuitively appealing.

Finally, we show two examples of how to use Theorem 10.1 to design the required fuzzy system.

Example 10.1. Design a fuzzy system $f(x)$ to uniformly approximate the continuous function $g(x) = sin(x)$ defined on $U = [-3, 3]$ with a required accuracy of $\epsilon = 0.2$; that is, $\sup_{x \in U} |g(x) - f(x)| < \epsilon$.

Since $||\frac{\partial g}{\partial x}||_\infty = ||cos(x)||_\infty = 1$, from (10.11) we see that the fuzzy system with $h = 0.2$ meets our requirement. Therefore, we define the following 31 fuzzy sets A^j in $U = [-3, 3]$ with the triangular membership functions

$$\mu_{A^1}(x) = \mu_{A^1}(x; -3, -3, -2.8) \tag{10.20}$$

$$\mu_{A^{31}}(x) = \mu_{A^{31}}(x; 2.8, 3, 3) \tag{10.21}$$

and

$$\mu_{A^j}(x) = \mu_{A^j}(x; e^{j-1}, e^j, e^{j+1}) \tag{10.22}$$

where $j = 2, 3, ..., 30$, and $e^j = -3 + 0.2(j - 1)$. These membership functions are shown in Fig. 10.4. According to (10.10), the designed fuzzy system is

$$f(x) = \frac{\sum_{j=1}^{31} sin(e^j) \mu_{A^j}(x)}{\sum_{j=1}^{31} \mu_{A^j}(x)} \tag{10.23}$$

which is plotted in Fig.10.5 against $g(x) = sin(x)$. We see from Fig.10.5 that $f(x)$ and $g(x)$ are almost identical. $\square$

Example 10.2. Design a fuzzy system to uniformly approximate the function $g(x) = 0.52 + 0.1x_1 + 0.28x_2 - 0.06x_1x_2$ defined on $U = [-1, 1] \times [-1, 1]$ with a required accuracy of $\epsilon = 0.1$.

Since $||\frac{\partial g}{\partial x_1}||_\infty = \sup_{x \in U} |0.1 - 0.06x_2| = 0.16$ and $||\frac{\partial g}{\partial x_2}||_\infty = \sup_{x \in U} |0.28 - 0.06x_1| = 0.34$, from (10.11) we see that $h_1 = h_2 = 0.2$ results in $||g - f||_\infty \leq 0.16 * 0.2 + 0.34 * 0.2 = 0.1$. Therefore, we define 11 fuzzy sets A^j $(j = 1, 2, ..., 11)$ in $[-1, 1]$ with the following triangular membership functions:

$$\mu_{A^1}(x) = \mu_{A^1}(x; -1, -1, -0.8) \tag{10.24}$$

$$\mu_{A^{11}}(x) = \mu_{A^{11}}(x; 0.8, 1, 1) \tag{10.25}$$

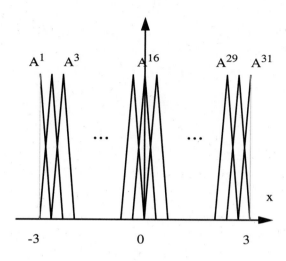

Figure 10.4. Membership functions in Example 10.1.

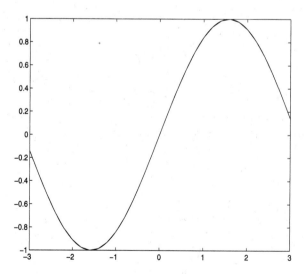

Figure 10.5. The designed fuzzy system $f(x)$ and the function $g(x) = six(x)$ (they are almost identical).

and

$$\mu_{A^j}(x) = \mu_{A^j}(x; e^{j-1}, e^j, e^{j+1}) \qquad (10.26)$$

for $j = 2, 3, ..., 10$, where $e^j = -1 + 0.2(j-1)$. The fuzzy system is constructed from the following $11 \times 11 = 121$ rules:

$$IF \; x_1 \; is \; A^{i_1} \; and \; x_2 \; is \; A^{i_2}, \; THEN \; y \; is \; B^{i_1 i_2} \qquad (10.27)$$

where $i_1, i_2 = 1, 2, ..., 11$, and the center of $B^{i_1 i_2}$ is $\bar{y}^{i_1 i_2} = g(e^{i_1}, e^{i_2})$. The final fuzzy system is

$$f(x) = \frac{\sum_{i_1=1}^{11} \sum_{i_2=1}^{11} g(e^{i_1}, e^{i_2})(\mu_{A^{i_1}}(x_1)\mu_{A^{i_2}}(x_2))}{\sum_{i_1=1}^{11} \sum_{i_2=1}^{11}(\mu_{A^{i_1}}(x_1)\mu_{A^{i_2}}(x_2))} \qquad (10.28)$$

□

From Example 10.2 we see that we need 121 rules to approximate the function $g(x)$. Are so many rules really necessary? In other words, can we improve the bound in (10.11) so that we can use less rules to approximate the same function to the same accuracy? The answer is yes and we will study the details in the next chapter.

10.4 Summary and Further Readings

In this chapter we have demonstrated the following:

- The three types of approximation problems (classified according to the information available).

- The concepts of completeness, consistency, and order of fuzzy sets, and their application to fuzzy sets with pseudo-trapezoid membership functions.

- For a given accuracy requirement, how to design a fuzzy system that can approximate a given function to the required accuracy.

- The idea of proving the approximation bound (10.11).

Approximation accuracies of fuzzy systems were analyzed in Ying [1994] and Zeng and Singh [1995]. This is a relatively new topic and very few references are available.

10.5 Exercises

Exercise 10.1. Let fuzzy sets A^j in $U = [a, b]$ ($j = 1, 2, ..., N$) be normal, consistent, and complete with pseudo-trapezoid membership functions $\mu_{A^j}(x) =$

$\mu_{A^j}(x; a_j, b_j, c_j, d_j)$. Suppose that $A^1 < A^2 < \cdots < A^N$. Define fuzzy sets B^j whose membership functions are given as

$$\mu_{B^j}(x) = \frac{\mu_{A^j}(x)}{\sum_{i=1}^{N} \mu_{A^i}(x)} \tag{10.29}$$

Show that:

(a) $\mu_{B^j}(x)$ $(j = 1, 2, ..., N)$ are also pesudo-trapezoid membership functions.

(b) The fuzzy sets B^j $(j = 1, 2, ..., N)$ are also normal, consistent, and complete.

(c) $B^1 < B^2 < \cdots < B^N$.

(d) If $\mu_{A^j}(x) = \mu_{A^j}(x; a_j, b_j, c_j, d_j)$ are trapezoid membership functions with $c_i = a_{i+1}, d_i = b_{i+1}$ for $i = 1, 2, ..., N-1$, then $\mu_{B^j}(x) = \mu_{A^j}(x)$ for $j = 1, 2, ..., N - 1$.

Exercise 10.2. Design a fuzzy system to uniformly approximate the function $g(x) = sin(x\pi) + cos(x\pi) + sin(x\pi)cos(x\pi)$ on $U = [-1, 1]$ with a required accuracy of $\epsilon = 0.1$.

Exercise 10.3. Design a fuzzy system to uniformly approximate the function $g(x) = sin(x_1\pi) + cos(x_2\pi) + sin(x_1\pi)cos(x_2\pi)$ on $U = [-1, 1] \times [-1, 1]$ with a required accuracy of $\epsilon = 0.1$.

Exercise 10.4. Extend the design method in Section 10.2 to n-input fuzzy systems.

Exercise 10.5. Let the function $g(x)$ on $U = [0, 1]^3$ be given by

$$g(x_1, x_2, x_3) = 1 + \sum_{k_1 k_2 k_3 \in K} x_1^{k_1} x_2^{k_2} x_3^{k_3} \tag{10.30}$$

where $K = \{k_1 k_2 k_3 | k_i = 0, 1; i = 1, 2, 3 \; and \; k_1 + k_2 + k_3 > 0\}$. Design a fuzzy system to uniformly approximate $g(x)$ with a required accuracy of $\epsilon = 0.05$.

Exercise 10.6. Show that if the fuzzy sets $A_i^1, A_i^2, ..., A_i^{N_i}$ in the design procedure of Section 10.2 are not complete, then the fuzzy system (10.10) is not well defined. If these fuzzy sets are not normal or not consistent, is the fuzzy system (10.10) well defined?

Exercise 10.7. Plot the fuzzy system (10.28) on $U = [-1, 1] \times [-1, 1]$ and compare it with $g(x) = 0.52 + 0.1x_1 + 0.28x_2 - 0.06x_1x_2$.

Chapter 11

Approximation Properties of
Fuzzy Systems II

In Chapter 10 we saw that by using the bound in Theorem 10.1 a large number of rules are usually required to approximate some simple functions. For example, Example 10.2 showed that we need 121 rules to approximate a two-dimensional quadratic function. Observing (10.11) we note that the bound is a linear function of h_i. Since h_i are usually small, if a bound could be a linear function of h_i^2, then this bound would be much smaller than the bound in (10.11). That is, if we can obtain tighter bound than that used in (10.11), we may use less rules to approximate the same function with the same accuracy.

In approximation theory (Powell [1981]), if $g(x)$ is a given function on U and $U^{i_1 i_2}$ ($i_1 = 1, 2, ..., N_1, i_2 = 1, 2, ..., N_2$) is a partition of U as in the proof of Theorem 10.1, then $f(x)$ is said to be the $k'th$ order accurate approximator for $g(x)$ if $\|g - f\|_\infty \leq M_g h^k$, where M_g is a constant that depends on the function g, and h is the module of the partition that in our case is $max(h_1, h_2)$. In this chapter, we first design a fuzzy system that is a second-order accurate approximator.

11.1 Fuzzy Systems with Second-Order Approximation Accuracy

We first design the fuzzy system in a step-by-step manner and then study its approximation accuracy. As in Chapter 10, we consider two-input fuzzy systems for notational simplicity; the approach and results are still valid for n-input fuzzy systems. The design problem is the same as in Section 10.2.

Design of Fuzzy System with Second-Order Accuracy:

- **Step 1.** Define N_i ($i = 1, 2$) fuzzy sets $A_i^1, A_i^2, ..., A_i^{N_i}$ in $[\alpha_i, \beta_i]$, which are normal, consistent, and complete with the triangular membership functions

$$\mu_{A_i^1}(x_i) = \mu_{A_i^1}(x_i; e_i^1, e_i^1, e_i^2) \tag{11.1}$$

$$\mu_{A_i^j}(x_i) = \mu_{A_i^j}(x_i; e_i^{j-1}, e_i^j, e_i^{j+1}) \tag{11.2}$$

for $j = 2, 3, ..., N_i - 1$, and

$$\mu_{A_i^{N_i}}(x_i) = \mu_{A_i^{N_i}}(x_i; e_i^{N_i-1}, e_i^{N_i}, e_i^{N_i}) \tag{11.3}$$

where $i = 1, 2$, and $\alpha_i = e_i^1 < e_i^2 < \cdots < e_i^{N_i} = \beta_i$. Fig. 11.1 shows an example with $N_1 = 4, N_2 = 5, \alpha_1 = \alpha_2 = 0$ and $\beta_1 = \beta_2 = 1$.

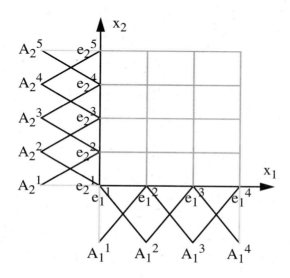

Figure 11.1. An example of fuzzy sets.

- **Steps 2 and 3**. The same as Steps 2 and 3 of the design procedure in Section 10.2. That is, the constructed fuzzy system is given by (10.10), where $\bar{y}^{i_1 i_2}$ are given by (10.9) and the $A_1^{i_1}$ and $A_2^{i_2}$ are given by (11.1)-(11.3).

Since the fuzzy system designed from the above steps is a special case of the fuzzy system designed in Section 10.2, Theorem 10.1 is still valid for this fuzzy system. The following theorem gives a stronger result.

Theorem 11.1. Let $f(x)$ be the fuzzy system designed through the above three steps. If the $g(x)$ is twice continuously differentiable on U, then

$$\|g - f\|_\infty \le \frac{1}{8}\left[\|\frac{\partial^2 g}{\partial x_1^2}\|_\infty h_1^2 + \|\frac{\partial^2 g}{\partial x_2^2}\|_\infty h_2^2\right] \tag{11.4}$$

where $h_i = \max_{1 \le j \le N_i - 1} |e_i^{j+1} - e_i^j|$ $(i = 1, 2)$.

Proof: As in the proof of Theorem 10.1, we partition U into $U = \bigcup_{i_1=1}^{N_1-1} \bigcup_{i_2=1}^{N_2-1} U^{i_1 i_2}$ where $U^{i_1 i_2} = [e_1^{i_1}, e_1^{i_1+1}] \times [e_2^{i_2}, e_2^{i_2+1}]$. So, for any $x \in U$, there exists $U^{i_1 i_2}$ such that $x \in U^{i_1 i_2}$. Now, suppose $x \in U^{i_1 i_2}$, then by the consistency and completeness of fuzzy sets $A_i^1, A_i^2, ..., A_i^{N_i}$ $(i = 1, 2)$, the fuzzy system can be simplified to (same as (10.13))

$$f(x) = \sum_{j_1=i_1}^{i_1+1} \sum_{j_2=i_2}^{i_2+1} \left[\frac{\mu_{A_1^{j_1}}(x_1) \mu_{A_2^{j_2}}(x_2)}{\sum_{j_1=i_1}^{i_1+1} \sum_{j_2=i_2}^{i_2+1} \mu_{A_1^{j_1}}(x_1) \mu_{A_2^{j_2}}(x_2)} \right] g(e_1^{i_1}, e_2^{i_2}) \tag{11.5}$$

Since $\mu_{A_i^{j_i}}(x_i)$ are the special triangular membership functions given by (11.1)-(11.3), we have

$$\mu_{A_i^{i_1}}(x_i) + \mu_{A_i^{i_1+1}}(x_i) = 1 \tag{11.6}$$

for $i = 1, 2$. Hence,

$$\sum_{j_1=i_1}^{i_1+1} \sum_{j_2=i_2}^{i_2+1} \mu_{A_1^{j_1}}(x_1) \mu_{A_2^{j_2}}(x_2) = \sum_{j_1=i_1}^{i_1+1} \mu_{A_1^{j_1}}(x_1) [\sum_{j_2=i_2}^{i_2+1} \mu_{A_2^{j_2}}(x_2)] = 1 \tag{11.7}$$

and the fuzzy system (11.5) is simplified to

$$f(x) = \sum_{j_1=i_1}^{i_1+1} \sum_{j_2=i_2}^{i_2+1} [\mu_{A_1^{j_1}}(x_1) \mu_{A_2^{j_2}}(x_2)] g(e_1^{i_1}, e_2^{i_2}) \tag{11.8}$$

Let $C^2(U^{i_1 i_2})$ be the set of all twice continuously differentiable functions on $U^{i_1 i_2}$ and define linear operators L_1 and L_2 on $C^2(U^{i_1 i_2})$ as follows:

$$(L_1 g)(x) = \sum_{j_1=i_1}^{i_1+1} (\mu_{A_1^{j_1}}(x_1)) g(e_1^{j_1}, x_2) \tag{11.9}$$

$$(L_2 g)(x) = \sum_{j_2=i_2}^{i_2+1} (\mu_{A_2^{j_2}}(x_2)) g(x_1, e_2^{j_2}) \tag{11.10}$$

Since $\mu_{A_1^{j_1}}(x_1)$ and $\mu_{A_2^{j_2}}(x_2)$ are linear functions in $U^{i_1 i_2}$, they are twice continuously differentiable. Hence, $g \in C^2(U^{i_1 i_2})$ implies $L_1 g \in C^2(U^{i_1 i_2})$ and $L_2 g \in C^2(U^{i_1 i_2})$. From (11.9) and (11.6) we have

$$\|L_1 g\|_\infty \le \|g\|_\infty |\sum_{j_1=i_1}^{i_1+1} \mu_{A_1^{j_1}}(x_1)| = \|g\|_\infty \tag{11.11}$$

Combining (11.9) and (11.10) and observing (11.8), we have

$$(L_1 L_2 g)(x) = \sum_{j_1=i_1}^{i_1+1} \mu_{A_1^{j_1}}(x_1) [\sum_{j_2=i_2}^{i_2+1} \mu_{A_2^{j_2}}(x_2) g(e_1^{j_1}, e_2^{j_2})] = f(x) \tag{11.12}$$

Therefore, from (11.11) and (11.12) we obtain

$$
\begin{aligned}
||g - f||_\infty &= ||g - L_1 L_2 g||_\infty \\
&\le ||g - L_1 g||_\infty + ||L_1(g - L_2 g)||_\infty \\
&\le ||g - L_1 g||_\infty + ||g - L_2 g||_\infty
\end{aligned}
\tag{11.13}
$$

Since $x \in U^{i_1 i_2} = [e_1^{i_1}, e_1^{i_1+1}] \times [e_2^{i_2}, e_2^{i_2+1}]$ and using the result in univariate linear interpolation (Powell [1981]), we obtain

$$
\begin{aligned}
||g - L_1 g||_\infty &= ||\sum_{j_1=i_1}^{i_1+1} \mu_{A_1^{j_1}}(x_1)[g(x_1, x_2) - g(e_1^{j_1}, x_2)]||_\infty \\
&\le ||g(x_1, x_2) - g(e_1^{j_1}, x_2)||_\infty \\
&\le \frac{1}{8}(e_1^{i_1+1} - e_1^{i_1})^2 ||\frac{\partial^2 g}{\partial x_1^2}||_\infty
\end{aligned}
\tag{11.14}
$$

Similarly, we have

$$
||g - L_2 g||_\infty \le \frac{1}{8}(e_2^{i_2+1} - e_2^{i_2})^2 ||\frac{\partial^2 g}{\partial x_2^2}||_\infty
\tag{11.15}
$$

Substituting (11.14) and (11.15) into (11.13) and noticing the definition of h_i, we obtain (11.4). □

From Theorem 11.1 we see that if we choose the particular triangular membership functions, a second-order accurate approximator can be obtained. We now design fuzzy systems to approximate the same functions $g(x)$ in Examples 10.1 and 10.2 using the new bound (11.4). We will see that we can achieve the same accuracy with fewer rules.

Example 11.1. Same as Example 10.1 except that we now use the bound (11.4). Since $||\frac{\partial^2 g}{\partial x^2}||_\infty = 1$, we have from (11.4) that if we choose $h = 1$, then we have $||g - f||_\infty \le \frac{1}{8} < \epsilon$. Therefore, we define 7 fuzzy sets A^j in the form of (11.1)-(11.3) with $e^j = -3 + (j - 1)$ for $j = 1, 2, ..., 7$. The designed fuzzy system is

$$
f(x) = \frac{\sum_{j=1}^{7} sin(e^j) \mu_{A^j}(x)}{\sum_{j=1}^{7} \mu_{A^j}(x)}
\tag{11.16}
$$

Comparing (11.16) with (10.23) we see that the number of rules is reduced from 31 to 7, but the accuracy remains the same. The $f(x)$ of (11.16) is plotted in Fig. 11.2 against $g(x) = sin(x)$. □

Example 11.2. Same as Example 10.2 except that we now use the bound (11.4). Since $\frac{\partial^2 g}{\partial x_i^2} = 0$ ($i = 1, 2$), we know from (11.4) that $f(x) = g(x)$ for all $x \in U$. In fact, choosing $h_i = 2, e_i^1 = -1$ and $e_i^2 = 1$ for $i = 1, 2$ (that is, $N_1 = N_2 = 2$), we

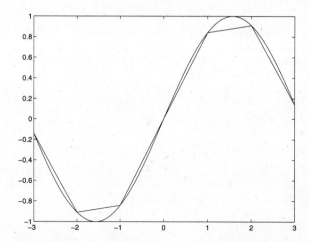

Figure 11.2. The designed fuzzy system $f(x)$ of (11.16) and the function $g(x) = sin(x)$ to be approximated.

obtain the designed fuzzy system as

$$f(x) = \frac{\sum_{i_1=1}^{2} \sum_{i_2=1}^{2} g(e_1^{i_1}, e_2^{i_2})(\mu_{A_1^{i_1}}(x_1)\mu_{A_2^{i_2}}(x_2))}{\sum_{i_1=1}^{2} \sum_{i_2=1}^{2} (\mu_{A_1^{i_1}}(x_1)\mu_{A_2^{i_2}}(x_2))} \tag{11.17}$$

where the membership functions are given by (11.1)-(11.3). For this particular case, we have for $i = 1, 2$ and $x \in U$ that

$$\mu_{A_i^1}(x_i) = \mu_{A_i^1}(x_i; -1, -1, 1) = \frac{1}{2}(1 - x_i) \tag{11.18}$$

$$\mu_{A_i^2}(x_i) = \mu_{A_i^2}(x_i; -1, 1, 1) = \frac{1}{2}(1 + x_i) \tag{11.19}$$

$g(e_1^1, e_2^1) = g(-1, -1) = 0.08, g(e_1^1, e_2^2) = g(-1, 1) = 0.76, g(e_1^2, e_2^1) = g(1, -1) = 0.4,$ and $g(e_1^2, e_2^2) = g(1, 1) = 0.84$. Substituting these results into (11.17), we obtain

$$\begin{aligned}
f(x) &= [\frac{0.08}{4}(1 - x_1)(1 - x_2) + \frac{0.76}{4}(1 - x_1)(1 + x_2) + \frac{0.4}{4}(1 + x_1)(1 - x_2) \\
&\quad + \frac{0.84}{4}(1 + x_1)(1 + x_2)]/[\frac{1}{4}(1 - x_1)(1 - x_2) + \frac{1}{4}(1 - x_1)(1 + x_2) \\
&\quad + \frac{1}{4}(1 + x_1)(1 - x_2) + \frac{1}{4}(1 + x_1)(1 + x_2)] \\
&= 0.52 + 0.1x_1 + 0.28x_2 - 0.06x_1x_2 \tag{11.20}
\end{aligned}$$

which is exactly the same as $g(x)$. This confirms our conclusion that $f(x)$ exactly reproduces $g(x)$. In Example 10.2 we used 121 rules to construct the fuzzy system, whereas in this example we only use 4 rules to achieve zero approximation error. $\square$

To generalize Example 11.2 , we observe from (11.4) that for any function with $||\frac{\partial^2 g}{\partial x_i^2}||_\infty = 0$, our fuzzy system $f(x)$ designed through the three steps reproduces the $g(x)$, that is, $f(x) = g(x)$ for all $x \in U$. This gives the following corollary of Theorem 11.1.

Corollary 11.1. Let $f(x)$ be the fuzzy system designed through the three steps in this section. If the function $g(x)$ is of the following form:

$$g(x) = \sum_{k_1=0}^{1} \sum_{k_2=0}^{1} a_{k_1 k_2} x_1^{k_1} x_2^{k_2} \tag{11.21}$$

where $a_{k_1 k_2}$ are constants, then $f(x) = g(x)$ for all $x \in U$.

Proof: Since $\frac{\partial^2 g}{\partial x_1^2} = \frac{\partial^2 g}{\partial x_2^2} = 0$ for this class of $g(x)$, the conclusion follows immediately from (11.4). $\square$

11.2 Approximation Accuracy of Fuzzy Systems with Maximum Defuzzifier

In Chapter 9 we learned that fuzzy systems with maximum defuzzifier are quite different from those with center average defuzzifier. In this section, we study the approximation properties of fuzzy systems with maximum defuzzifier.

Similar to the approach in Sections 11.1 and 10.2, we first design a particular fuzzy system and then study its properties.

Design of Fuzzy System with Maximum Defuzzifier:

- **Step 1.** Same as Step 1 of the design procedure in Section 11.1.

- **Step 2.** Same as Step 2 of the design procedure in Section 10.2.

- **Step 3.** Construct a fuzzy system $f(x)$ from the $N_1 \times N_2$ rules in the form of (10.8) using product inference engine (7.23), singleton fuzzifier (8.1), and maximum defuzzifier (8.23). According to Lemma 9.4, this fuzzy system is

$$f(x) = \bar{y}^{i_1^* i_2^*} = g(e_1^{i_1^*}, e_2^{i_2^*}) \tag{11.22}$$

where $i_1^* i_2^*$ is such that

$$\mu_{A_1^{i_1^*}}(x_1)\mu_{A_2^{i_2^*}}(x_2) \geq \mu_{A_1^{i_1}}(x_1)\mu_{A_2^{i_2}}(x_2) \tag{11.23}$$

for all $i_1 = 1, 2, ..., N_1$ and $i_2 = 1, 2, ..., N_2$.

The following theorem shows that the fuzzy system designed above is a first-order accurate approximator of $g(x)$.

Theorem 11.2. Let $f(x)$ be the fuzzy system (11.22) designed from the three steps above. If $g(x)$ is continuously differentiable on $U = [\alpha_1, \beta_1] \times [\alpha_2, \beta_2]$, then

$$\|g - f\|_\infty \leq \|\frac{\partial g}{\partial x_1}\|_\infty h_1 + \|\frac{\partial g}{\partial x_2}\|_\infty h_2 \qquad (11.24)$$

where $h_i = \max_{1 \leq j \leq N_i - 1} |e_i^{j+1} - e_i^j|$, $i = 1, 2$.

Proof: As in the proof of Theorem 11.1, we partition U into $U = \bigcup_{i_1=1}^{N_1-1} \bigcup_{i_2=1}^{N_2-1} U^{i_1 i_2}$ where $U^{i_1 i_2} = [e_1^{i_1}, e_1^{i_1+1}] \times [e_2^{i_2}, e_2^{i_2+1}]$. We now further partition $U^{i_1 i_2}$ into $U^{i_1 i_2} = U_{00}^{i_1 i_2} \cup U_{01}^{i_1 i_2} \cup U_{10}^{i_1 i_2} \cup U_{11}^{i_1 i_2}$, where $U_{00}^{i_1 i_2} = [e_1^{i_1}, \frac{1}{2}(e_1^{i_1} + e_1^{i_1+1})] \times [e_2^{i_2}, \frac{1}{2}(e_2^{i_2} + e_2^{i_2+1})]$, $U_{01}^{i_1 i_2} = [e_1^{i_1}, \frac{1}{2}(e_1^{i_1} + e_1^{i_1+1})] \times [\frac{1}{2}(e_2^{i_2} + e_2^{i_2+1}), e_2^{i_2+1}]$, $U_{10}^{i_1 i_2} = [\frac{1}{2}(e_1^{i_1} + e_1^{i_1+1}), e_1^{i_1+1}] \times [e_2^{i_2}, \frac{1}{2}(e_2^{i_2} + e_2^{i_2+1})]$, and $U_{11}^{i_1 i_2} = [\frac{1}{2}(e_1^{i_1} + e_1^{i_1+1}), e_1^{i_1+1}] \times [\frac{1}{2}(e_2^{i_2} + e_2^{i_2+1}), e_2^{i_2+1}]$; see Fig. 11.3 for an illustration. So for any $x \in U$, there exist $U_{pq}^{i_1 i_2}$ $(p, q = 0 \text{ or } 1)$ such that $x \in U_{pq}^{i_1 i_2}$. If x is in the interior of $U_{pq}^{i_1 i_2}$, then with the help of Fig. 11.3 we see that $\mu_{A_1^{i_1+p}}(x_1) > 0.5$, $\mu_{A_2^{i_2+q}}(x_2) > 0.5$, and all other membership values are less than 0.5. Hence, from (11.22) and (11.23) we obtain

$$f(x) = g(e_1^{i_1+p}, e_2^{i_2+q}) \qquad (11.25)$$

Using the Mean Value Theorem and the fact that $x \in U^{i_1 i_2}$, we have

$$|g(x) - f(x)| = |g(x_1, x_2) - g(e_1^{i_1+p}, e_2^{i_2+q})|$$
$$\leq \|\frac{\partial g}{\partial x_1}\|_\infty |x_1 - e_1^{i_1+p}| + \|\frac{\partial g}{\partial x_2}\|_\infty |x_2 - e_2^{i_2+q}|$$
$$\leq \|\frac{\partial g}{\partial x_1}\|_\infty |e_1^{i_1+1} - e_1^{i_1}| + \|\frac{\partial g}{\partial x_2}\|_\infty |e_2^{i_2+1} - e_2^{i_2}| \qquad (11.26)$$

If x is on the boundary of $U_{pq}^{i_1 i_2}$, then with the help of Fig. 11.3 we see that $f(x)$ may take any value from a set of at most the four elements $\{g(e_1^{i_1}, e_2^{i_2}), g(e_1^{i_1}, e_2^{i_2+1}), g(e_1^{i_1+1}, e_2^{i_2}), g(e_1^{i_1+1}, e_2^{i_2+1})\}$; so (11.26) is still true in this case. Finally, (11.24) follows from (11.26). $\square$

From Theorem 11.2 we immediately see that the fuzzy systems with product inference engine, singleton fuzzifier, and maximum defuzzifier are universal approximators. In fact, by choosing the h_1 and h_2 sufficiently small, we can make $\|g - f\|_\infty < \epsilon$ for arbitrary $\epsilon > 0$ according to (11.24).

We now approximate the functions $g(x)$ in Examples 11.1 and 11.2 using the fuzzy system (11.22).

Example 11.3. Same as Example 10.1 except that we use the fuzzy system (11.22). Since $\|\frac{\partial g}{\partial x}\|_\infty = 1$, we choose $h = 0.2$ and define 31 fuzzy sets A^j in the form of (10.20)-(10.22) (Fig. 10.4). Let $e^j = -3 + 0.2(j-1)$, then the fuzzy system

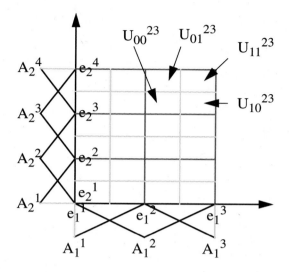

Figure 11.3. An example of partition of $U^{i_1 i_2}$ into subsets.

(11.22) becomes

$$f(x) = \begin{cases} sin(-3), \ x \in [-3, -2.8] \\ sin(e^j), \ x \in (e^j - 0.1, e^j + 0.1], \ j = 2, 3, ..., 30 \\ sin(3), \ x \in (2.8, 3] \end{cases} \tag{11.27}$$

which is plotted in Fig. 11.4. □

Example 11.4. Same as Example 10.2 except that we use the fuzzy system (11.22). As in Example 10.2, we choose $h_1 = h_2 = 0.2$ and define 11 fuzzy sets A^j on $[-1, 1]$ given by (10.24)-(10.26). We construct the fuzzy system using the 121 rules in the form of (10.27). For this example, $e^j = -1 + 0.2(j - 1)$ $(j = 1, 2, ..., 11)$ and $U^{i_1 i_2} = [e^{i_1}, e^{i_1+1}] \times [e^{i_1}, e^{i_1+1}]$ $(i_1, i_2 = 1, 2, ..., 10)$. As shown in Fig. 11.3, we further decompose $U^{i_1 i_2}$ into $U^{i_1 i_2} = \bigcup_{p=0}^{1} \bigcup_{q=0}^{1} U_{pq}^{i_1 i_2}$, where $U_{00}^{i_1 i_2} = [e_1^{i_1}, \frac{1}{2}(e_1^{i_1} + e_1^{i_1+1})] \times [e_2^{i_2}, \frac{1}{2}(e_2^{i_2} + e_2^{i_2+1})]$, $U_{01}^{i_1 i_2} = [e_1^{i_1}, \frac{1}{2}(e_1^{i_1} + e_1^{i_1+1})] \times [\frac{1}{2}(e_2^{i_2} + e_2^{i_2+1}), e_2^{i_2+1}]$, $U_{10}^{i_1 i_2} = [\frac{1}{2}(e_1^{i_1} + e_1^{i_1+1}), e_1^{i_1+1}] \times [e_2^{i_2}, \frac{1}{2}(e_2^{i_2} + e_2^{i_2+1})]$, and $U_{11}^{i_1 i_2} = [\frac{1}{2}(e_1^{i_1} + e_1^{i_1+1}), e_1^{i_1+1}] \times [\frac{1}{2}(e_2^{i_2} + e_2^{i_2+1}), e_2^{i_2+1}]$. Then the fuzzy system (11.22) becomes

$$f(x) = g(e^{i_1+p}, e^{i_2+q}), \ x \in U_{pq}^{i_1 i_2} \tag{11.28}$$

which is computed through the following two steps: (i) for given $x \in U$, determine i_1, i_2, p, q such that $x \in U_{pq}^{i_1 i_2}$, and (ii) the $f(x)$ equals $g(e^{i_1+p}, e^{i_2+q})$. □

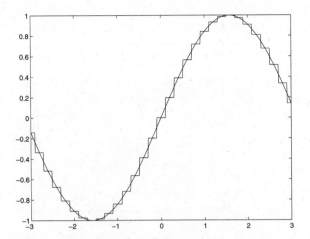

Figure 11.4. The designed fuzzy system (11.22) and the function $g(x) = sin(x)$ to be approximated in Example 11.3.

In Section 11.1 we showed that the fuzzy systems with center average defuzzifier are second-order accurate approximators. Theorem 11.2 shows that the fuzzy systems with maximum defuzzifier are first-order accurate approximators. So it is natural to ask whether they are second-order accurate approximators also? The following example shows, unfortunately, that they cannot be second-order accurate approximators.

Example 11.5. Let $g(x) = x$ on $U = [0, 1]$ and $\mu_{A^i}(x) = \mu_{A^i}(x; e^{i-1}, e^i, e^{i+1})$, where $e^0 = 0, e^{N+1} = 1, e^i = \frac{i-1}{N-1}, i = 1, 2, ..., N$, and N can be any positive integer. So, in this case we have N rules and $h = \frac{1}{N-1}$. Let $U^i = [e^i, e^{i+1}]$ ($i = 1, 2, ..., N-1$) and $f(x)$ be the fuzzy system (11.22). If $x \in U^i$, then

$$\max_{x \in U^i} |g(x) - f(x)| = \max_{x \in [e^i, e^{i+1}]} |x - g(e^i) \text{ or } g(e^{i+1})| = \frac{1}{2}(e^{i+1} - e^i) = \frac{1}{2}h \quad (11.29)$$

Since $h(= \frac{1}{N-1}) \geq h^2$ for any positive integer N, the fuzzy system (11.22) cannot approximate the simple function $g(x) = x$ to second-order accuracy. Because of this counter-example, we conclude that fuzzy systems with maximum defuzzifier cannot be second-order accurate approximators. Therefore, fuzzy systems with center average defuzzifier are better approximators than fuzzy systems with maximum defuzzifier.

11.3 Summary and Further Readings

In this chapter we have demonstrated the following:

- Using the second-order bound (11.4) to design fuzzy systems with required accuracy.

- Designing fuzzy systems with maximum defuzzifier to approximate functions with required accuracy.

- The ideas of proving the second-order bound for fuzzy systems with center average defuzzifier (Theorem 11.1) and the first-order bound for fuzzy systems with maximum defuzzifier (Theorem 11.2).

Again, very few references are available on the topic of this chapter. The most relevant papers are Ying [1994] and Zeng and Singh [1995].

11.4 Exercises

Exercise 11.1. Use the first-order bound (10.11) and the second-order bound (11.4) to design two fuzzy systems with center average defuzzifier to uniformly approximate the function $g(x_1, x_2) = \frac{1}{3+x_1+x_2}$ on $U = [-1, 1] \times [-1, 1]$ to the accuracy of $\epsilon = 0.1$. Plot the designed fuzzy systems and compare them.

Exercise 11.2. Repeat Exercise 11.1 with $g(x_1, x_2) = \frac{1}{1+x_1^2+x_2^2}$.

Exercise 11.3. Design a fuzzy system with maximum defuzzifier to uniformly approximate the $g(x_1, x_2)$ in Exercise 11.1 on the same U to the accuracy of $\epsilon = 0.1$. Plot the designed fuzzy system.

Exercise 11.4. Repeat Exercise 11.3 with the $g(x_1, x_2)$ in Exercise 11.2.

Exercise 11.5. Generalize the design procedure in Section 11.1 to n-input fuzzy systems and prove that the designed fuzzy system $f(x)$ satisfies

$$||g - f||_\infty \leq \frac{1}{8}(\sum_{i=1}^{n} ||\frac{\partial^2 g}{\partial x_i^2}||_\infty h_i^2) \tag{11.30}$$

Exercise 11.6. Verify graphically that (11.29) is true.

Part III

Design of Fuzzy Systems from Input-Output Data

Fuzzy systems are used to formulate human knowledge. Therefore, an important question is: What forms does human knowledge usually take? Roughly speaking, human knowledge about a particular engineering problem may be classified into two categories: conscious knowledge and subconscious knowledge. By *conscious knowledge* we mean the knowledge that can be *explicitly* expressed in words, and by *subconscious knowledge* we refer to the situations where the human experts know what to do but cannot express exactly in words how to do it. For example, the experienced truck drivers know how to drive the truck in very difficult situations (they have subconscious knowledge), but it is difficult for them to express their actions in precise words. Even if they can express the actions in words, the description is usually incomplete and insufficient for accomplishing the task.

For conscious knowledge, we can simply ask the human experts to express it in terms of fuzzy IF-THEN rules and put them into fuzzy systems. For subconscious knowledge, what we can do is to ask the human experts to *demonstrate*, that is, to show what they do in some typical situations. When the expert is demonstrating, we view him/her as a black box and measure the inputs and the outputs; that is, we can collect a set of input-output data pairs. In this way, the subconscious knowledge is transformed into a set of input-output pairs; see Fig. 12.1. Therefore, a problem of fundamental importance is to construct fuzzy systems from input-output pairs.

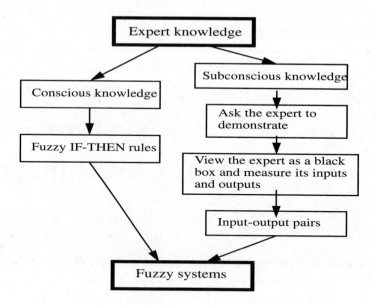

Figure 12.1. Converting expert knowledge into fuzzy systems.

In this part (Chapters 12-15), we will develop a number of methods for constructing fuzzy systems from input-output pairs. Because in many practical situations we are only provided with a limited number of input-output pairs and cannot obtain the outputs for arbitrary inputs, the design methods in Chapters 10 and 11 are not applicable (recall that the methods in Chapters 10 and 11 require that we can determine the output $g(x)$ for arbitrary input $x \in U$). That is, we are now considering the third case discussed in the beginning of Chapter 10. Our task is to design a fuzzy system that characterizes the input-output behavior represented by the input-output pairs.

In Chapter 12, we will develop a simple heuristic method for designing fuzzy systems from input-output pairs and apply the method to the truck backer-upper control and time series prediction problems. In Chapter 13, we will design the fuzzy system by first specifying its structure and then adjusting its parameters using a gradient descent training algorithm; we will use the designed fuzzy systems to identify nonlinear dynamic systems. In Chapter 14, the recursive least squares algorithm will be used to design the parameters of the fuzzy system and the designed fuzzy system will be used as an equalizer for nonlinear communication channels. Finally, Chapter 15 will show how to use clustering ideas to design fuzzy systems.

Chapter 12

Design of Fuzzy Systems Using A Table Look-Up Scheme

12.1 A Table Look-Up Scheme for Designing Fuzzy Systems from Input-Output Pairs

Suppose that we are given the following input-output pairs:

$$(x_0^p; y_0^p), \ p = 1, 2, ..., N \tag{12.1}$$

where $x_0^p \in U = [\alpha_1, \beta_1] \times \cdots \times [\alpha_n, \beta_n] \subset R^n$ and $y_0^p \in V = [\alpha_y, \beta_y] \subset R$. Our objective is to design a fuzzy system $f(x)$ based on these N input-output pairs. We now propose the following five step table look-up scheme to design the fuzzy system:

Step 1. Define fuzzy sets to cover the input and output spaces.

Specifically, for each $[\alpha_i, \beta_i]$, $i = 1, 2, ..., n$, define N_i fuzzy sets A_i^j ($j = 1, 2, ..., N_i$), which are required to be complete in $[\alpha_i, \beta_i]$; that is, for any $x_i \in [\alpha_i, \beta_i]$, there exists A_i^j such that $\mu_{A_i^j}(x_i) \neq 0$. For example, we may choose $\mu_{A_i^j}(x_i)$ to be the pseudo-trapezoid membership functions: $\mu_{A_i^j}(x_i) = \mu_{A_i^j}(x_i; a_i^j, b_i^j, c_i^j, d_i^j)$, where $a_i^1 = b_i^1 = \alpha_i, c_i^j = a_i^{j+1} < b_i^{j+1} = d_i^j$ ($j = 1, 2, ..., N_i - 1$), and $c_i^{N_i} = d_i^{N_i} = \beta_i$. Similarly, define N_y fuzzy sets B^j, $j = 1, 2, ..., N_y$, which are complete in $[\alpha_y, \beta_y]$. We also may choose $\mu_{B^j}(y)$ to be the pseudo-trapezoid membership functions: $\mu_{B^j}(y) = \mu_{B^j}(y; a^j, b^j, c^j, d^j)$, where $a^1 = b^1 = \alpha_y, c^j = a^{j+1} < b^{j+1} = d^j$ ($j = 1, 2, ..., N_y - 1$), and $c^{N_y} = d^{N_y} = \beta_y$. Fig.12.2 shows an example for the $n = 2$ case, where $N_1 = 5, N_2 = 7, N_y = 5$, and the membership functions are all triangular.

Step 2. Generate one rule from one input-output pair.

First, for each input-output pair $(x_{01}^p, ..., x_{0n}^p; y_0^p)$, determine the membership values of x_{0i}^p ($i = 1, 2, ..., n$) in fuzzy sets A_i^j ($j = 1, 2, ..., N_i$) and the membership values of y_0^p in fuzzy sets B^l ($l = 1, 2, ..., N_y$). That is, compute the following:

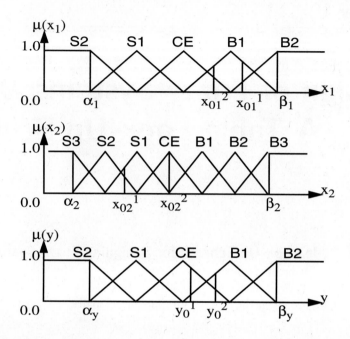

Figure 12.2. An example of membership functions and input-output pairs for the two-input case.

$\mu_{A_i^j}(x_{0i}^p)$ for $j = 1, 2, ..., N_i, i = 1, 2, ..., n$, and $\mu_{B^l}(y_0^p)$ for $l = 1, 2, ..., N_y$. For the example in Fig. 12.2, we have approximately that: x_{01}^1 has membership value 0.8 in B1, 0.2 in B2, and zero in other fuzzy sets; x_{02}^1 has membership value 0.6 in S1, 0.4 in S2, and zero in other fuzzy sets; and, y_0^1 has membership value 0.8 in CE, 0.2 in B1, and zero in other fuzzy sets.

Then, for each input variable x_i $(i = 1, 2, ..., n)$, determine the fuzzy set in which x_{0i}^p has the largest membership value, that is, determine A_i^{j*} such that $\mu_{A_i^{j*}}(x_{0i}^p) \geq \mu_{A_i^j}(x_{0i}^p)$ for $j = 1, 2, ..., N_i$. Similarly, determine B^{l*} such that $\mu_{B^{l*}}(y_0^p) \geq \mu_{B^l}(y_0^p)$ for $l = 1, 2, ..., N_y$. For the example in Fig. 12.2, the input-output pair $(x_{01}^1, x_{02}^1; y_0^1)$ gives $A_1^{j*} = B1, A_2^{j*} = S1$ and $B^{l*} = CE$, and the pair $(x_{01}^2, x_{02}^2; y_0^2)$ gives $A_1^{j*} = B1, A_2^{j*} = CE$ and $B^{l*} = B1$.

Finally, obtain a fuzzy IF-THEN rule as

$$IF\ x_1\ is\ A_1^{j*}\ and\ \cdots\ and\ x_n\ is\ A_n^{j*},\ THEN\ y\ is\ B^{l*} \tag{12.2}$$

For the example in Fig. 12.2, the pair $(x_{01}^1, x_{02}^1; y_0^1)$ gives the rule: *IF x_1 is B1* and x_2 is $S1$, $THEN$ y is CE; and the pair $(x_{01}^2, x_{02}^2; y_0^2)$ gives the rule: IF x_1 is $B1$ and x_2 is CE, $THEN$ y is $B1$.

Step 3. Assign a degree to each rule generated in Step 2.

Since the number of input-output pairs is usually large and with each pair generating one rule, it is highly likely that there are conflicting rules, that is, rules with the same IF parts but different THEN parts. To resolve this conflict, we assign a degree to each generated rule in Step 2 and keep only one rule from a conflicting group that has the maximum degree. In this way not only is the conflict problem resolved, but also the number of rules is greatly reduced.

The degree of a rule is defined as follows: suppose that the rule (12.2) is generated from the input-output pair $(x_0^p; y_0^p)$, then its degree is defined as

$$D(rule) = \prod_{i=1}^{n} \mu_{A_i^{j*}}(x_{0i}^p)\mu_{B^{l*}}(y_0^p) \tag{12.3}$$

For the example in Fig. 12.2, the rule generated by $(x_{01}^1, x_{02}^1; y_0^1)$ has degree

$$D(rule1) = \mu_{B1}(x_{01}^1)\mu_{S1}(x_{02}^1)\mu_{CE}(y_0^1)$$
$$= 0.8 * 0.6 * 0.8 = 0.384 \tag{12.4}$$

and the rule generated by $(x_{01}^2, x_{02}^2; y_0^2)$ has degree

$$D(rule2) = \mu_{B1}(x_{01}^2)\mu_{CE}(x_{02}^2)\mu_{B1}(y_0^2)$$
$$= 0.6 * 1 * 0.7 = 0.42 \tag{12.5}$$

If the input-output pairs have different reliability and we can determine a number to assess it, we may incorporate this information into the degrees of the rules. Specifically, suppose the input-output pair $(x_0^p; y_0^p)$ has reliable degree μ^p ($\in [0,1]$), then the degree of the rule generated by $(x_0^p; y_0^p)$ is redefined as

$$D(rule) = \prod_{i=1}^{n} \mu_{A_i^{j*}}(x_{0i}^p)\mu_{B^{l*}}(y_0^p)\mu^p \tag{12.6}$$

In practice, we may ask an expert to check the data (if the number of input-output pairs is small) and estimate the degree μ^p. Or, if we know the characteristics of the noise in the data pair, we may choose μ^p to reflect the strength of the noise. If we cannot tell the difference among the input-output pairs, we simply choose all $\mu^p = 1$ so that (12.6) reduces to (12.3).

Step 4. Create the fuzzy rule base.

The fuzzy rule base consists of the following three sets of rules:

- The rules generated in Step 2 that do not conflict with any other rules.

- The rule from a conflicting group that has the maximum degree, where a group of conflicting rules consists of rules with the same IF parts.

- Linguistic rules from human experts (due to conscious knowledge).

Since the first two sets of rules are obtained from subconscious knowledge, the final fuzzy rule base combines both conscious and subconscious knowledge.

Intuitively, we can illustrate a fuzzy rule base as a look-up table in the two-input case. For example, Fig. 12.3 demonstrates a table-lookup representation of the fuzzy rule base corresponding to the fuzzy sets in Fig. 12.2. Each box represents a combination of fuzzy sets in $[\alpha_1, \beta_1]$ and fuzzy sets in $[\alpha_2, \beta_2]$ and thus a possible rule. A conflicting group consists of rules in the same box. This method can be viewed as filling up the boxes with appropriate rules; this is why we call this method a table look-up scheme.

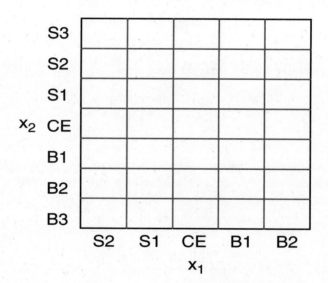

Figure 12.3. Table look-up illustration of the fuzzy rule base.

Step 5. Construct the fuzzy system based on the fuzzy rule base.

We can use any scheme in Chapter 9 to construct the fuzzy system based on the fuzzy rule base created in Step 4. For example, we may choose fuzzy systems with product inference engine, singleton fuzzifier, and center average defuzzifier (Lemma 9.1).

We now make a few remarks on this five step procedure of designing the fuzzy system from the input-output pairs.

- A fundamental difference between this method and the methods in Chapters 10 and 11 is that the methods in Chapters 10 and 11 require that we are able to determine the exact output $g(x)$ for any input $x \in U$, whereas in this method we cannot freely choose the input points in the given input-output pairs. Also, in order to design a fuzzy system with the required accuracy, the methods in Chapters 10 and 11 need to know the bounds of the first or second order derivatives of the function to be approximated, whereas this method does not require this information.

- If the input-output pairs happen to be the $(e_1^{i_1}, e_2^{i_2}; \bar{y}^{i_1 i_2})$ in (10.9), then it is easy to verify that the fuzzy system designed through these five steps is the same fuzzy system as designed in Section 10.2. Therefore, this method can be viewed as a generalization of the design method in Section 10.2 to the case where the input-output pairs cannot be arbitrarily chosen.

- The number of rules in the final fuzzy rule base is bounded by two numbers: N, the number of input-output pairs, and $\prod_{i=1}^{n} N_i$, the number of all possible combinations of the fuzzy sets defined for the input variables. If the dimension of the input space n is large, $\prod_{i=1}^{n} N_i$ will be a huge number and may be larger than N. Also, some input-output pairs may correspond to the same box in Fig. 12.3 and thus can contribute only one rule to the fuzzy rule base. Therefore, the number of rules in the fuzzy rule base may be much less than both $\prod_{i=1}^{n} N_i$ and N. Consequently, the fuzzy rule base generated by this method may not be complete. To make the fuzzy rule base complete, we may fill in the empty boxes in the fuzzy rule base by interpolating the existing rules; we leave the details to the reader to think about.

Next, we apply this method to a control problem and a time series prediction problem.

12.2 Application to Truck Backer-Upper Control

Backing up a truck to a loading dock is a nonlinear control problem. Using conventional control approach, we can first develop a mathematical model of the system and then design a controller based on nonlinear control theory (Walsh, Tilbury, Sastry, Murray, and Laumond [1994]). Another approach is to design a controller to emulate the human driver. We adapt the second approach. Assume that an experienced human driver is available and we can measure the truck's states and the corresponding control action of the human driver while he/she is backing the truck to the dock; that is, we can collect a set of input-output (state-control) pairs. We will design a fuzzy system based on these input-output pairs using the table look-up scheme in the last section, and replace the human driver by the designed fuzzy system.

The simulated truck and loading zone are shown in Fig. 12.4. The truck position is determined by three state variables ϕ, x and y, where ϕ is the angle of the truck with respect to the horizontal line as shown in Fig. 12.4. Control to the truck is the steeling angle θ. Only backing up is permitted. The truck moves backward by a fixed unit distance every stage. For simplicity, we assume enough clearance between the truck and the loading dock such that y does not have to be considered as a state variable. The task is to design a controller whose inputs are (x, ϕ) and whose output is θ, such that the final state will be $(x_f, \phi_f) = (10, 90^o)$. We assume that $x \in [0, 20], \phi \in [-90^o, 270^o]$ and $\theta \in [-40^o, 40^o]$; that is, $U = [0, 20] \times [-90^o, 270^o]$ and $V = [-40^o, 40^o]$.

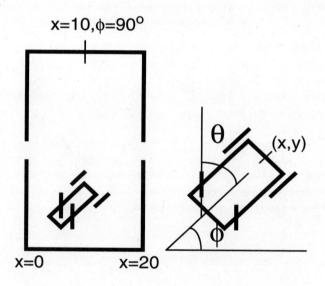

Figure 12.4. The simulated truck and loading zone.

First, we generate the input-output pairs $(x^p, \phi^p; \theta^p)$. We do this by trial and error: at every stage (given x and ϕ) starting from an initial state, we determine a control θ based on common sense (that is, our own experience of how to control the steering angle in the situation); after some trials, we choose the input-output pairs corresponding to the smoothest successful trajectory. The following 14 initial states were used to generate the desired input-output pairs: $(x_0, \phi_0^o) = (1, 0), (1, 90), (1, 270); (7, 0), (7, 90), (7, 180), (7, 270); (13, 0), (13, 90), (13, 180), (13, 270);$

$(19, 90), (19, 180), (19, 270)$. Table 12.1 shows the input-output pairs starting from the initial state $(x_0, \phi_0) = (1, 0^o)$. The input-output pairs starting from the other 13 initial states can be obtained in a similar manner. Totally, we have about 250 input-output pairs. We now design a fuzzy system based on these input-output pairs using the table look-up scheme developed in the last section.

Table 12.1. Ideal trajectory (x_t, ϕ_t^o) and the corresponding control θ_t^o starting from $(x_0, \phi_0) = (1, 0^o)$.

t	x_t	ϕ_t^o	θ_t^o
0	1.00	0.00	-19.00
1	1.95	9.37	-17.95
2	2.88	18.23	-16.90
3	3.79	26.57	-15.85
4	4.65	34.44	-14.80
5	5.45	41.78	-13.75
6	6.18	48.60	-12.70
7	7.48	54.91	-11.65
8	7.99	60.71	-10.60
9	8.72	65.99	-9.55
10	9.01	70.75	-8.50
11	9.28	74.98	-7.45
12	9.46	78.70	-6.40
13	9.59	81.90	-5.34
14	9.72	84.57	-4.30
15	9.81	86.72	-3.25
16	9.88	88.34	-2.20
17	9.91	89.44	0.00

In Step 1, we define 7 fuzzy sets in $[-90^o, 270^o]$, 5 fuzzy sets in $[0, 20]$ and 7 fuzzy sets in $[-40^o, 40^o]$, where the membership functions are shown in Fig.12.5. In Steps 2 and 3, we generate one rule from one input-output pair and compute the corresponding degrees of the rules. Table 12.2 shows the rules and their degrees generated by the corresponding input-output pairs in Table 12.1. The final fuzzy rule base generated in Step 4 is shown in Fig.12.6 (we see that some boxes are empty, so the input-output pairs do not cover all the state space; however, we will see that the rules in Fig. 12.6 are sufficient for controlling the truck to the desired position starting from a wide range of initial positions). Finally, in Step 5 we use the fuzzy system with product inference engine, singleton fuzzifier, and center average defuzzifier; that is, the designed fuzzy system is in the form of (9.1) with the rules in Fig. 12.6.

We now use the fuzzy system designed above as a controller for the truck. To simulate the control system, we need a mathematical model of the truck. We use

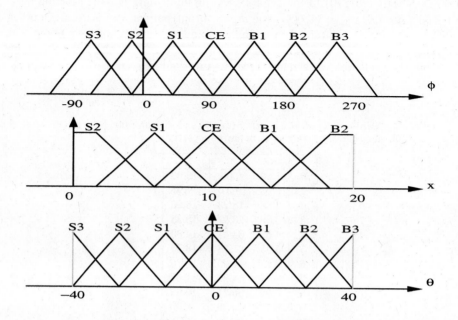

Figure 12.5. Membership functions for the truck backer-upper control problem.

ϕ	S2	S1	CE	B1	B2
S3	S2	S3			
S2	S2	S3	S3	S3	
S1	B1	S1	S2	S3	S2
CE	B2	B2	CE	S2	S2
B1	B2	B3	B2	B1	S1
B2		B3	B3	B3	B2
B3				B3	B2

X

Figure 12.6. The final fuzzy rule base for the truck backer-upper control problem.

Table 12.2. Fuzzy IF-THEN rules generated from the input-output pairs in Table 12.1 and their degrees.

x is	ϕ is	θ is	degree
S2	S2	S2	1.00
S2	S2	S2	0.92
S2	S2	S2	0.35
S2	S2	S2	0.12
S2	S2	S2	0.07
S1	S2	S1	0.08
S1	S1	S1	0.18
S1	S1	S1	0.53
S1	S1	S1	0.56
S1	S1	S1	0.60
CE	S1	S1	0.35
CE	S1	S1	0.21
CE	S1	CE	0.16
CE	CE	CE	0.32
CE	CE	CE	0.45
CE	CE	CE	0.54
CE	CE	CE	0.88
CE	CE	CE	0.92

the following approximate model (Wang and Mendel [1992b]):

$$x(t+1) = x(t) + cos[\phi(t) + \theta(t)] + sin[\theta(t)]sin[\phi(t)] \qquad (12.7)$$

$$y(t+1) = y(t) + sin[\phi(t) + \theta(t)] - sin[\theta(t)]cos[\phi(t)] \qquad (12.8)$$

$$\phi(t+1) = \phi(t) - sin^{-1}[\frac{2sin(\theta(t))}{b}] \qquad (12.9)$$

where b is the length of the truck and we assume $b = 4$ in our simulations. Fig. 12.7 shows the truck trajectory using the designed fuzzy system as the controller for two initial conditions: $(x_0, \phi_0) = (3, -30^\circ)$ and $(13, 30^\circ)$. We see that the fuzzy controller can successfully control the truck to the desired position.

12.3   Application to Time Series Prediction

Time series prediction is an important practical problem. Applications of time series prediction can be found in the areas of economic and business planning, inventory and production control, weather forecasting, signal processing, control, and many other fields. In this section, we use the fuzzy system designed by the table look-up scheme to predict the Mackey-Glass chaotic time series that is generated by the

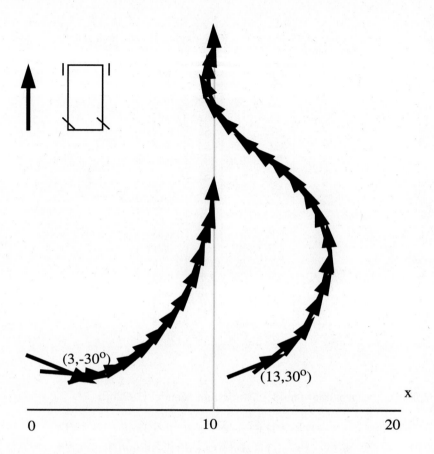

Figure 12.7. Truck trajectories using the fuzzy controller.

following delay differential equation:

$$\frac{dx(t)}{dt} = \frac{0.2x(t-\tau)}{1+x^{10}(t-\tau)} - 0.1x(t) \qquad (12.10)$$

When $\tau > 17$, (12.10) shows chaotic behavior. We choose $\tau = 30$.

Let $x(k)$ $(k = 1, 2, 3, ...)$ be the time series generated by (12.10) (sampling the continuous curve $x(t)$ generated by (12.10) with an interval of 1 sec.). Fig.12.8 shows 600 points of $x(k)$. The problem of time series prediction can be formulated as follows: given $x(k-n+1), x(k-n+2), ..., x(k)$, estimate $x(k+1)$, where n is a positive integer. That is, the task is to determine a mapping from $[x(k-n+1), x(k-n+2), ..., x(k)] \in R^n$ to $[x(k+1)] \in R$, and this mapping in our case is the designed fuzzy system based on the input-output pairs. Assuming that $x(1), x(2), ..., x(k)$ are given with $k > n$, we can form $k - n$ input-output pairs as follows:

$$[x(k - n), ..., x(k - 1); x(k)]$$
$$[x(k - n - 1), ..., x(k - 2); x(k - 1)]$$
$$...$$ (12.11)
$$[x(1), ..., x(n); x(n + 1)]$$

These input-output pairs are used to design a fuzzy system $f(x)$ using the table lookup scheme in Section 12.2, and this $f(x)$ is then used to predict $x(k + l)$ for $l = 1, 2, ...$, where the input to $f(x)$ is $[x(k - n + l), ..., x(k - 1 + l)]$ when predicting $x(k + l)$.

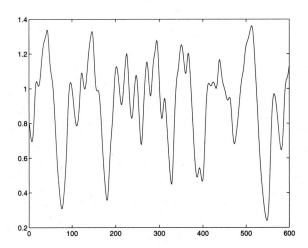

Figure 12.8. A section of the Mackey-Glass chaotic time series.

We now use the first 300 points in Fig. 12.8 to construct the input-output pairs and the designed fuzzy system is then used to predict the remaining 300 points. We consider two cases: (i) $n = 4$ and the 7 fuzzy sets in Fig.12.9 are defined for each input variable, and (ii) $n = 4$ and the 15 fuzzy sets in Fig.12.10 are used. The prediction results for these two cases are shown in Figs.12.11 and 12.12, respectively. Comparing Figs. 12.11 and 12.12, we see that the prediction accuracy is improved by defining more fuzzy sets for each input variable.

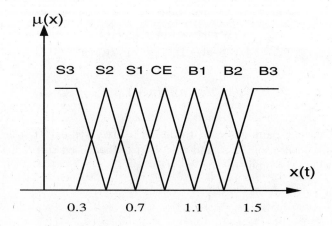

Figure 12.9. The first choice of membership functions for each input variable.

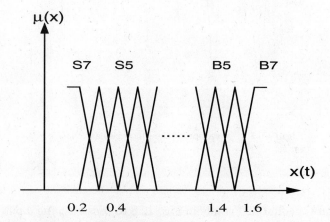

Figure 12.10. The second choice of membership functions for each input variable.

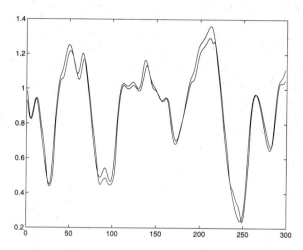

Figure 12.11. Prediction and the true values of the time series using the membership functions in Fig. 12.9.

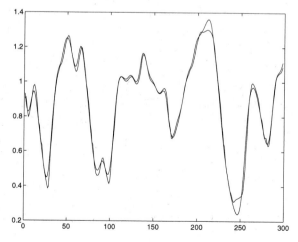

Figure 12.12. Prediction and the true values of the time series using the membership functions in Fig. 12.10.

12.4 Summary and Further Readings

In this chapter we have demonstrated the following:

- The details of the table look-up method for designing fuzzy systems from input-output pairs.

- How to apply the method to the truck backer-upper control and the time series prediction problems.

- How to combine conscious and subconscious knowledge into fuzzy systems using this table look-up scheme.

This table look-up scheme is taken from Wang and Mendel [1992b] and Wang [1994], which discussed more details about this method and gave more examples. Application of the method to financial data prediction can be found in Cox [1994].

12.5 Exercises and Projects

Exercise 12.1. Consider the design of a 2-input-1-output fuzzy system using the table look-up scheme. Suppose that in Step 1 we define the fuzzy sets as shown in Fig. 12.2, where $\alpha_1 = \alpha_2 = \alpha_y = 0$ and $\beta_1 = \beta_2 = \beta_y = 1$, and the membership functions are triangular and equally spaced.

(a) What is the minimum number of input-output pairs such that every fuzzy sets in Fig. 12.2 will appear at least once in the generated rules? Give an example of this minimum set of input-output pairs.

(b) What is the minimum number of input-output pairs such that the generated fuzzy rule base is complete? Give an example of this minimum set of input-output pairs.

Exercise 12.2. Consider the truck backer-upper control problem in Section 12.3.

(a) Generate a set of input-output pairs by driving the truck from the initial state $(x_0, \phi_0) = (1, 90^o)$ to the final state $(x_f, \phi_f) = (10, 90^o)$ using common sense.

(b) Use the table look-up scheme to create a fuzzy rule base from the input-output pairs generated in (a), where the membership functions in Step 1 are given in Fig. 12.5.

(c) Construct a fuzzy system based on the fuzzy rule base in (b) and use it to control the truck from $(x_0, \phi_0) = (0, 90^o)$ and $(x_0, \phi_0) = (-3, 90^o)$. Comment on the simulation results.

Exercise 12.3. Let $f(x)$ be the fuzzy system designed using the table look-up scheme. Can you determine an error bound for $|f(x_0^p) - y_0^p|$? Explain your answer.

Exercise 12.4. Propose a method to fill up the empty boxes in the fuzzy rule base generated by the table look-up scheme. Justify your method and test it through examples.

Exercise 12.5. We are given 10 points $x(1), x(2), ..., x(10)$ of a time series and we want to predict $x(12)$.

(a) If we use the fuzzy system $f[x(10), x(8)]$ to predict $x(12)$, list all the input-output pairs for constructing this fuzzy system.

(b) If we use the fuzzy system $f[x(10), x(9), x(8)]$ to predict $x(12)$, list all the input-output pairs for constructing this fuzzy system.

12.6 (Project). Write a computer program to implement the table look-up scheme and apply your program to the time series prediction problem in Section 12.4. To make your codes generally applicable, you may have to include a method to fill up the empty boxes.

Chapter 13

Design of Fuzzy Systems Using Gradient Descent Training

13.1 Choosing the Structure of Fuzzy Systems

In the table look-up scheme of Chapter 12, the membership functions are fixed in the first step and do not depend on the input-output pairs; that is, the membership functions are not optimized according to the input-output pairs. In this chapter, we propose another approach to designing fuzzy systems where the membership functions are chosen in such a way that certain criterion is optimized.

From a conceptual point of view, the design of fuzzy systems from input-output pairs may be classified into two types of approaches. In the first approach, fuzzy IF-THEN rules are first generated from input-output pairs, and the fuzzy system is then constructed from these rules according to certain choices of fuzzy inference engine, fuzzifier, and defuzzifier. The table look-up scheme of Chapter 12 belongs to this approach. In the second approach, the structure of the fuzzy system is specified first and some parameters in the structure are free to change, then these free parameters are determined according to the input-output pairs. In this chapter, we adapt this second approach.

First, we specify the structure of the fuzzy system to be designed. Here we choose the fuzzy system with product inference engine, singleton fuzzifier, center average defuzzifier, and Gaussian membership function, given by (9.6). That is, we assume that the fuzzy system we are going to design is of the following form:

$$f(x) = \frac{\sum_{l=1}^{M} \bar{y}^l [\prod_{i=1}^{n} exp(-(\frac{x_i - \bar{x}_i^l}{\sigma_i^l})^2)]}{\sum_{l=1}^{M} [\prod_{i=1}^{n} exp(-(\frac{x_i - \bar{x}_i^l}{\sigma_i^l})^2)]} \tag{13.1}$$

where M is fixed, and $\bar{y}^l$, $\bar{x}_i^l$ and σ_i^l are free parameters (we choose $a_i^l = 1$). Although the structure of the fuzzy system is chosen as (13.1), the fuzzy system has not been designed because the parameters $\bar{y}^l$, $\bar{x}_i^l$ and σ_i^l are not specified. Once we specify the parameters $\bar{y}^l, \bar{x}_i^l$ and σ_i^l, we obtain the designed fuzzy system; that is, designing the fuzzy system is now equivalent to determining the parameters $\bar{y}^l, \bar{x}_i^l$ and σ_i^l.

To determine these parameters in some optimal fashion, it is helpful to represent the fuzzy system $f(x)$ of (13.1) as a feedforward network. Specifically, the mapping from the input $x \in U \subset R^n$ to the output $f(x) \in V \subset R$ can be implemented according to the following operations: first, the input x is passed through a product Gaussian operator to become $z^l = \prod_{i=1}^{n} exp(-(\frac{x_i - \bar{x}_i^l}{\sigma_i^l})^2)$; then, the z^l are passed through a summation operator and a weighted summation operator to obtain $b = \sum_{l=1}^{M} z^l$ and $a = \sum_{l=1}^{M} \bar{y}^l z^l$; finally, the output of the fuzzy system is computed as $f(x) = a/b$. This three-stage operation is shown in Fig. 13.1 as a three-layer feedforward network.

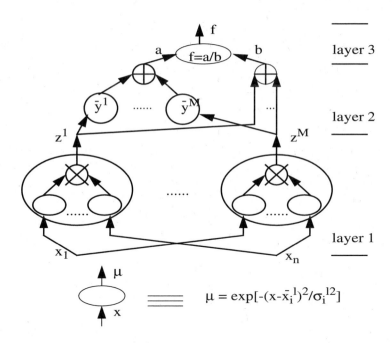

Figure 13.1. Network representation of the fuzzy system.

13.2   Designing the Parameters by Gradient Descent

As in Chapter 12, the data we have are the input-output pairs given by (12.1). Our task is to design a fuzzy system $f(x)$ in the form of (13.1) such that the matching error

$$e^p = \frac{1}{2}[f(x_0^p) - y_0^p]^2 \tag{13.2}$$

is minimized. That is, the task is to determine the parameters $\bar{y}^l, \bar{x}_i^l$ and σ_i^l such that e^p of (13.2) is minimized. In the sequel, we use e, f and y to denote $e^p, f(x_0^p)$ and y_0^p, respectively.

We use the gradient descent algorithm to determine the parameters. Specifically, to determine $\bar{y}^l$, we use the algorithm

$$\bar{y}^l(q+1) = \bar{y}^l(q) - \alpha \frac{\partial e}{\partial \bar{y}^l}|_q \tag{13.3}$$

where $l = 1, 2, ..., M, q = 0, 1, 2, ...,$ and α is a constant stepsize. If $\bar{y}^l(q)$ converges as q goes to infinity, then from (13.3) we have $\frac{\partial e}{\partial \bar{y}^l} = 0$ at the converged $\bar{y}^l$, which means that the converged $\bar{y}^l$ is a local minimum of e. From Fig. 13.1 we see that f (and hence e) depends on $\bar{y}^l$ only through a, where $f = a/b$, $a = \sum_{l=1}^{M}(\bar{y}^l z^l)$, $b = \sum_{l=1}^{M} z^l$, and $z^l = \prod_{i=1}^{n} exp(-(\frac{x_i - \bar{x}_i^l}{\sigma_i^l})^2)$; hence, using the Chain Rule, we have

$$\frac{\partial e}{\partial \bar{y}^l} = (f - y)\frac{\partial f}{\partial a}\frac{\partial a}{\partial \bar{y}^l} = (f - y)\frac{1}{b}z^l \tag{13.4}$$

Substituting (13.4) into (13.3), we obtain the training algorithm for $\bar{y}^l$:

$$\bar{y}^l(q+1) = \bar{y}^l(q) - \alpha \frac{f - y}{b} z^l \tag{13.5}$$

where $l = 1, 2, ..., M$, and $q = 0, 1, 2, ...$

To determine $\bar{x}_i^l$, we use

$$\bar{x}_i^l(q+1) = \bar{x}_i^l(q) - \alpha \frac{\partial e}{\partial \bar{x}_i^l}|_q \tag{13.6}$$

where $i = 1, 2, ..., n, l = 1, 2, ..., M$, and $q = 0, 1, 2, ...$ We see from Fig. 13.1 that f (and hence e) depends on $\bar{x}_i^l$ only through z^l; hence, using the Chain Rule, we have

$$\frac{\partial e}{\partial \bar{x}_i^l} = (f - y)\frac{\partial f}{\partial z^l}\frac{\partial z^l}{\partial \bar{x}_i^l} = (f - y)\frac{\bar{y}^l - f}{b} z^l \frac{2(x_{0i}^p - \bar{x}_i^l)}{\sigma_i^{l2}} \tag{13.7}$$

Substituting (13.7) into (13.6), we obtain the training algorithm for $\bar{x}_i^l$:

$$\bar{x}_i^l(q+1) = \bar{x}_i^l(q) - \alpha \frac{f - y}{b}(\bar{y}^l(q) - f)z^l \frac{2(x_{0i}^p - \bar{x}_i^l(q))}{\sigma_i^{l2}(q)} \tag{13.8}$$

where $i = 1, 2, ..., n, l = 1, 2, ..., M$, and $q = 0, 1, 2, ...$

Using the same procedure, we obtain the training algorithm for σ_i^l:

$$\sigma_i^l(q+1) = \sigma_i^l(q) - \alpha \frac{\partial e}{\partial \sigma_i^l}|_q$$

$$= \sigma_i^l(q) - \alpha \frac{f-y}{b}(\bar{y}^l(q) - f)z^l \frac{2(x_{0i}^p - \bar{x}_i^l(q))^2}{\sigma_i^{l3}(q)} \tag{13.9}$$

where $i = 1, 2, ..., n$, $l = 1, 2, ..., M$, and $q = 0, 1, 2, ...$

The training algorithm (13.5), (13.8), and (13.9) performs an error back-propagation procedure. To train $\bar{y}^l$, the "normalized" error $(f-y)/b$ is back-propagated to the layer of $\bar{y}^l$; then $\bar{y}^l$ is updated using (13.5) in which z^l is the input to $\bar{y}^l$ (see Fig. 13.1). To train $\bar{x}_i^l$ and σ_i^l, the "normalized" error $(f-y)/b$ times $(\bar{y}^l - f)$ and z^l is back-propagated to the processing unit of Layer 1 whose output is z^l; then $\bar{x}_i^l$ and σ_i^l are updated using (13.8) and (13.9), respectively, in which the remaining variables $\bar{x}_i^l, x_{0i}^p$, and σ_i^l (that is, the variables on the right-hand sides of (13.8) and (13.9), except the back-propagated error $\frac{f-y}{b}(\bar{y}^l - f)z^l$) can be obtained locally. Therefore, this algorithm is also called the *error back-propagation training algorithm*.

We now summarize this design method.

Design of Fuzzy Systems Using Gradient Descent Training:

- **Step 1. Structure determination and initial parameter setting.** Choose the fuzzy system in the form of (13.1) and determine the M. Larger M results in more parameters and more computation, but gives better approximation accuracy. Specify the initial parameters $\bar{y}^l(0), \bar{x}_i^l(0)$ and $\sigma_i^l(0)$. These initial parameters may be determined according to the linguistic rules from human experts, or be chosen in such a way that the corresponding membership functions uniformly cover the input and output spaces. For particular applications, we may use special methods; see Section 13.3 for an example.

- **Step 2. Present input and calculate the output of the fuzzy system.** For a given input-output pair $(x_0^p; y_0^p)$, $p = 1, 2, ...$, and at the $q'th$ stage of training, $q = 0, 1, 2, ...$, present x_0^p to the input layer of the fuzzy system in Fig. 13.1 and compute the outputs of Layers 1-3. That is, compute

$$z^l = \prod_{i=1}^{n} exp(-(\frac{x_{0i}^p - \bar{x}_i^l(q)}{\sigma_i^l(q)})^2) \tag{13.10}$$

$$b = \sum_{l=1}^{M} z^l \tag{13.11}$$

$$a = \sum_{l=1}^{M} \bar{y}^l(q)z^l \tag{13.12}$$

$$f = a/b \tag{13.13}$$

- **Step 3. Update the parameters**. Use the training algorithm (13.5), (13.8) and (13.9) to compute the updated parameters $\bar{y}^l(q+1), \bar{x}_i^l(q+1)$ and $\sigma_i^l(q+1)$, where $y = y_0^p$, and z^l, b, a and f equal those computed in Step 2.

- **Step 4**. Repeat by going to Step 2 with $q = q + 1$, until the error $|f - y_0^p|$ is less than a prespecified number ϵ, or until the q equals a prespecified number.

- **Step 5**. Repeat by going to Step 2 with $p = p + 1$; that is, update the paramters using the next input-output pair $(x_0^{p+1}; y_0^{p+1})$.

- **Step 6**. If desirable and feasible, set $p = 1$ and do Steps 2-5 again until the designed fuzzy system is satisfactory. For on-line control and dynamic system identification, this step is not feasible because the input-output pairs are provided one-by-one in a real-time fashion. For pattern recognition problems where the input-output pairs are provided off-line, this step is usually desirable.

Because the training algorithm (13.5), (13.8) and (13.9) is a gradient descent algorithm, the choice of the initial parameters is crucial to the success of the algorithm. If the initial parameters are close to the optimal parameters, the algorithm has a good chance to converge to the optimal solution; otherwise, the algorithm may converge to a nonoptimal solution or even diverge. The advantage of using the fuzzy system is that the parameters $\bar{y}^l, \bar{x}_i^l$ and σ_i^l have clear physical meanings and we have methods to choose good initial values for them. Keep in mind that the parameters $\bar{y}^l$ are the centers of the fuzzy sets in the THEN parts of the rules, and the parameters $\bar{x}_i^l$ and σ_i^l are the centers and widths of the Gaussian fuzzy sets in the IF parts of the rules. Therefore, given a designed fuzzy system in the form of (13.1), we can recover the fuzzy IF-THEN rules that constitute the fuzzy system. These recovered fuzzy IF-THEN rules may help to explain the designed fuzzy system in a user-friendly manner.

Next, we apply this method to the problem of nonlinear dynamic system identification.

13.3   Application to Nonlinear Dynamic System Identification

13.3.1   Design of the Identifier

System identification is a process of determining an appropriate model for the system based on measurements from sensors. It is an important process because many approaches in engineering depend on the model of the system. Because the fuzzy systems are powerful universal approximators, it is reasonable to use them as identification models for nonlinear systems. In this section, we use the fuzzy system

(13.1) equipped with the training algorithm (13.5), (13.8) and (13.9) to approximate unknown nonlinear components in dynamic systems.

Consider the discrete time nonlinear dynamic system

$$y(k + 1) = f(y(k), ..., y(k - n + 1); u(k), ..., u(k - m + 1)) \qquad (13.14)$$

where f is an unknown function we want to identify, u and y are the input and output of the system, respectively, and n and m are positive integers. Our task is to identify the unknown function f based on fuzzy systems.

Let $\hat{f}(x)$ be the fuzzy system in the form of (13.1). We replace the $f(x)$ in (13.14) by $\hat{f}(x)$ and obtain the following identification model:

$$\hat{y}(k + 1) = \hat{f}(y(k), ..., y(k - n + 1); u(k), ..., u(k - m + 1)) \qquad (13.15)$$

Our task is to adjust the parameters in $\hat{f}(x)$ such that the output of the identification model $\hat{y}(k + 1)$ converges to the output of the true system $y(k + 1)$ as k goes to infinity. Fig. 13.2 shows this identification scheme.

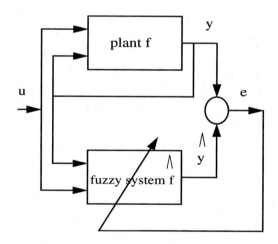

Figure 13.2. Basic scheme of identification model for the nonlinear dynamic system using the fuzzy system.

The input-output pairs in this problem are $(x_0^{k+1}; y_0^{k+1})$, where $x_0^{k+1} = (y(k), ..., y(k - n + 1); u(k), ..., u(k - m + 1))$, $y_0^{k+1} = y(k + 1)$, and $k = 0, 1, 2, ...$ Because the system is dynamic, these input-output pairs are collected one at a time. The operation of the identification process is the same as the Steps 1-5 in Section 13.2. Note that the p there is the k in (13.14) and (13.15) and the n in (13.1) equals $n + m$.

13.3.2 Initial Parameter Choosing

As we discussed in Section 13.2, a good initial $\hat{f}$ is crucial for the success of the approach. For this particular identification problem, we propose the following method for on-line initial parameter choosing and provide a theoretical justification (Lemma 13.1) to explain why this is a good method.

An on-line initial parameter choosing method: Collect the input-output pairs $(x_0^{k+1}; y_0^{k+1}) = (y(k), ..., y(k-n+1), u(k), ..., u(k-m+1); y(k+1))$ for the first M time points $k = 0, 1, ..., M-1$, and do not start the training algorithm until $k = M-1$ (that is, set $q = k-M$ in (13.5), (13.8) and (13.9)). The initial parameters are chosen as: $\bar{y}^l(0) = y_0^l$, $\bar{x}_i^l(0) = x_{0i}^l$, and $\sigma_i^l(0)$ equals small numbers (see Lemma 13.1) or $\sigma_i^l(0) = [max(x_{oi}^l : l = 1, 2, ..., M) - min(x_{oi}^l : l = 1, 2, ..., M)]/M$, where $l = 1, 2, ..., M$ and $i = 1, 2, ..., n+m$. The second choice of $\sigma_i^l(0)$ makes the membership functions uniformly cover the range of x_{0i}^l from $l = 1$ to $l = M$.

We now show that by choosing the $\sigma_i^l(0)$ sufficiently small, the fuzzy system with the preceding initial parameters can match all the M input-output pairs $(x_0^{k+1}; y_0^{k+1})$, $k = 0, 1, ..., M-1$, to arbitrary accuracy.

Lemma 13.1. For arbitrary $\epsilon > 0$, there exists $\sigma^* > 0$ such that the fuzzy system $\hat{f}(x)$ of (13.1) with the preceding initial parameters $\bar{y}^l$ and $\bar{x}_i^l$ and $\sigma_i^l = \sigma^*$ has the property that

$$|\hat{f}(x_0^{k+1}) - y_0^{k+1}| < \epsilon \tag{13.16}$$

for $k = 0, 1, ..., M-1$.

Proof: Substituting the initial parameters $\bar{y}^l(0)$ and $\bar{x}_i^l(0)$ into (13.1) and setting $\sigma_i^l = \sigma^*$, we have

$$\hat{f}(x) = \frac{\sum_{l=1}^M y_0^l [\prod_{i=1}^{n+m} exp(-(\frac{x_i - x_{oi}^l}{\sigma^*})^2)]}{\sum_{l=1}^M [\prod_{i=1}^{n+m} exp(-(\frac{x_i - x_{oi}^l}{\sigma^*})^2)]} \tag{13.17}$$

Setting $x = x_0^{k+1}$ in (13.17) and noticing $1 \leq k+1 \leq M$, we have

$$\hat{f}(x_0^{k+1}) = \frac{\sum_{l=1}^M y_0^l [\prod_{i=1}^{n+m} exp(-(\frac{x_{oi}^{k+1} - x_{oi}^l}{\sigma^*})^2)]}{\sum_{l=1}^M [\prod_{i=1}^{n+m} exp(-(\frac{x_{oi}^{k+1} - x_{oi}^l}{\sigma^*})^2)]}$$

$$= \frac{y_0^{k+1} + \sum_{l=1, l \neq k+1}^M y_0^l [\prod_{i=1}^{n+m} exp(-(\frac{x_{oi}^{k+1} - x_{oi}^l}{\sigma^*})^2)]}{1 + \sum_{l=1, l \neq k+1}^M [\prod_{i=1}^{n+m} exp(-(\frac{x_{oi}^{k+1} - x_{oi}^l}{\sigma^*})^2)]} \tag{13.18}$$

Hence,

$$|\hat{f}(x_0^{k+1}) - y_0^{k+1}| = \left| \frac{\sum_{l=1, l \neq k+1}^M (y_0^l - y_0^{k+1}) [\prod_{i=1}^{n+m} exp(-(\frac{x_{oi}^{k+1} - x_{oi}^l}{\sigma^*})^2)]}{1 + \sum_{l=1, l \neq k+1}^M [\prod_{i=1}^{n+m} exp(-(\frac{x_{oi}^{k+1} - x_{oi}^l}{\sigma^*})^2)]} \right| \tag{13.19}$$

If $x_0^{l_1} \neq x_0^{l_2}$ for $l_1 \neq l_2$, then we have $\prod_{i=1}^{n+m} exp(-(\frac{x_{oi}^{k+1}-x_{oi}^l}{\sigma^*})^2) \to 0$ as $\sigma^* \to 0$ for $l \neq k+1$. Therefore, by choosing σ^* sufficiently small, we can make $|\hat{f}(x_0^{k+1})-y_0^{k+1}| < \epsilon$. Similarly, we can show that (13.16) is true in the case where $x_0^l = x_0^{k+1}$ for some $l \neq k+1$; this is left as an exercise. $\square$

Based on Lemma 13.1, we can say that the initial parameter choosing method is a good one because the fuzzy system with these initial parameters can at least match the first M input-output pairs arbitrarily well. If these first M input-output pairs contain important features of the unknown nonlinear function $f(x)$, we can hope that after the training starts from time point M, the fuzzy identifier will converge to the unknown nonlinear system very quickly. In fact, based on our simulation results in the next subsection, this is indeed true. However, we cannot choose σ_i^l too small because, although a fuzzy system with small σ_i^l matches the first M pairs quite well, it may give large approximation errors for other input-output pairs. Therefore, in our following simulations we will use the second choice of σ_i^l described in the on-line initial parameter choosing method.

13.3.3 Simulations

Example 13.1. The plant to be identified is governed by the difference equation

$$y(k + 1) = 0.3y(k) + 0.6y(k - 1) + g[u(k)] \qquad (13.20)$$

where the unknown function has the form $g(u) = 0.6sin(\pi u) + 0.3sin(3\pi u) + 0.1sin(5\pi u)$. From (13.15), the identification model is governed by the difference equation

$$\hat{y}(k + 1) = 0.3y(k) + 0.6y(k - 1) + \hat{f}[u(k)] \qquad (13.21)$$

where $\hat{f}[*]$ is in the form of (13.1) with $M = 10$. We choose $\alpha = 0.5$ in the training algorithm (13.5), (13.8) and (13.9) and use the on-line parameter choosing method. We start the training from time point $k = 10$, and adjust the parameters $\bar{y}^l, \bar{x}_i^l$, and σ_i^l for one cycle at each time point. That is, we use (13.5), (13.8) and (13.9) once at each time point. Fig. 13.3 shows the outputs of the plant and the identification model when the training stops at $k = 200$, where the input $u(k) = sin(2\pi k/200)$. We see from Fig. 13.3 that the output of the identification model follows the output of the plant almost immediately and still does so when the training stops at $k = 200$. $\square$

Example 13.2. In this example, we show how the fuzzy identifier works for a multi-input-multi-output plant that is described by the equations

$$\begin{bmatrix} y_1(k + 1) \\ y_2(k + 1) \end{bmatrix} = \begin{bmatrix} \frac{y_1(k)}{1+y_2^2(k)} \\ \frac{y_1(k)y_2(k)}{1+y_2^2(k)} \end{bmatrix} + \begin{bmatrix} u_1(k) \\ u_2(k) \end{bmatrix} \qquad (13.22)$$

The identification model consists of two fuzzy systems, $\hat{f}^1$ and $\hat{f}^2$, and is described

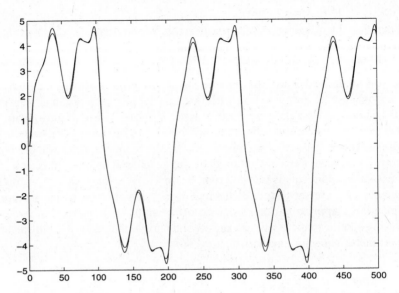

Figure 13.3. Outputs of the plant and the identification model for Example 13.1 when the training stops at $k = 200$.

by the equations

$$\begin{bmatrix} \hat{y}_1(k+1) \\ \hat{y}_2(k+1) \end{bmatrix} = \begin{bmatrix} \hat{f}^1(y_1(k), y_2(k)) \\ \hat{f}^2(y_1(k), y_2(k)) \end{bmatrix} + \begin{bmatrix} u_1(k) \\ u_2(k) \end{bmatrix} \qquad (13.23)$$

Both $\hat{f}^1$ and $\hat{f}^2$ are in the form of (13.1) with $M = 121$. The identification procedure is carried out for 5,000 time steps using random inputs $u_1(k)$ and $u_2(k)$ whose magnitudes are uniformly distributed over [-1,1], where we choose $\alpha = 0.5$, use the on-line initial parameter choosing method, and train the parameters for one cycle at each time point. The responses of the plant and the trained identification model for a vector input $[u_1(k), u_2(k)] = [sin(2\pi k/25), cos(2\pi k/25)]$ are shown in Figs. 13.4 and 13.5 for $y_1(k)$-$\hat{y}_1(k)$ and $y_2(k)$-$\hat{y}_2(k)$, respectively. We see that the fuzzy identifier follows the true system almost perfectly. □

13.4   Summary and Further Readings

In this chapter we have demonstrated the following:

- The derivation of the gradient descent training algorithm.

- The method for choosing the initial parameters of the fuzzy identification model and its justification (Lemma 13.1).

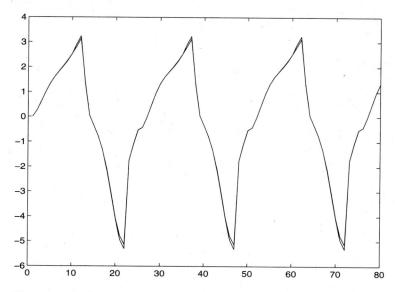

Figure 13.4. Outputs of the plant $y_1(k)$ and the identification model $\hat{y}_1(k)$ for Example 13.2 after 5,000 steps of training.

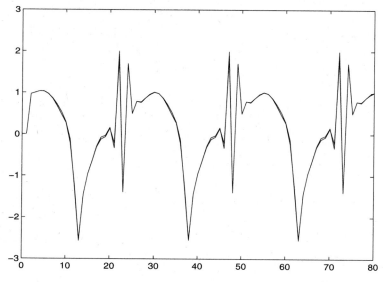

Figure 13.5. Outputs of the plant $y_2(k)$ and the identification model $\hat{y}_2(k)$ for Example 13.2 after 5,000 steps of training.

- Design of fuzzy systems based on the gradient descent training algorithm and the initial parameter choosing method.

- Application of this approach to nonlinear dynamic system identification and other problems.

The method in this chapter was inspired by the error back-propagation algorithm for neural networks (Werbos [1974]). Many similar methods were proposed in the literature; see Jang [1993] and Lin [1994]. More simulation results can be found in Wang [1994]. Narendra and Parthasarathy [1990] gave extensive simulation results of neural network identifiers for the nonlinear systems such as those in Examples 13.1 and 13.2.

13.5   Exercises and Projects

Exercise 13.1. Suppose that the parameters $\bar{x}_i^l$ and σ_i^l in (13.1) are fixed and only the $\bar{y}^l$ are free to change.

(a) Show that if the training algorithm (13.5) converges, then it converges to the global minimum of e^p of (13.2).

(b) Let $e^p(\bar{y}^l)$ be the e^p of (13.2). Find the optimal stepsize α by minimizing $e^p[\bar{y}^l(q) - \alpha \frac{f-y}{b} z^l]$ with respect to α.

Exercise 13.2. Why is the proof of Lemma 13.1 not valid if $x_0^l = x_0^{k+1}$ for some $l \neq k+1$? Prove Lemma 13.1 for this case.

Exercise 13.3. Suppose that we are given K input-output pairs $(x_0^p; y_0^p), p = 1, 2, ..., K$, and we want to design a fuzzy system $f(x)$ in the form of (13.1) such that the summation of squared errors

$$J = \sum_{p=1}^{K} [f(x_0^p) - y_0^p]^2 \tag{13.24}$$

is minimized. Let the parameters $\bar{x}_i^l$ and σ_i^l be fixed and only the $\bar{y}^l$ be free to change. Determine the optimal $\bar{y}^l$ $(l = 1, 2, ..., M)$ such that J is minimized (write the optimal $\bar{y}^l$ in a closed-form expression).

Exercise 13.4. Let $[x(k)]$ be a sequence of real-valued vectors generated by the gradient descent algorithm

$$x(k + 1) = x(k) - \alpha \nabla e(x(k)), \tag{13.25}$$

where $e : R^n \to R$ is a cost function and $e \in C^2$ (i.e., e has continuous second derivative). Assume that all $x(k) \in D \subset R^n$ for some compact D, then there exist $\epsilon > 0$ and $L > 0$ such that if

$$0 < \epsilon \leq \alpha \leq \frac{2 - \epsilon}{L}, \tag{13.26}$$

then:

(a) $e(x(k+1)) < e(x(k))$ if $\nabla e(x(k)) \neq 0$;

(b) $x(k) \to x^* \in D$ as $k \to \infty$ if $e(*)$ is bounded from below; and

(c) x^* is a local minimum of $e(x)$.

Exercise 13.5. It is a common perception that the gradient decent algorithm in Section 13.2 converges to local minima. Create a set of training samples that numerically shows this phenomenon; that is, show that a different initial condition may lead to a different final convergent solution.

Exercise 13.6. Explain how conscious and subconscious knowledge are combined into the fuzzy system using the design method in Section 13.2. Is it possible that conscious knowledge disappears in the final designed fuzzy system? Why? If we want to preserve conscious knowledge in the final fuzzy system, how to modify the design procedure in Section 13.2?

13.7 (Project). Write a computer program to implement the training algorithm (13.5), (13.8), and (13.9), and apply your codes to the time series prediction problem in Chapter 12.

Chapter 14

Design of Fuzzy Systems Using Recursive Least Squares

14.1 Design of the Fuzzy System

The gradient descent algorithm in Chapter 13 tries to minimum the criterion e^p of (13.2), which accounts for the matching error of only one input-output pair $(x_0^p; y_0^p)$. That is, the training algorithm updates the parameters to match one input-output pair at a time. In this chapter, we develop a training algorithm that minimizes the summation of the matching errors for all the input-output pairs up to p, that is, the objective now is to design a fuzzy system $f(x)$ such that

$$J_p = \sum_{j=1}^{p} [f(x_0^j) - y_0^j]^2 \tag{14.1}$$

is minimized. Additionally, we want to design the fuzzy system recursively; that is, if f_p is the fuzzy system designed to minimize J_p, then f_p should be represented as a function of f_{p-1}. We now use the recursive least squares algorithm to design the fuzzy system.

Design of the Fuzzy System by Recursive Least Squares:

- **Step 1.** Suppose that $U = [\alpha_1, \beta_1] \times \cdots \times [\alpha_n, \beta_n] \subset R^n$. For each $[\alpha_i, \beta_i]$ $(i = 1, 2, ..., n)$, define N_i fuzzy sets $A_i^{l_i}$ $(l_i = 1, 2, ..., N_i)$, which are complete in $[\alpha_i, \beta_i]$. For example, we may choose $A_i^{l_i}$ to be the pseudo-trapezoid fuzzy sets: $\mu_{A_i^{l_i}}(x_i) = \mu_{A_i^{l_i}}(x_i; a_i^{l_i}, b_i^{l_i}, c_i^{l_i}, d_i^{l_i})$, where $a_i^1 = b_i^1 = \alpha_i, c_i^j \leq a_i^{j+1} < d_i^j \leq b_i^{j+1}$ for $j = 1, 2, ..., N_i - 1$, and $c_i^{N_i} = d_i^{N_i} = \beta_i$.

- **Step 2.** Construct the fuzzy system from the following $\prod_{i=1}^{n} N_i$ fuzzy IF-THEN rules:

$$IF\ x_1\ is\ A_1^{l_1}\ and\ \cdots\ and\ x_n\ is\ A_n^{j_n},\ THEN\ y\ is\ B^{l_1 \cdots l_n} \tag{14.2}$$

where $l_i = 1, 2, ..., N_i$, $i = 1, 2, ..., n$ and $B^{l_1 \cdots l_n}$ is any fuzzy set with center at $\bar{y}^{l_1 \cdots l_n}$ which is free to change. Specifically, we choose the fuzzy system with product inference engine, singleton fuzzifier, and center average defuzzifier; that is, the designed fuzzy system is

$$f(x) = \frac{\sum_{l_1=1}^{N_1} \cdots \sum_{l_n=1}^{N_n} \bar{y}^{l_1 \cdots l_n} [\prod_{i=1}^{n} \mu_{A_i^{l_i}}(x_i)]}{\sum_{l_1=1}^{N_1} \cdots \sum_{l_n=1}^{N_n} [\prod_{i=1}^{n} \mu_{A_i^{l_i}}(x_i)]} \qquad (14.3)$$

where $\bar{y}^{l_1 \cdots l_n}$ are free parameters to be designed, and $A_i^{l_i}$ are designed in Step 1. Collect the free parameters $\bar{y}^{l_1 \cdots l_n}$ into the $\prod_{i=1}^{n} N_i$-dimensional vector

$$\theta = (\bar{y}^{1 \cdots 1}, ..., \bar{y}^{N_1 1 \cdots 1}, \bar{y}^{121 \cdots 1}, ..., \bar{y}^{N_1 21 \cdots 1}, ..., \bar{y}^{1 N_2 \cdots N_n}, ..., \bar{y}^{N_1 N_2 \cdots N_n})^T \qquad (14.4)$$

and rewrite (14.3) as

$$f(x) = b^T(x)\theta \qquad (14.5)$$

where

$$b(x) = (b^{1 \cdots 1}(x), ..., b^{N_1 1 \cdots 1}(x), b^{121 \cdots 1}(x), ..., b^{N_1 21 \cdots 1}(x), ...,$$
$$b^{1 N_2 \cdots N_n}(x), ..., b^{N_1 N_2 \cdots N_n}(x))^T \qquad (14.6)$$

$$b^{l_1 \cdots l_n}(x) = \frac{\prod_{i=1}^{n} \mu_{A_i^{l_i}}(x_i)}{\sum_{l_1=1}^{N_1} \cdots \sum_{l_n=1}^{N_n} [\prod_{i=1}^{n} \mu_{A_i^{l_i}}(x_i)]} \qquad (14.7)$$

- **Step 3.** Choose the initial parameters $\theta(0)$ as follows: if there are linguistic rules from human experts (conscious knowledge) whose IF parts agree with the IF parts of (14.2), then choose $\bar{y}^{l_1 \cdots l_n}(0)$ to be the centers of the THEN part fuzzy sets in these linguistic rules; otherwise, choose $\theta(0)$ arbitrarily in the output space $V \subset R$ (for example, choose $\theta(0) = 0$ or the elements of $\theta(0)$ uniformly distributed over V). In this way, we can say that the initial fuzzy system is constructed from conscious human knowledge.

- **Step 4.** For $p = 1, 2, ...$, compute the parameters θ using the following recursive least squares algorithm:

$$\theta(p) = \theta(p-1) + K(p)[y_0^p - b^T(x_0^p)\theta(p-1)] \qquad (14.8)$$
$$K(p) = P(p-1)b(x_0^p)[b^T(x_0^p)P(p-1)b(x_0^p) + 1]^{-1} \qquad (14.9)$$
$$P(p) = P(p-1) - P(p-1)b(x_0^p)$$
$$[b^T(x_0^p)P(p-1)b(x_0^p) + 1]^{-1}b^T(x_0^p)P(p-1) \qquad (14.10)$$

where $\theta(0)$ is chosen as in Step 3, and $P(0) = \sigma I$ where σ is a large constant. The designed fuzzy system is in the form of (14.3) with the parameters $\bar{y}^{l_1 \cdots l_n}$ equal to the corresponding elements in $\theta(p)$.

The recursive least squares algorithm (14.8)-(14-10) is obtained by minimizing J_p of (14.1) with $f(x_0^j)$ in the form of (14.3); its derivation is given next.

14.2   Derivation of the Recursive Least Squares Algorithm

Let $Y_0^{p-1} = (y_0^1, ..., y_0^{p-1})^T$ and $B_{p-1} = (b(x_0^1), ..., b(x_0^{p-1}))^T$, then from (14.5) we can rewrite J_{p-1} as

$$J_{p-1} = \sum_{j=1}^{p-1} [f(x_0^j) - y_0^j]^2$$

$$= (B_{p-1}\theta - Y_0^{p-1})^T (B_{p-1}\theta - Y_0^{p-1}) \qquad (14.11)$$

Since J_{p-1} is a quadratic function of θ, the optimal θ that minimizes J_{p-1}, denoted by $\theta(p-1)$, is

$$\theta(p-1) = (B_{p-1}^T B_{p-1})^{-1} B_{p-1}^T Y_0^{p-1} \qquad (14.12)$$

When the input-output pair $(x_0^p; y_0^p)$ becomes available, the criterion changes to J_p of (14.1) which can be rewritten as

$$J_p = \left[\begin{pmatrix} B_{p-1} \\ b^T(x_0^p) \end{pmatrix} \theta - \begin{pmatrix} Y_0^{p-1} \\ y_0^p \end{pmatrix} \right]^T \left[\begin{pmatrix} B_{p-1} \\ b^T(x_0^p) \end{pmatrix} \theta - \begin{pmatrix} Y_0^{p-1} \\ y_0^p \end{pmatrix} \right] \qquad (14.13)$$

Similar to (14.12), the optimal θ which minimizes J_p, denoted by $\theta(p)$, is obtained as

$$\theta(p) = [(B_{p-1}^T b(x_0^p)) \begin{pmatrix} B_{p-1} \\ b^T(x_0^p) \end{pmatrix}]^{-1} (B_{p-1}^T b(x_0^p)) \begin{pmatrix} Y_0^{p-1} \\ y_0^p \end{pmatrix}$$

$$= [B_{p-1}^T B_{p-1} + b(x_0^p)b^T(x_0^p)]^{-1} [B_{p-1}^T Y_0^{p-1} + b(x_0^p)y_0^p] \qquad (14.14)$$

To further simplify (14.14), we need to use the matrix identity

$$(P^{-1} + bb^T)^{-1} = P - Pb(b^T Pb + 1)^{-1} b^T P \qquad (14.15)$$

Defining $P(p-1) = (B_{p-1}^T B_{p-1})^{-1}$ and using (14.15), we can rewrite (14.14) as

$$\theta(p) = \{P(p-1) - P(p-1)b(x_0^p)[b^T(x_0^p)P(p-1)b(x_0^p) + 1]^{-1} b^T(x_0^p)P(p-1)\}$$
$$[B_{p-1}^T Y_0^{p-1} + b(x_0^p)y_0^p] \qquad (14.16)$$

Since $P(p-1)B_{p-1}^T Y_0^{p-1} = (B_{p-1}^T B_{p-1})^{-1} B_{p-1}^T Y_0^{p-1} = \theta(p-1)$ (see (14.12)), we can simplify (14.16) to

$$\theta(p) = \theta(p-1) - P(p-1)b(x_0^p)[b^T(x_0^p)P(p-1)b(x_0^p) + 1]^{-1} b^T(x_0^p)\theta(p-1)$$
$$+ P(p-1)b(x_0^p)[1 - (b^T(x_0^p)P(p-1)b(x_0^p) + 1)^{-1} b^T(x_0^p)P(p-1)b(x_0^p)]y_0^p$$
$$= \theta(p-1) + P(p-1)b(x_0^p)[b^T(x_0^p)P(p-1)b(x_0^p) + 1]^{-1}$$
$$(y_0^p - b^T(x_0^p)\theta(p-1)) \qquad (14.17)$$

Defining $K(p) = P(p-1)b(x_0^p)[b^T(x_0^p)P(p-1)b(x_0^p)+1]^{-1}$, we obtain (14.8) and (14.9).

Finally, we derive (14.10). By definition, we have

$$P(p) = [(B_{p-1}^T b(x_0^p))\begin{pmatrix} B_{p-1} \\ b^T(x_0^p) \end{pmatrix})]^{-1}$$
$$= [B_{p-1}^T B_{p-1} + b(x_0^p)b^T(x_0^p)]^{-1} \qquad (14.18)$$

Using the matrix identity (14.15) and the fact that $B_{p-1}^T B_{p-1} = P^{-1}(p-1)$, we obtain (14.10) from (14.18).

14.3 Application to Equalization of Nonlinear Communication Channels

14.3.1 The Equalization Problem and Its Geometric Formulation

Nonlinear distortion over a communication channel is now a significant factor hindering further increase in the attainable data rate in high-speed data transmission (Biglieri, E., A. Gersho, R.D. Gitlin, and T.L. Lim [1984]). Because the received signal over a nonlinear channel is a nonlinear function of the past values of the transmitted symbols and the nonlinear distortion varies with time and from place to place, effective equalizers for nonlinear channels should be nonlinear and adaptive. In this section, we use the fuzzy system designed from the recursive least squares algorithm as an equalizer for nonlinear channels.

The digital communication system considered here is shown in Fig. 14.1, where the *channel* includes the effects of the transmitter filter, the transmission medium, the receiver matched filter, and other components. The transmitted data sequence $s(k)$ is assumed to be an independent sequence taking values from $\{-1, 1\}$ with equal probability. The inputs to the equalizer, $x(k), x(k-1), \cdots, x(k-n+1)$, are the channel outputs corrupted by an additive noise $e(k)$. The task of the equalizer at the sampling instant k is to produce an estimate of the transmitted symbol $s(k-d)$ using the information contained in $x(k), x(k-1), \cdots, x(k-n+1)$, where the integers n and d are known as the order and the lag of the equalizer, respectively.

We use the geometric formulation of the equalization problem due to Chen, S., G.J. Gibson, C.F.N. Cowan and P.M. Grand [1990]. Define

$$P_{n,d}(1) = \{\hat{\mathbf{x}}(k) \in R^n | s(k-d) = 1\} \qquad (14.19)$$
$$P_{n,d}(-1) = \{\hat{\mathbf{x}}(k) \in R^n | s(k-d) = -1\} \qquad (14.20)$$

where

$$\hat{\mathbf{x}}(k) = [\hat{x}(k), \hat{x}(k-1), \cdots, \hat{x}(k-n+1)]^T, \qquad (14.21)$$

$\hat{x}(k)$ is the noise-free output of the channel (see Fig. 14.1), and $P_{n,d}(1)$ and $P_{n,d}(-1)$ represent the two sets of possible channel noise-free output vectors $\hat{\mathbf{x}}(k)$ that can

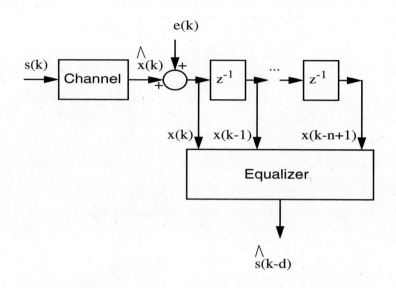

Figure 14.1. Schematic of data transmission system.

be produced from sequences of the channel inputs containing $s(k - d) = 1$ and $s(k - d) = -1$, respectively. The equalizer can be characterized by the function

$$g_k : R^n \to \{-1, 1\} \tag{14.22}$$

with

$$\hat{s}(k - d) = g_k(\mathbf{x}(k)), \tag{14.23}$$

where

$$\mathbf{x}(k) = [x(k), x(k - 1), \cdots, x(k - n + 1)]^T \tag{14.24}$$

is the observed channel output vector. Let $p_1[\mathbf{x}(k)|\hat{\mathbf{x}}(k) \in P_{n,d}(1)]$ and $p_{-1}[\mathbf{x}(k)|\hat{\mathbf{x}}(k) \in P_{n,d}(-1)]$ be the conditional probability density functions of $\mathbf{x}(k)$ given $\hat{\mathbf{x}}(k) \in P_{n,d}(1)$ and $\hat{\mathbf{x}}(k) \in P_{n,d}(-1)$, respectively. It was shown in Chen, S., G.J. Gibson, C.F.N. Cowan and P.M. Grand [1990] that the equalizer that is defined by

$$f_{opt}(\mathbf{x}(k)) = sgn[p_1(\mathbf{x}(k)|\hat{\mathbf{x}}(k) \in P_{n,d}(1)) - p_{-1}(\mathbf{x}(k)|\hat{\mathbf{x}}(k) \in P_{n,d}(-1))] \tag{14.25}$$

achieves the minimum bit error rate for the given order n and lag d, where $sgn(y) = 1(-1)$ if $y \geq 0$ ($y < 0$). If the noise $e(k)$ is zero-mean and Gaussian with covariance matrix

$$Q = E[(e(k), ..., e(k - n + 1))(e(k), ..., e(k - n + 1))^T] \tag{14.26}$$

then from $x(k) = \hat{x}(k) + e(k)$ we have that

$$p_1[\mathbf{x}(k)|\hat{\mathbf{x}}(k) \in P_{n,d}(1)] - p_{-1}[\mathbf{x}(k)|\hat{\mathbf{x}}(k) \in P_{n,d}(-1)]$$
$$= \sum exp[-\frac{1}{2}(\mathbf{x}(k) - \hat{\mathbf{x}}_+)^T Q^{-1}(\mathbf{x}(k) - \hat{\mathbf{x}}_+)]$$
$$- \sum exp[-\frac{1}{2}(\mathbf{x}(k) - \hat{\mathbf{x}}_-)^T Q^{-1}(\mathbf{x}(k) - \hat{\mathbf{x}}_-)] \qquad (14.27)$$

where the first (second) sum is over all the points $\hat{\mathbf{x}}_+ \in P_{n,d}(1)$ ($\hat{\mathbf{x}}_- \in P_{n,d}(-1)$).

Now consider the nonlinear channel

$$\hat{x}(k) = s(k) + 0.5s(k-1) - 0.9[s(k) + 0.5s(k-1)]^3 \qquad (14.28)$$

and white Gaussian noise $e(k)$ with $E[e^2(k)] = 0.2$. For this case, the optimal decision region for $n = 2$ and $d = 0$,

$$[\mathbf{x}(k) \in R^2 | p_1[\mathbf{x}(k)|\hat{\mathbf{x}}(k) \in P_{2,0}(1)] - p_{-1}[\mathbf{x}(k)|\hat{\mathbf{x}}(k) \in P_{2,0}(-1)] \geq 0] \qquad (14.29)$$

is shown in Fig. 14.2 as the shaded area. The elements of the sets $P_{2,0}(1)$ and $P_{2,0}(-1)$ are illustrated in Fig. 14.2 by the "o" and "*", respectively. From Fig. 14.2 we see that the optimal decision boundary for this case is severely nonlinear.

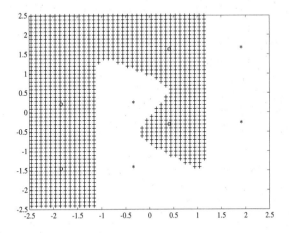

Figure 14.2. Optimal decision region for the channel (14.28), Gaussian white noise with variance $\sigma_e^2 = 0.2$, and equalizer order $n = 2$ and lag $d = 0$, where the horizontal axis denotes $x(k)$ and the vertical axis denotes $x(k-1)$.

14.3.2   Application of the Fuzzy System to the Equalization Problem

We use the fuzzy system (14.3) as the equalizer in Fig. 14.1. The operation consists of the following two phases:

- **Training Phase**: In this phase, the transmitted signal $s(k)$ is known and the task is to design the equalizer (that is, the fuzzy system (14.3)). We use the design method in Section 14.1 to specify the structure and the parameters of the equalizer. The input-output pairs for this problem are: $x_0^k = (x(k), ..., x(k-n+1))^T$ and $y_0^k = s(k-d)$ (the index p in Section 14.1 becomes the time index k here).

- **Application Phase**: In this phase, the transmitted signal $s(k)$ is unknown and the designed equalizer (the fuzzy system (14.3)) is used to estimate $s(k-d)$. Specifically, if the output of the fuzzy system is greater than or equal to zero, the estimate $\hat{s}(k-d) = 1$; otherwise, $\hat{s}(k-d) = -1$.

Example 14.1. Consider the nonlinear channel (14.28). Suppose that $n = 2$ and $d = 0$, so that the optimal decision region is shown in Fig. 14.2. Our task is to design a fuzzy system whose input-output behavior approximates that in Fig. 14.2, where the output of the fuzzy system is quantized as in the Application Phase. We use the design procedure in Section 14.1. In Step 1, we choose $N_1 = N_2 = 9$ and $\mu_{A_i^l}(x_i) = exp(-(\frac{x_i - \bar{x}_i^l}{0.3})^2)$, where $i = 1, 2$ and $\bar{x}_i^l = -2 + 0.5(l-1)$ for $l = 1, 2, ..., 9$. In Step 3, we choose the initial parameters $\theta(0)$ randomly in the interval $[-0.3, 0.3]$. In Step 4, we choose $\sigma = 0.1$. Figs.14.3-14.5 show the decision regions resulting from the designed fuzzy system when the training in Step 4 stops at $k = 30, 50$ and 100 (that is, when the p in (14.8)-(14.10) equals $30, 50$ and 100), respectively. From Figs.14.3-14.5 we see that the decision regions tend to converge to the optimal decision region as more training is performed. □

Example 14.2. In this example, we consider the same situation as in Example 14.1 except that we choose $d = 1$ rather than $d = 0$. The optimal decision region for this case is shown in Fig.14.6. Figs.14.7 and 14.8 show the decision regions resulting from the fuzzy system equalizer when the training in Step 4 stops at $k = 20$ and $k = 50$, respectively. We see, again, that the decision regions tend to converge to the optimal decision region. □

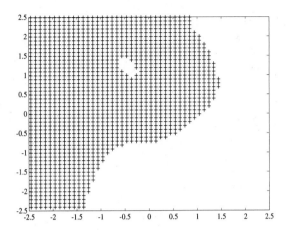

Figure 14.3. Decision region of the fuzzy system equalizer when the training stops at $k = 30$, where the horizontal axis denotes $x(k)$ and the vertical axis denotes $x(k - 1)$.

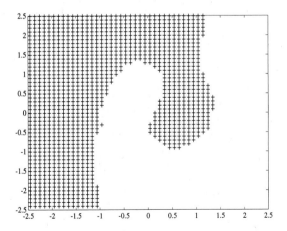

Figure 14.4. Decision region of the fuzzy system equalizer when the training stops at $k = 50$, where the horizontal axis denotes $x(k)$ and the vertical axis denotes $x(k - 1)$.

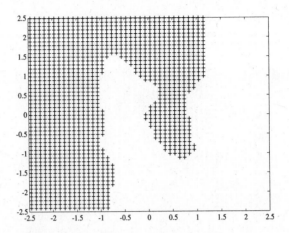

Figure 14.5. Decision region of the fuzzy system equalizer when the training stops at $k = 100$, where the horizontal axis denotes $x(k)$ and the vertical axis denotes $x(k - 1)$.

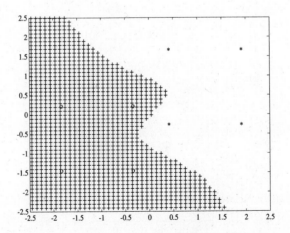

Figure 14.6. Optimal decision region for the channel (14.28), Gaussian white noise with variance $\sigma_e^2 = 0.2$, and equalizer order $n = 2$ and lag $d = 1$, where the horizontal axis denotes $x(k)$ and the vertical axis denotes $x(k - 1)$.

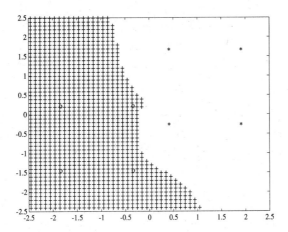

Figure 14.7. Decision region of the fuzzy system equalizer in Example 14.2 when the training stops at $k = 20$, where the horizontal axis denotes $x(k)$ and the vertical axis denotes $x(k-1)$.

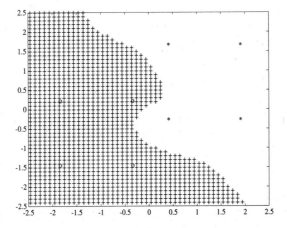

Figure 14.8. Decision region of the fuzzy system equalizer in Example 14.2 when the training stops at $k = 50$, where the horizontal axis denotes $x(k)$ and the vertical axis denotes $x(k-1)$.

14.4   Summary and Further Readings

In this chapter we have demonstrated the following:

- Using the recursive least squares algorithm to design the parameters of the fuzzy system.

- The techniques used in the derivation of the recursive least squares algorithm.

- Formulation of the channel equalization problem as a pattern recognition problem and the application of the fuzzy system with the recursive least squares algorithm to the equalization and similar pattern recognition problems.

The recursive least squares algorithm was studied in detail in many standard textbooks on estimation theory and adaptive filters; for example, Mendel [1994] and Cowan and Grant [1985]. The method in this chapter is taken from Wang [1994a] and Wang and Mendel [1993] where more simulation results can be found. For a similar approach using neural networks, see Chen, S., G.J. Gibson, C.F.N. Cowan and P.M. Grand [1990].

14.5   Exercises and Projects

Exercise 14.1. The XoR function is defined as

x_1	x_2	$x_1 XoR x_2$
-1	-1	1
-1	1	-1
1	-1	-1
1	1	1

(a) Design a fuzzy system $f(x_1, x_2)$ in the form of (14.3) such that $sgn[f(x_1, x_2)]$ implements the XoR function, where $sgn(f) = 1$ if $f \geq 0$ and $sgn(f) = -1$ if $f < 0$.

(b) Plot the decision region $\{x \in U | sgn[f(x)] \geq 0\}$, where $U = [-2, 2] \times [-2, 2]$ and $f(x)$ is the fuzzy system you designed in (a).

Exercise 14.2. Discuss the physical meaning of each of (14.8)-(14.10). Explain why the initial $P(0) = \sigma I$ should be large.

Exercise 14.3. Prove the matrix identity (14.15).

Exercise 14.4. Suppose that we change the criterion J_k of (14.1) to

$$J'_k = \sum_{j=1}^{k} \lambda^{k-j} [f(x_0^j) - y_0^j]^2 \tag{14.30}$$

where $\lambda \in (0, 1]$ is a forgetting factor, and that we still use the fuzzy system in the form of (14.5). Derive the recursive least squares algorithm similar to (14.8)-(14.10) for this new criterion J_k'.

Exercise 14.5. The objective to use the J_k' of (14.30) is to discount old data by putting smaller weights for them. Another way is to consider only the most recent N data pairs, that is, the criterion now is

$$J_k'' = \sum_{j=k-N+1}^{k} [f(x_0^j) - y_0^j]^2 \qquad (14.31)$$

Let $f(x_0^j)$ be the fuzzy system in the form of (14.5). Derive the recursive least squares algorithm similar to (14.8)-(14.10) which minimizes J_k''.

Exercise 14.6. Determine the exact locations of the sets of points: (a) $P_{2,0}(1)$ and $P_{2,0}(-1)$ in Fig. 14.2, and (b) $P_{2,1}(1)$ and $P_{2,1}(-1)$ in Fig. 14.6.

14.7 (Project). Write a computer program to implement the design method in Section 14.1 and apply your program to the nonlinear system identification problems in Chapter 13.

Chapter 15

Design of Fuzzy Systems Using Clustering

In Chapters 12-14, we proposed three methods for designing fuzzy systems. In all these methods, we did not propose a systematic procedure for determining the number of rules in the fuzzy systems. More specifically, the gradient descent method of Chapter 13 fixes the number of rules before training, while the table look-up scheme of Chapter 12 and the recursive least squares method of Chapter 14 fix the IF-part fuzzy sets, which in turn sets a bound for the number of rules. Choosing an appropriate number of rules is important in designing the fuzzy systems, because too many rules result in a complex fuzzy system that may be unnecessary for the problem, whereas too few rules produce a less powerful fuzzy system that may be insufficient to achieve the objective.

In this chapter, we view the number of rules in the fuzzy system as a design parameter and determine it based on the input-output pairs. The basic idea is to group the input-output pairs into clusters and use one rule for one cluster; that is, the number of rules equals the number of clusters. We first construct a fuzzy system that is optimal in the sense that it can match all the input-output pairs to arbitrary accuracy; this optimal fuzzy system is useful if the number of input-output pairs is small. Then, we determine clusters of the input-output pairs using the nearest neighborhood clustering algorithm, view the clusters as input-output pairs, and use the optimal fuzzy system to match them.

15.1 An Optimal Fuzzy System

Suppose that we are given N input-output pairs $(x_0^l; y_0^l), l = 1, 2, ..., N$, and N is small, say, $N = 20$. Our task is to construct a fuzzy system $f(x)$ that can match all the N pairs to any given accuracy; that is, for any given $\epsilon > 0$, we require that $|f(x_0^l) - y_0^l| < \epsilon$ for all $l = 1, 2, ..., N$.

This optimal fuzzy system is constructed as

$$f(x) = \frac{\sum_{l=1}^{N} y_0^l exp(-\frac{|x-x_0^l|^2}{\sigma^2})}{\sum_{l=1}^{N} exp(-\frac{|x-x_0^l|^2}{\sigma^2})} \tag{15.1}$$

Clearly, the fuzzy system (15.1) is constructed from the N rules in the form of (7.1) with $\mu_{A_i^l}(x_i) = exp(-\frac{|x_i-x_{0i}^l|^2}{\sigma^2})$ and the center of B^l equal to y_0^l, and using the product inference engine, singleton fuzzifier, and center average defuzzifier. The following theorem shows that by properly choosing the parameter σ, the fuzzy system (15.1) can match all the N input-output pairs to any given accuracy.

Theorem 15.1: For arbitrary $\epsilon > 0$, there exists $\sigma^* > 0$ such that the fuzzy system (15.1) with $\sigma = \sigma^*$ has the property that

$$|f(x_0^l) - y_0^l| < \epsilon \tag{15.2}$$

for all $l = 1, 2, ..., N$.

Proof: Viewing x_0^l and y_0^l as the x_0^{k+1} and y_0^{k+1} in Lemma 13.1 and using exactly the same method as in the proof of Lemma 13.1, we can prove this theorem. □

The σ is a smoothing parameter: the smaller the σ, the smaller the matching error $|f(x_0^l) - y_0^l|$, but the less smooth the $f(x)$ becomes. We know that if $f(x)$ is not smooth, it may not generalize well for the data points not in the training set. Thus, the σ should be properly chosen to provide a balance between matching and generalization. Because the σ is a one-dimensional parameter, it is usually not difficult to determine an appropriate σ for a practical problem. Sometimes, a few trial-and-error procedures may determine a good σ. As a general rule, large σ can smooth out noisy data, while small σ can make $f(x)$ as nonlinear as is required to approximate closely the training data.

The $f(x)$ is a general nonlinear regression that provides a smooth interpolation between the observed points $(x_0^l; y_0^l)$. It is well behaved even for very small σ.

Example 15.1. In this example, we would like to see the influence of the parameter σ on the smoothness and the matching accuracy of the optimal fuzzy system. We consider a simple single-input case. Suppose that we are given five input-output pairs: $(-2, 1), (-1, 0), (0, 2), (1, 2)$ and $(2, 1)$. The optimal fuzzy system $f(x)$ is in the form of (15.1) with $(x_0^l; y_0^l) = (-2, 1), (-1, 0), (0, 2), (1, 2), (2, 1)$ for $l = 1, 2, ..., 5$, respectively. Figs. 15.1-15.3 plot the $f(x)$ for $\sigma = 0.1, 0.3$ and 0.5, respectively. These plots confirm our early comment that smaller σ gives smaller matching errors but less smoothing functions. □

15.2   Design of Fuzzy Systems By Clustering

The optimal fuzzy system (15.1) uses one rule for one input-output pair, thus it is no longer a practical system if the number of input-output pairs is large. For

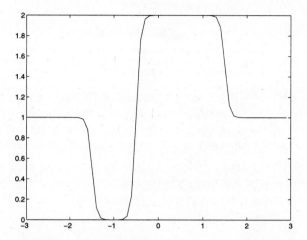

Figure 15.1. The optimal fuzzy system in Example 15.1 with $\sigma = 0.1$.

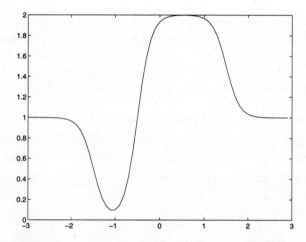

Figure 15.2. The optimal fuzzy system in Example 15.1 with $\sigma = 0.3$.

these large sample problems, various clustering techniques can be used to group the input-output pairs so that a group can be represented by one rule.

From a general conceptual point of view, clustering means partitioning of a collection of data into disjoint subsets or clusters, with the data in a cluster having

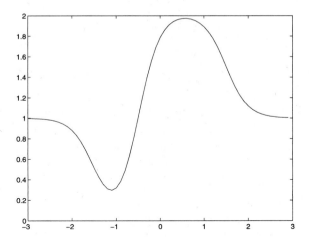

Figure 15.3. The optimal fuzzy system in Example 15.1 with $\sigma = 0.5$.

some properties that distinguish them from the data in the other clusters. For our problem, we first group the input-output pairs into clusters according to the distribution of the input points, and then use one rule for one cluster. Fig. 15.4 illustrates an example where six input-output pairs are grouped into two clusters and the two rules in the figure are used to construct the fuzzy system. The detailed algorithm is given next.

One of the simplest clustering algorithms is the nearest neighborhood clustering algorithm. In this algorithm, we first put the first datum as the center of the first cluster. Then, if the distances of a datum to the cluster centers are less than a prespecified value, put this datum into the cluster whose center is the closest to this datum; otherwise, set this datum as a new cluster center. The details are given as follows.

Design of the Fuzzy System Using Nearest Neighborhood Clustering:

- **Step 1.** Starting with the first input-output pair $(x_0^1; y_0^1)$, establish a cluster center x_c^1 at x_0^1, and set $A^1(1) = y_0^1$, $B^1(1) = 1$. Select a radius r.

- **Step 2.** Suppose that when we consider the $k'th$ input-output pair $(x_0^k; y_0^k)$, $k = 2, 3, ...$, there are M clusters with centers at $x_c^1, x_c^2, ..., x_c^M$. Compute the distances of x_0^k to these M cluster centers, $|x_0^k - x_c^l|$, $l = 1, 2, ..., M$, and let the smallest distances be $|x_0^k - x_c^{l_k}|$, that is, the nearest cluster to x_0^k is $x_c^{l_k}$. Then:

a) If $|x_0^k - x_c^{l_k}| > r$, establish x_0^k as a new cluster center $x_c^{M+1} = x_0^k$, set

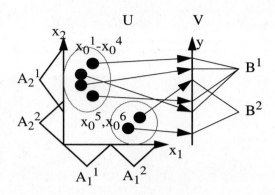

IF x_1 is $A_1{}^1$ and x_2 is $A_2{}^1$, THEN y is B^1

IF x_1 is $A_1{}^2$ and x_2 is $A_2{}^2$, THEN y is B^2

Figure 15.4. An example of constructing fuzzy IF-THEN rules from input-output pairs, where six input-output pairs $(x_0^1; y_0^1), ..., (x_0^6; y_0^6)$ are grouped into two clusters from which the two rules are generated.

$A^{M+1}(k) = y_0^k, B^{M+1}(k) = 1$, and keep $A^l(k) = A^l(k-1), B^l(k) = B^l(k-1)$ for $l = 1, 2, ..., M$.

b) If $|x_0^k - x_c^{l_k}| \leq r$, do the following:

$$A^{l_k}(k) = A^{l_k}(k-1) + y_0^k \tag{15.3}$$

$$B^{l_k}(k) = B^{l_k}(k-1) + 1 \tag{15.4}$$

and set

$$A^l(k) = A^l(k-1) \tag{15.5}$$

$$B^l(k) = B^l(k-1) \tag{15.6}$$

for $l = 1, 2, ..., M$ with $l \neq l_k$.

- **Step 3**. If x_0^k does not establish a new cluster, then the designed fuzzy system based on the k input-output pairs $(x_0^j; y_0^j), j = 1, 2, ..., k$, is

$$f_k(x) = \frac{\sum_{l=1}^M A^l(k) exp(-\frac{|x - x_c^l|^2}{\sigma})}{\sum_{l=1}^M B^l(k) exp(-\frac{|x - x_c^l|^2}{\sigma})} \tag{15.7}$$

If x_0^k establishes a new cluster, then the designed fuzzy system is

$$f_k(x) = \frac{\sum_{l=1}^{M+1} A^l(k)exp(-\frac{|x-x_c^l|^2}{\sigma})}{\sum_{l=1}^{M+1} B^l(k)exp(-\frac{|x-x_c^l|^2}{\sigma})} \tag{15.8}$$

- **Step 4.** Repeat by going to Step 2 with $k = k + 1$.

From (15.3)-(15.6) we see that the variable $B^l(k)$ equals the number of input-output pairs in the $l'th$ cluster after k input-output pairs have been used, and $A^l(k)$ equals the summation of the output values of the input-output pairs in the $l'th$ cluster. Therefore, if each input-output pair establishes a cluster center, then the designed fuzzy system (15.8) becomes the optimal fuzzy system (15.1). Because the optimal fuzzy system (15.1) can be viewed as using one rule to match one input-output pair, the fuzzy system (15.7) or (15.8) can be viewed as using one rule to match one cluster of input-output pairs. Since a new cluster may be introduced whenever a new input-output pair is used, the number of rules in the designed fuzzy system also is changing during the design process. The number of clusters (or rules) depends on the distribution of the input points in the input-output pairs and the radius r.

The radius r determines the complexity of the designed fuzzy system. For smaller r, we have more clusters, which result in a more sophisticated fuzzy system. For larger r, the designed fuzzy system is simpler but less powerful. In practice, a good radius r may be obtained by trials and errors.

Example 15.2. Consider the five input-output pairs in Example 15.1. Our task now is to design a fuzzy system using the design procedure in this section. If $r < 1$, then each of the five input-output pair establishes a cluster center and the designed fuzzy system $f_5(x)$ is the same as in Example 15.1. We now design the fuzzy system with $r = 1.5$.

In Step 1, we establish the center of the first cluster $x_c^1 = -2$ and set $A^1(1) = y_0^1 = 1, B^1(1) = 1$. In Step 2 with $k = 2$, since $|x_0^2 - x_c^{l_2}| = |x_0^2 - x_c^1| = |-1-(-2)| = 1 < r = 1.5$, we have $A^1(2) = A^1(1) + y_0^2 = 1 + 0 = 1$ and $B^1(2) = B^1(1) + 1 = 2$. For $k = 3$, since $|x_0^3 - x_c^{l_3}| = |x_0^3 - x_c^1| = |0-(-2)| = 2 > r$, we establish a new cluster center $x_c^2 = x^3 = 0$ together with $A^2(3) = y_0^3 = 2$ and $B^2(3) = 1$. The A^1 and B^1 remain the same, that is, $A^1(3) = A^1(2) = 1$ and $B^1(3) = B^1(2) = 2$. For $k = 4$, since $|x_0^4 - x_c^{l_4}| = |x_0^4 - x_c^2| = |1-0| = 1 < r$, we have $A^2(4) = A^2(3) + y_0^4 = 2 + 2 = 4, B^2(4) = B^2(3) + 1 = 2, A^1(4) = A^1(3) = 1$ and $B^1(4) = B^1(3) = 2$. Finally, for $k = 5$, since $|x_0^5 - x_c^{l_5}| = |x_0^5 - x_c^2| = |2-0| = 2 > r$, a new cluster center $x_c^3 = x_0^5 = 2$ is established with $A^3(5) = y_0^5 = 1$ and $B^3(5) = 1$. The other variables remain the same, that is, $A^1(5) = A^1(4) = 1, B^1(5) = B^1(4) = 2, A^2(5) = A^2(4) = 4$ and $B^2(5) = B^2(4) = 2$. The final fuzzy system is

$$f_5(x) = \frac{\sum_{l=1}^{3} A^l(5)exp(-\frac{(x-x_c^l)^2}{\sigma})}{\sum_{l=1}^{3} B^l(5)exp(-\frac{(x-x_c^l)^2}{\sigma})}$$

$$= \frac{exp(-\frac{(x+2)^2}{\sigma}) + 4exp(-\frac{x^2}{\sigma}) + exp(-\frac{(x-2)^2}{\sigma})}{2exp(-\frac{(x+2)^2}{\sigma}) + 2exp(-\frac{x^2}{\sigma}) + exp(-\frac{(x-2)^2}{\sigma})} \tag{15.9}$$

which is plotted in Fig. 15.5 with $\sigma = 0.3$. Comparing Figs. 15.5 with 15.2 we see, as expected, that the matching errors of the fuzzy system (15.9) at the five input-output pairs are larger than those of the optimal fuzzy system. $\square$

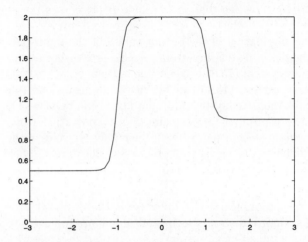

Figure 15.5. The designed fuzzy system $f_5(x)$ (15.9) in Example 15.2 with $\sigma = 0.3$.

Since the $A^l(k)$ and $B^l(k)$ coefficients in (15.7) and (15.8) are determined using the recursive equations (15.3)-(15.6), it is easy to add a forgetting factor to (15.3)-(15.6). This is desirable if the fuzzy system is being used to model systems with changing characteristics. For these cases, we replace (15.3) and (15.4) with

$$A^{l_k}(k) = \frac{\tau - 1}{\tau} A^{l_k}(k - 1) + \frac{1}{\tau} y_0^k \tag{15.10}$$

$$B^{l_k}(k) = \frac{\tau - 1}{\tau} B^{l_k}(k - 1) + \frac{1}{\tau} \tag{15.11}$$

and replace (15.5) and (15.6) with

$$A^l(k) = \frac{\tau - 1}{\tau} A^l(k - 1) \tag{15.12}$$

$$B^l(k) = \frac{\tau - 1}{\tau} B^l(k - 1) \tag{15.13}$$

where τ can be considered as a time constant of an exponential decay function. For practical considerations, there should be a lower threshold for $B^l(k)$ so that when sufficient time has elapsed without update for a particular cluster (which results in the $B^l(k)$ to be smaller than the threshold), that cluster would be eliminated.

15.3   Application to Adaptive Control of Nonlinear Systems

In this section, we use the designed fuzzy system (15.7) or (15.8) as a basic component to construct adaptive fuzzy controllers for discrete-time nonlinear dynamic systems. We consider two examples, but the approach can be generalized to other cases.

Example 15.3. Consider the discrete-time nonlinear system described by the difference equation

$$y(k+1) = g[y(k), y(k-1)] + u(k) \qquad (15.14)$$

where the nonlinear function

$$g[y(k), y(k-1)] = \frac{y(k)y(k-1)[y(k)+2.5]}{1+y^2(k)+y^2(k-1)} \qquad (15.15)$$

is assumed to be unknown. The objective here is to design a controller $u(k)$ (based on the fuzzy system (15.7) or (15.8)) such that the output $y(k)$ of the closed-loop system follows the output $y_m(k)$ of the reference model

$$y_m(k+1) = 0.6y_m(k) + 0.2y_m(k-1) + r(k) \qquad (15.16)$$

where $r(k) = sin(2\pi k/25)$. That is, we want $e(k) = y(k) - y_m(k)$ converge to zero as k goes to infinity.

If the function $g[y(k), y(k-1)]$ is known, we can construct the controller as

$$u(k) = -g[y(k), y(k-1)] + 0.6y(k) + 0.2y(k-1) + r(k) \qquad (15.17)$$

which, when applied to (15.14), results in

$$y(k+1) = 0.6y(k) + 0.2y(k-1) + r(k) \qquad (15.18)$$

Combining (15.16) and (15.18), we have

$$e(k+1) = 0.6e(k) + 0.2e(k-1) \qquad (15.19)$$

from which it follows that $lim_{k\to\infty} e(k) = 0$. However, since $g[y(k), y(k-1)]$ is unknown, the controller (15.17) cannot be implemented. To solve this problem, we replace the $g[y(k), y(k-1)]$ in (15.17) by the fuzzy system (15.7) or (15.8); that is, we use the following controller

$$u(k) = -f_k[y(k), y(k-1)] + 0.6y(k) + 0.2y(k-1) + r(k) \qquad (15.20)$$

where $f_k[y(k), y(k-1)]$ is in the form of (15.7) or (15.8) with $x = (y(k), y(k-1))^T$. This results in the nonlinear difference equation

$$y(k+1) = g[y(k), y(k-1)] - f_k[y(k), y(k-1)] + 0.6y(k) + 0.2y(k-1) + r(k) \quad (15.21)$$

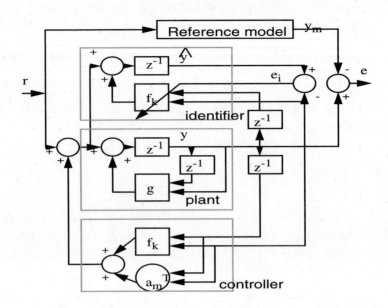

Figure 15.6. Overall adaptive fuzzy control system for Example 15.3.

governing the behavior of the closed-loop system. The overall control system is
shown in Fig. 15.6. From Fig. 15.6 we see that the controller consists of two parts:
an identifier and a controller. The identifier uses the fuzzy system f_k to approximate
the unknown nonlinear function g, and this f_k is then copied to the controller.

We simulated the following two cases for this example:

- *Case 1*: The controller in Fig. 15.6 was first disconnected and only the identifier was operating to identify the unknown plant. In this identification phase,
  we chose the input $u(k)$ to be an i.i.d. random signal uniformly distributed in
  the interval $[-3, 3]$. After the identification procedure was terminated, (15.20)
  was used to generate the control input; that is, the controller in Fig. 15.6 began operating with f_k copied from the final f_k in the identifier. Figs.15.7
  and 15.8 show the output $y(k)$ of the closed-loop system with this controller
  together with the reference model output $y_m(k)$ for the cases where the identification procedure was terminated at $k = 100$ and $k = 500$, respectively.
  In these simulations, we chose $\sigma = 0.3$ and $r = 0.3$. From these simulation
  results we see that: (i) with only 100 steps of training the identifier could
  produce an accurate model that resulted in a good tracking performance, and
  (ii) with more steps of training the control performance was improved.

- *Case 2*: The identifier and the controller operated simultaneously (as shown

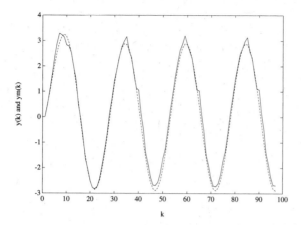

Figure 15.7. The output $y(k)$ (solid line) of the closed-loop system and the reference trajectory $y_m(k)$ (dashed line) for Case 1 in Example 15.3 when the identification procedure was terminated at $k = 100$.

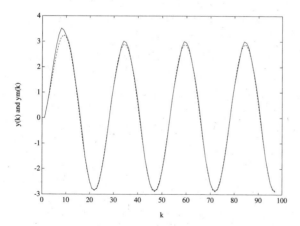

Figure 15.8. The output $y(k)$ (solid line) of the closed-loop system and the reference trajectory $y_m(k)$ (dashed line) for Case 1 in Example 15.3 when the identification procedure was terminated at $k = 500$.

in Fig. 15.6) from $k = 0$. We still chose $\sigma = 0.3$ and $r = 0.3$. Fig. 15.9 shows $y(k)$ and $y_m(k)$ for this simulation. $\square$

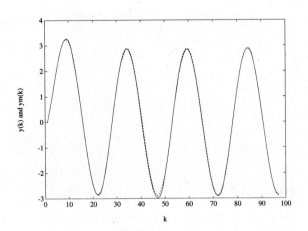

Figure 15.9. The output $y(k)$ (solid line) of the closed-loop system and the reference trajectory $y_m(k)$ (dashed line) for Case 2 in Example 15.3.

Example 15.4. In this example we consider the plant

$$y(k + 1) = \frac{5y(k)y(k - 1)}{1 + y^2(k) + y^2(k - 1) + y^2(k - 2)} + u(k) + 0.8u(k - 1) \quad (15.22)$$

where the nonlinear function is assumed to be unknown. The aim is to design a controller $u(k)$ such that $y(k)$ will follow the reference model

$$y_m(k + 1) = 0.32y_m(k) + 0.64y_m(k - 1) - 0.5y_m(k - 2) + sin(2\pi k/25) \quad (15.23)$$

Using the same idea as in Example 15.3, we choose

$$u(k) = -f_k[y(k), y(k - 1), y(k - 2)] - 0.8u(k - 1) + 0.32y(k) + 0.64y(k - 1)$$
$$-0.5y(k - 2) + sin(2\pi k/25) \quad (15.24)$$

where $f_k[y(k), y(k - 1), y(k - 2)]$ is in the form of (15.7) or (15.8). The basic configuration of the overall control scheme is the same as Fig.15.6. Fig.15.10 shows $y(k)$ and $y_m(k)$ when both the identifier and the controller began operating from $k = 0$. We chose $\sigma = 0.3$ and $r = 0.3$ in this simulation. $\square$

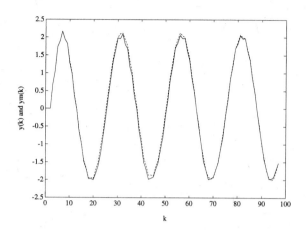

Figure 15.10. The output $y(k)$ (solid line) of the closed-loop system and the reference trajectory $y_m(k)$ (dashed line) for Example 15.4.

15.4   Summary and Further Readings

In this chapter we have demonstrated the following:

- The idea and construction of the optimal fuzzy system.

- The detailed steps of using the nearest neighborhood clustering algorithm to design the fuzzy systems from input-output pairs.

- Applications of the designed fuzzy system to the adaptive control of discrete-time dynamic systems and other problems.

Various clustering algorithms can be found in the textbooks on pattern recognition, among which Duda and Hart [1973] is still one of the best. The method in this chapter is taken from Wang [1994a], where more examples can be found.

15.5   Exercises and Projects

Exercise 15.1. Repeat Example 15.2 with $r = 2.2$.

Exercise 15.2. Modify the design method in Section 15.2 such that a cluster center is the average of inputs of the points in the cluster, the $A^l(k)$ parameter is the average of outputs of the points in the cluster, and the $B^l(k)$ parameter is deleted.

Exercise 15.3. Create an example to show that even with the same set of input-output pairs, the clustering method in Section 15.2 may create different fuzzy systems if the ordering of the input-output pairs used is different.

Exercise 15.4. The basic idea of *hierarchical clustering* is illustrated in Fig. 15.11. Propose a method to design fuzzy systems using the hierarchical clustering idea. Show your method in detail in a step-by-step manner and demonstrate it through a simple example.

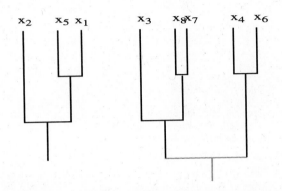

Figure 15.11. The basic idea of hierarchical clustering.

Exercise 15.5. Consider the two-input-two-output system

$$\left[\begin{array}{c} y_1(k+1) \\ y_2(k+1) \end{array} \right] = \left[\begin{array}{c} \frac{y_1(k)}{1+y_2^2(k)} \\ \frac{y_1(k)y_2(k)}{1+y_2^2(k)} \end{array} \right] + \left[\begin{array}{c} u_1(k) \\ u_2(k) \end{array} \right] \tag{15.25}$$

where the nonlinear functions are assumed to be unknown.

(a) Design an identifier for the system using the fuzzy system (15.7) or (15.8) as basic block. Explain the working procedure of the identifier.

(b) Design a controller for the system such that the closed-loop system outputs follow the reference model

$$\left[\begin{array}{c} y_{m1}(k+1) \\ y_{m2}(k+1) \end{array} \right] = \left[\begin{array}{c} y_{m1}(k) + y_{m2}(k) \\ y_{m2}(k) \end{array} \right] + \left[\begin{array}{c} r_1(k) \\ r_2(k) \end{array} \right] \tag{15.26}$$

where $r_1(k)$ and $r_2(k)$ are known reference signals. Under what conditions will the tracking error converge to zero?

15.6 (Project). Write a computer program to implement the design method in Section 15.2 and apply your program to the time series prediction and nonlinear system identification problems in Chapters 12 and 13.

Part IV

Nonadaptive Fuzzy Control

When fuzzy systems are used as controllers, they are called *fuzzy controllers*. If fuzzy systems are used to model the process and controllers are designed based on the model, then the resulting controllers also are called *fuzzy controllers*. Therefore, fuzzy controllers are nonlinear controllers with a special structure. Fuzzy control has represented the most successful applications of fuzzy theory to practical problems.

Fuzzy control can be classified into nonadaptive fuzzy control and adaptive fuzzy control. In nonadaptive fuzzy control, the structure and parameters of the fuzzy controller are fixed and do not change during real-time operation. In adaptive fuzzy control, the structure or/and parameters of the fuzzy controller change during real-time operation. Nonadaptive fuzzy control is simpler than adaptive fuzzy control, but requires more knowledge of the process model or heuristic rules. Adaptive fuzzy control, on the other hand, is more expensive to implement, but requires less information and may perform better. In this part (Chapters 16-22), we will study nonadaptive fuzzy control.

In Chapter 16, we will exam the trial-and-error approach to fuzzy controller design through two case studies: fuzzy control of a cement kiln and fuzzy control of a wastewater treatment process. In Chapters 17 and 18, stable and optimal fuzzy controllers for linear plants will be designed, respectively. In Chapters 19 and 20, fuzzy controllers for nonlinear plants will be developed in such a way that stability is guaranteed by using the ideas of sliding control and supervisory control. A fuzzy gain scheduling for PID controllers also will be studied in Chapter 20. In Chapter 21, both the plant and the controller will be modeled by the Takagi-Sugeno-Kang fuzzy systems and we will show how to choose the parameters such that the closed-loop system is guaranteed to be stable. Finally, Chapter 22 will introduce a few robustness indices for fuzzy control systems and show the basics of the hierarchical fuzzy systems.

Chapter 16

The Trial-and-Error Approach to Fuzzy Controller Design

16.1 Fuzzy Control Versus Conventional Control

Fuzzy control and conventional control have similarities and differences. They are similar in the following aspects:

- They try to solve the same kind of problems, that is, control problems. Therefore, they must address the same issues that are common to any control problem, for example, stability and performance.

- The mathematical tools used to analyze the designed control systems are similar, because they are studying the same issues (stability, convergence, etc.) for the same kind of systems.

However, there is a fundamental difference between fuzzy control and conventional control:

- Conventional control starts with a mathematical model of the process and controllers are designed for the model; fuzzy control, on the other hand, starts with heuristics and human expertise (in terms of fuzzy IF-THEN rules) and controllers are designed by synthesizing these rules. That is, the information used to construct the two types of controllers are different; see Fig.16.1. Advanced fuzzy controllers can make use of both heuristics and mathematical models; see Chapter 24.

For many practical control problems (for example, industrial process control), it is difficult to obtain an accurate yet simple mathematical model, but there are experienced human experts who can provide heuristics and rule-of-thumb that are very useful for controlling the process. Fuzzy control is most useful for these kinds

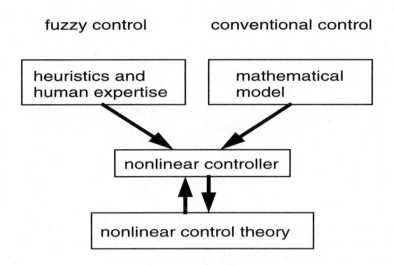

Figure 16.1. Fuzzy control versus conventional control.

of problems. As we will learn in this and the next few chapters, if the mathematical model of the process is unknown or partially unknown, we can design fuzzy controllers in a systematic manner that guarantee certain key performance criteria.

We classify the design methodologies for fuzzy controllers into two categories: the trial-and-error approach and the theoretical approach. In the trial-and-error approach, a set of fuzzy IF-THEN rules are collected from an introspective verbalization of experience-based knowledge (for example, an operating manual) and by asking the domain experts to answer a carefully organized questionnaire; then, fuzzy controllers are constructed from these fuzzy IF-THEN rules; finally, the fuzzy controllers are tested in the real system and if the performance is not satisfactory, the rules are fine-tuned or redesigned in a number of trial-and-error cycles until the performance is satisfactory. In the theoretical approach, the structure and parameters of the fuzzy controller are designed in such a way that certain performance criteria (for example, stability) are guaranteed. Of course, in designing fuzzy controllers for practical systems we should combine both approaches whenever possible to get the best fuzzy controllers. In this chapter, we will illustrate the trial-and-error approach through two practical examples—a cement kiln fuzzy control system, and a wastewater treatment fuzzy control system. The theoretical approaches will be studied in Chapters 17-21.

16.2   The Trial-and-Error Approach to Fuzzy Controller Design

The trial-and-error approach to fuzzy controller design can be roughly summarized in the following three steps:

- **Step 1: Analyze the real system and choose state and control variables.** The state variables should characterize the key features of the system and the control variables should be able to influence the states of the system. The state variables are the inputs to the fuzzy controller and the control variables are the outputs of the fuzzy controller. Essentially, this step defines the domain in which the fuzzy controller is going to operate.

- **Step 2. Derive fuzzy IF-THEN rules that relate the state variables with the control variables.** The formulation of these rules can be achieved by means of two heuristic approaches. The most common approach involves an introspective verbalization of human expertise. A typical example of such verbalization is the operating manual for the cement kiln, which we will show in the next section. Another approach includes an interrogation of experienced experts or operators using a carefully organized questionnaire. In these ways, we can obtain a prototype of fuzzy control rules.

- **Step 3. Combine these derived fuzzy IF-THEN rules into a fuzzy system and test the closed-loop system with this fuzzy system as the controller.** That is, run the closed-loop system with the fuzzy controller and if the performance is not satisfactory, fine-tune or redesign the fuzzy controller by trial and error and repeat the procedure until the performance is satisfactory.

We now show how to design fuzzy controllers for two practical systems using this approach—a cement kiln system and a wastewater treatment process.

16.3   Case Study I: Fuzzy Control of Cement Kiln

As we mentioned in Chapter 1, fuzzy control of cement kiln was one of the first successful applications of fuzzy control to full-scale industrial systems. In this section, we summarize the cement kiln fuzzy control system developed by Holmblad and Østerguard [1982] in the late '70s.

16.3.1   The Cement Kiln Process

Cement is manufactured by fine grinding of cement clinkers. The clinkers are produced in the cement kiln by heating a mixture of limestone, clay, and sand components. For a wet process cement kiln, the raw material mixture is prepared in a slurry; see Fig.16.2. Then four processing stages follow. In the first stage, the water is driven off; this is called preheating. In the second stage, the raw mix is heated up and calcination (CO_2 is driven off) takes place. The third stage comprises burning of the material at a temperature of approximately $1430^\circ C$, where free lime (CaO) combines with the other components to form the clinker minerals. In the final stage, the clinkers are cooled by air.

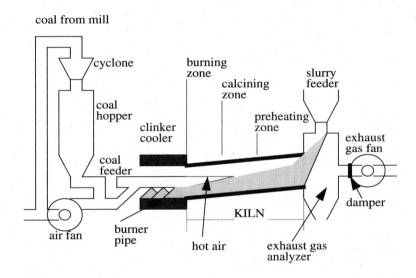

Figure 16.2. The cement kiln process.

The kiln is a steel tube about 165 meters long and 5 meters in diameter. The kiln tube is mounted slightly inclined from horizontal and rotates 1-2 rev/min. The clinker production is a continuous process. The slurry is fed to the upper (back) end of the kiln, whereas heat is provided to the lower (front) end of the kiln; see Fig. 16.2. Due to the inclination and the rotation of the kiln tube, the material is transported slowly through the kiln in 3-4 hours and heated with hot gases. The hot combustion gases are pulled through the kiln by an exhaust gas fan and controlled by a damper that is in the upper end of the kiln as shown in Fig. 16.2.

Cement kilns exhibit time-varying nonlinear behavior and experience indicates that mathematical models for the process become either too simple to be of any practical value, or too comprehensive and needled into the specific process to possess any general applicability. However, humans can be trained and in a relatively short time become skilled kiln operators. Consequently, cement kilns are specially suitable for fuzzy control.

16.3.2 Fuzzy Controller Design for the Cement Kiln Process

First, we determine the state and control variables for this system. The state variables must characterize the main functioning of the cement kiln process and their values can be determined from sensory measurements. The control variables must be able to influence the values of the state variables. By analyzing the system, the following three state variables were chosen:

- Temperature in the burning zone, denoted by BZ.

- Oxygen percentage in the exhaust gases, denoted by OX.

- Temperature at the back end of the kiln, denoted by BE.

The values of BZ can be obtained from the liter weight of clinkers, which is measureable. The values of OX and BE are obtained from the exhaust gas analyzer shown at the back end of the kiln in Fig.16.2. Two control variables were chosen as follows:

- Coal feed rate, denoted by CR.

- Exhaust gas damper position, denoted by DP.

The exhaust gas damper position influences the air volume through the kiln. Each of the control variables influences the various stages of preheating, calcining, clinker formation and cooling with different delays and time constants ranging from minutes to hours. Looking at kiln control in general, we see that kiln speed and slurry fed rate can also be used as control variables, but as kilns are normally operated at constant production, the adjustment on feed rate and kiln speed are seldom used for regulatory control.

In the second step, we derive fuzzy IF-THEN rules relating the state variables BZ, OX and BE with the control variables CR and DP. We derive these rules from the operator's manual for cement kiln control. Fig. 16.3 illustrates an extract from an American textbook for cement kiln operators. This section describes how an operator must adjust fuel rate (coal feed rate), kiln speed, and air volumn through the kiln under various conditions characterized by the temperature in the burning zone BZ, oxygen percentages in the exhaust gases OX, and temperature at the back end of the kiln BE. We see that the conditions and control actions are described in qualitative terms such as "high," "ok," "low," "slightly increase," etc.

In order to convert the instructions in Fig. 16.3 into fuzzy IF-THEN rules, we first define membership functions for the terms "low," "ok," and "high" for different

Case	Condition	Action to be taken	Reason
10	BZ ok	a. Increase air fan speed	To raise back-end temperature and increase oxygen for action 'b'
	OX low		
	BE low	b. Increase fuel rate	To maintain burning zone temperature
11	BZ ok	a. Decrease fuel rate speed slightly	To raise oxygen
	OX low		
	BE ok		
12	BZ ok	a. Reduce fuel rate	To increase oxygen for action 'b'
	OX low	b. Reduce air fan speed	To lower back-end temperature and maintain burning zone temperature
	BE high		
13	BZ ok	a. Increase air fan speed	To raise back-end temperature
	OX ok	b. Increase fuel rate	To maintain burning zone temperature
	BE ok		
14	BZ ok	None. However, do not get overconfident	
	OX ok	and keep all conditions under observation	
	BE ok		
15	BZ ok	When oxygen is in upper part of range	
	OX ok	a. Reduce air fan speed	To reduce back-end temperature
	BE high	When oxygen is in lower part of range	
		b. Reduce fuel rate	To raise oxygen for action 'c'
		c. Reduce air fan speed	To lower back-end temperature and maintain burning zone temperatur
16	BZ ok	a. Increase air fan speed	To raise back-end temperature
	OX high	b. Increase fuel rate	To maintain burning zone temperature and reduce oxygen
	BE low		
17	BZ ok	a. Reduce air fan speed slightly	To lower oxygen
	OX high		
	BE ok		

Figure 16.3. Extract from an American textbook for cement kiln operators.

variables. Fig. 16.4 shows the membership functions for these three terms for the variable OX. Others are similar and we omit the details. A set of fuzzy IF-THEN rules can be derived from the instructions in Fig. 16.3. For example, from item 10 in Fig. 16.3, we obtain the following fuzzy IF-THEN rule:

$$IF\ BZ\ is\ OK\ and\ OX\ is\ low\ and\ BE\ is\ low,$$
$$THEN\ CR\ is\ large,\ DP\ is\ large \qquad (16.1)$$

Similarly, item 11 gives the rule

$$IF\ BZ\ is\ OK\ and\ OX\ is\ low\ and\ BE\ is\ OK,\ THEN\ CR\ is\ small \qquad (16.2)$$

In real implementation, 27 rules were derived to form a complete set of fuzzy IF-THEN rules.

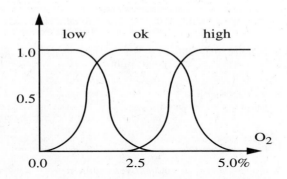

Figure 16.4. Membership functions for "low," "ok," and "high" for the variable OX.

16.3.3 Implementation

To develop a fuzzy control system for the real cement kiln process, it is not enough to have only one fuzzy controller. A number of fuzzy controllers were developed to operate in different modes. Specifically, the following two operating modes were considered:

- The kiln is in a reasonably stable operation, measured by the kiln drive torque showing only small variations.

- The kiln is running unstable, characterized by the kiln drive torque showing large and oscillating changes.

In the cement kiln fuzzy control system developed by Holmblad and Østerguard [1982], eight operation subroutines were developed, in which each subroutine was represented by a fuzzy controller or some supporting operations. Fig. 16.5 shows these subroutines. Whether or not the kiln is in stable operation is determined by Subroutine A, which monitors the variations of the torque during a period of 8 hours. If the operation is unstable, control is taken over by Subroutine B, which adjusts only the amount of coal until the kiln is in stable operation again.

During stable operation, the desired values of the state variables were: liter weight about $1350g/liter$, the oxygen about 1.8%, and the back-end temperature about 197^oC. To approach and maintain this desired state, Subroutine C adjusts the coal feed rate and the exhaust gas damper position. This subroutine is the fuzzy controller described in the last subsection; it performs the main control actions.

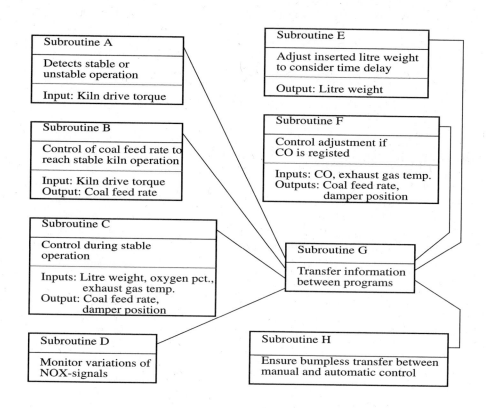

Figure 16.5. Structure of kiln control program.

Since the liter weight was determined manually from samples of the clinker and inserted manually via the operator's console, when the litre weight is inserted it represents a state that is between 1 and 2 hours old, compared to the actual state of the kiln. This time delay is considered in the Subroutines D and E. Subroutine D monitors how the NO-content in exhaust gases is varying as an increasing NO-content indicates increasing burning zone temperature and with that a higher liter weight. Information on increasing or decreasing NO-content is used in Subroutine E to adjust the inserted liter weight. If, for example, the inserted liter weight is high and the NO-content is decreasing, the liter weight will be adjusted toward lower values depending on how much the NO-content is decreasing. Subroutine E also considers that the process needs reasonable time before a response can be expected following a control adjustment.

During stable as well as unstable operations, Subroutine F measures the CO-content of the exhaust gases and executes appropriate control. Subroutine G transfers information between the subroutines and, finally, Subroutine H ensures that the human operator can perform bumpless transfers between manual and automatic kiln control.

16.4 Case Study II: Fuzzy Control of Wastewater Treatment Process

16.4.1 The Activated Sludge Wastewater Treatment Process

The activated sludge process is a commonly used method for treating sewage and wastewater. Fig. 16.6 shows the schematic of the system. The process (the part within the dotted lines) consists of an aeration tank and a clarifier. The wastewater entering the process is first mixed with the recycled sludge. Then, air is blown into the mixed liquor through diffusers placed along the base of the aeration tank. Complex biological and chemical reactions take place within the aeration tank, such that water and waste materials are separated. Finally, the processed mixed liquor enters the clarifier where the waste materials are settled out and the clear water is discharged.

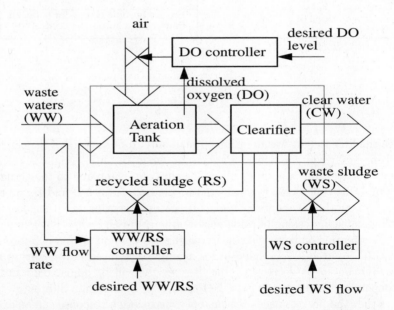

Figure 16.6. Schematic of the activated sludge wastewater treatment process.

There are three low-level controllers as shown in Fig.16.6. The WW/RS controller regulates the ratio of the wastewater flow rate to the recycled sludge flow rate to the desired value. The objectives of this controller are to maintain a desirable ratio of substrate/organism concentrations and to manipulate the distribution of sludge between the aeration tank and the clarifier. The DO controller controls the air flow rate to maintain a desired dissolved oxygen level in the aeration tank. A higher dissolved oxygen level is necessary to oxidize nitrogen-bearing waste materials (this is called nitrification). Finally, the WS controller is intended to regulate the total amount and average retention time of the sludge in both the aeration tank and the clarifier by controlling the waste sludge flow rate. Higher retention times of sludge are generally necessary to achieve nitrification.

Because the basic biological mechanism within the process is poorly understood, a usable mathematical model is difficult to obtain. In practice, the desired values of the WW/RS ratio, the DO level and the WS flow are set and adjusted by human operators. Our objective is to summarize the expertise of the human operators into a fuzzy system so that human operators can be released from the on-line operations. That is, we will design a fuzzy controller that gives the desired values of the WW/RS ratio, the DO level, and the WS flow. This fuzzy controller is therefore a higher level decision maker; the lower level direct control is performed by the WW/RS, DO, and WS controllers.

16.4.2 Design of the Fuzzy Controller

First, we specify the state and control variables. Clearly, we have the following three control variables:

- Δ WW/RS: change in the desired WW/RS ratio.

- Δ DO: change in the desired DO level.

- Δ WS: change in the desired WS flow.

The state variables should characterize the essential features of the system. Since the overall objectives of the wastewater treatment process are to control the total amounts of biochemical oxygen and suspended solid materials in the output clear water to below certain levels, these two variables should be chosen as states. Additionally, the suspended solids in the mixed liquor leaving the aeration tank and in the recycled sludge also are important and should be chosen as state variables. Finally, the ammonia-nitrogen content of the output clear water ($NH_3 - N$) and the waste sludge flow rate also are chosen as state variables. In summary, we have the following six state variables:

- TBO: the total amount of biochemical oxygen in the output clear water.

- TSS: the total amount of suspended solid in the output clear water.

- MSS: the suspended solid in the mixed liquor leaving the aeration tank.

- RSS: the suspended solid in the recycled sludge.

- NH_3-N: the ammonia-nitrogen content of the output clear water.

- WSR: the waste sludge flow rate.

The next task is to derive fuzzy IF-THEN rules relating the state variables to the control variables. Based on the expertise of human operators, the 15 rules in Table 16.1 were proposed in Tong, Beck, and Latten [1980], where S, M, L, SN, LN, SP, LP, VS and NL correspond to fuzzy sets "small," "medium," "large," "small negative," "large negative," "small positive," "large positive," "very small," and "not large," respectively. Rules 1-2 are resetting rules in that, if the process is in a satisfactory state but WSR is at abnormal levels, then the WSR is adjusted accordingly. Rules 3-6 deal with high NH_3-N levels in the output water. Rules 7-8 cater for high output water solids. Rules 9-13 describe the required control actions if MSS is outside its normal range. Finally, Rules 14-15 deal with the problem of high biochemical oxygen in the output clear water.

Table 16.1. Fuzzy IF-THEN rules for the wastewater treatment fuzzy controller.

Rule No.	TBO	TSS	MSS	RSS	NH_3-N	WSR	Δ WW/RS	Δ DO	Δ WS
1	S	S	M	M	S	S			SP
2	S	S	M	M	S	L			SN
3		S			M		SP		
4		S			M				SN
5		S			L		LP		
6		S			L				LN
7	NL	M						SP	
8	NL	L						LP	
9			L						LP
10			S						SN
11			VS						LN
12			VS			S		SP	
13			L			L		SN	
14	M	S			S				SN
15	L	S			S				LN

The detailed operation of the system was described in Tong, Beck, and Latten [1980].

16.5 Summary and Further Readings

In this chapter we have demonstrated the following:

- The general steps of the trial-and-error approach to designing fuzzy controllers.

- The cement kiln system and how to design fuzzy controllers for it.

- Practical considerations in the real implementation of the cement kiln fuzzy control system.

- How to design the fuzzy controller for the activated sludge wastewater treatment process.

   Various case studies of fuzzy control can be found in the books Sugeno and Nishida [1985] for earlier applications and Terano, Asai, and Sugeno [1994] for more recent applications. The two case studies in this chapter are taken from Holmblad and Østergaard [1982] and Tong, Beck, and Latten [1980] where more details can be found.

16.6 Exercises

Exercise 16.1. Consider the inverted pendulum system in Fig. 1.9. Let $x_1 = \theta$ and $x_2 = \dot{\theta}$ be the state variables. The dynamic equations governing the inverted pendulum system are

$$\dot{x}_1 = x_2$$

$$\dot{x}_2 = \frac{gsinx_1 - \frac{mlx_2^2cosx_1 sinx_1}{m_c+m}}{l(\frac{4}{3} - \frac{mcos^2x_1}{m_c+m})} + \frac{\frac{cosx_1}{m_c+m}}{l(\frac{4}{3} - \frac{mcos^2x_1}{m_c+m})}u \qquad (16.3)$$

   where $g = 9.8m/s^2$ is the acceleration due to gravity, m_c is the mass of the cart, m is the mass of the pole, l is the half length of the pole, and u is the applied force (control). Design a fuzzy controller to balance the pendulum using the trial-and-error approach. Test your fuzzy controller by computer simulations with $m_c = 1kg$, $m = 0.1kg$, and $l = 0.5m$.

Exercise 16.2. Consider the ball-and-beam system in Fig.16.7. Let $\mathbf{x} = (r, \dot{r}, \theta, \dot{\theta})^T$ be the state vector of the system and $y = r$ be the output of the system. Then the system can be represented by the state-space model

$$\begin{bmatrix} \dot{x}_1 \\ \dot{x}_2 \\ \dot{x}_3 \\ \dot{x}_4 \end{bmatrix} = \begin{bmatrix} x_2 \\ \alpha(x_1x_4^2 - \beta sinx_3) \\ x_4 \\ 0 \end{bmatrix} + \begin{bmatrix} 0 \\ 0 \\ 0 \\ 1 \end{bmatrix} u \qquad (16.4)$$

$$y = x_1 \tag{16.5}$$

where the control u is the acceleration of θ, and α, β are parameters. Design a fuzzy controller to control the ball to stay at the origin from a number of initial positions using the trial-and-error approach. Test your fuzzy controller by computer simulations with $\alpha = 0.7143$ and $\beta = 9.81$.

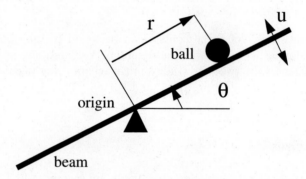

Figure 16.7. The ball-and-beam system.

Exercise 16.3. Find more similarities and differences between fuzzy control and conventional control other than whose discussed in Section 16.1.

Exercise 16.4. Find an application of fuzzy control from the books Sugeno and Nishida [1985] or Terano, Asai, and Sugeno [1994] and describe the working principle of the fuzzy control system in some detail.

Chapter 17

Fuzzy Control of Linear Systems I: Stable Controllers

From Chapter 16 we see that the starting point of designing a fuzzy controller is to get a collection of fuzzy IF-THEN rules from human experts and operational manuals. That is, knowing the mathematical model of the process under control is not a necessary condition for *designing* fuzzy controllers. However, in order to *analyze* the designed closed-loop fuzzy control system theoretically, we must have at least a rough model of the process. Additionally, if we want to design a fuzzy controller that guarantees some performance criterion, for example, globally exponential stability, then we must assume a mathematical model for the process, so that mathematical analysis can be done to establish the properties of the designed system. In this chapter, we consider the case where the process is a linear system and the controller is a fuzzy system. Our goal is to establish conditions on the fuzzy controller such that the closed-loop fuzzy control system is stable.

17.1 Stable Fuzzy Control of Single-Input-Single-Output Systems

For any control systems (including fuzzy control systems), stability is the most important requirement, because an unstable control system is typically useless and potentially dangerous. Conceptually, there are two classes of stability: Lyapunov stability and input-output stability. Within each class of stability, there are a number of specific stability concepts. We now briefly review these concepts.

For Lyapunov stability, let us consider the autonomous system

$$\dot{x}(t) = g[x(t)] \tag{17.1}$$

where $x \in R^n$ and $g(x)$ is a $n \times 1$ vector function. Assume that $g(0) = 0$, thus $x = 0$ is an equilibrium point of the system.

Definition 17.1. (Lyapunov stability) The equilibrium point $x = 0$ is said to be *stable* if for any $\epsilon > 0$ there exists $\delta > 0$ such that $||x(0)|| < \delta$ implies $||x(t)|| < \epsilon$

for all $t \geq 0$. It is said to be *asymptotically stable* if it is stable and additionally, there exists $\delta' > 0$ such that $||x(0)|| < \delta'$ implies $x(t) \to 0$ as $t \to \infty$. Finally, the equilibrium point $x = 0$ is said to be *exponentially stable* if there exist positive numbers α, λ and r such that

$$||x(t)|| \leq \alpha ||x(0)|| e^{-\lambda t} \tag{17.2}$$

for all $t \geq 0$ and $||x(0)|| \leq r$. If asymptotic or exponential stability holds for any initial state $x(0)$, then the equilibrium point $x = 0$ is said to be *globally* asymptotic or exponential stable, respectively.

For input-output stability, we consider any system that maps input $u(t) \in R^r$ to output $y(t) \in R^m$.

Definition 17.2. (Input-output stability) Let L_p^n be the set of all vector functions $g(t) = (g_1(t), ..., g_n(t))^T : [0, \infty) \to R^n$ such that $||g||_p = (\sum_{i=1}^n ||g_i||_p^2)^{1/2} < \infty$, where $||g_i||_p = (\int_0^\infty |g_i(t)|^p dt)^{1/p}$, $p \in [1, \infty]$ and $||g_i||_\infty = \sup_{t \in [0,\infty)} |g_i(t)|$. A system with input $u(t) \in R^r$ and output $y(t) \in R^m$ is said to be L_p-stable if

$$u(t) \in L_p^r \ implies \ y(t) \in L_p^m \tag{17.3}$$

where $p \in [0, \infty]$. In particular, a system is L_∞-*stable* (or *bounded-input-bounded-output stable*) if $u(t) \in L_\infty^r$ implies $y(t) \in L_\infty^m$.

Now assume that the process under control is a single-input-single-output (SISO) time-invariant linear system represented by the following state variable model:

$$\dot{x}(t) = Ax(t) + bu(t) \tag{17.4}$$
$$y(t) = cx(t) \tag{17.5}$$

where $u \in R$ is the control, $y \in R$ is the output, and $x \in R^n$ is the state vector. A number of important concepts were introduced for this linear system.

Definition 17.3. (Controllability, Observability, and Positive Real) The system (17.4)-(17.5) is said to be *controllable* if

$$rank[b \ AB \ \cdots \ A^{n-1}b] = n \tag{17.6}$$

and *observable* if

$$rank \begin{bmatrix} c \\ cA \\ ... \\ cA^{n-1} \end{bmatrix} = n \tag{17.7}$$

The transfer function of the system $h(s) = c(sI - A)^{-1}b$ is said to be *strictly positive real* if

$$\inf_{w \in R} Re[h(jw)] > 0 \tag{17.8}$$

We are now ready to apply some relevant results in control theory to fuzzy control systems.

17.1.1   Exponential Stability of Fuzzy Control Systems

Suppose that the control $u(t)$ in the system (17.4)-(17.5) is a fuzzy system whose input is $y(t)$, that is,

$$u(t) = -f[y(t)] \tag{17.9}$$

where f is a fuzzy system. Substituting (17.9) into (17.4)-(17.5), we obtain the closed-loop fuzzy control system that is shown in Fig.17.1. We now cite a famous result in control theory (its proof can be found in Vidyasagar [1993]).

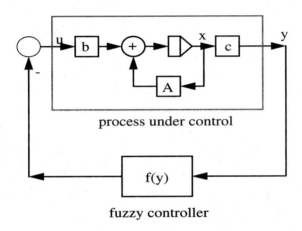

process under control

fuzzy controller

Figure 17.1. Closed-loop fuzzy control system.

Proposition 17.1. Consider the closed-loop control system (17.4), (17.5), and (17.9), and suppose that (a) all eigenvalues of A lie in the open left half of the complex plane, (b) the system (17.4)-(17.5) is controllable and observable, and (c) the transfer function of the system (17.4)-(17.5) is strictly positive real. If the nonlinear function f satisfies $f(0) = 0$ and

$$yf(y) \geq 0, \ \forall y \in R \tag{17.10}$$

then the equilibrium point $x = 0$ of the closed-loop system (17.4), (17.5) and (17.9) is globally exponentially stable.

Conditions (a)-(c) in Proposition 17.1 are imposed on the process under control, not on the controller $f(y)$. They are strong conditions, essentially requiring that the open-loop system is stable and well-behaved. Conceptually, these systems are not difficult to control, thus the conditions on the fuzzy controller, $f(0) = 0$ and (17.10), are not very strong. Proposition 17.1 guarantees that if we design a fuzzy controller

$f(y)$ that satisfies $f(0) = 0$ and (17.10), then the closed-loop system is globally exponentially stable, provided that the process under control is linear and satisfies conditions (a)-(c) in Proposition 17.1. We now design such a fuzzy controller.

Design of Stable Fuzzy Controller:

- **Step 1**. Suppose that the output $y(t)$ takes values in the interval $U = [\alpha, \beta] \subset R$. Define $2N+1$ fuzzy sets A^l in U that are normal, consistent, and complete with the triangular membership functions shown in Fig.17.2. That is, we use the N fuzzy sets $A^1, ..., A^N$ to cover the negative interval $[\alpha, 0)$, the other N fuzzy sets $A^{N+2}, ..., A^{2N+1}$ to cover the positive interval $(0, \beta]$, and choose the center $\bar{x}^{N+1}$ of fuzzy set A^{N+1} at zero.

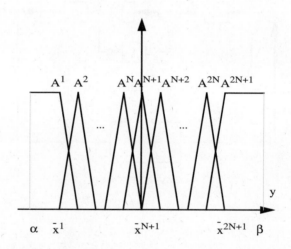

Figure 17.2. Membership functions for the fuzzy controller.

- **Step 2**. Consider the following $2N + 1$ fuzzy IF-THEN rules:

$$IF\ y\ is\ A^l,\ THEN\ u\ is\ B^l \tag{17.11}$$

where $l = 1, 2, ..., 2N + 1$, and the centers $\bar{y}^l$ of fuzzy sets B^l are chosen such that

$$\bar{y}^l \begin{cases} \leq 0 & for \quad l = 1, ..., N \\ = 0 & for \quad l = N + 1 \\ \geq 0 & for \quad l = N + 2, ..., 2N + 1 \end{cases} \tag{17.12}$$

- **Step 3**. Design the fuzzy controller from the $2N + 1$ fuzzy IF-THEN rules (17.11) using product inference engine, singleton fuzzifier, and center average defuzzifier; that is, the designed fuzzy controller is

$$u = -f(y) = -\frac{\sum_{l=1}^{2N+1} \bar{y}^l \mu_{A^l}(y)}{\sum_{l=1}^{2N+1} \mu_{A^l}(y)} \tag{17.13}$$

where $\mu_{A^l}(y)$ are shown in Fig. 17.2 and $\bar{y}^l$ satisfy (17.12).

We now prove that the fuzzy controller designed from the three steps above produces a stable closed-loop system.

Theorem 17.1. Consider the closed-loop fuzzy control system in Fig. 17.1. If the fuzzy controller $u = -f(y)$ is designed through the above three steps (that is, u is given by (17.13)) and the process under control satisfies conditions (a)-(c) in Proposition 17.1, then the equilibrium point $x = 0$ of the closed-loop fuzzy control system is guaranteed to be globally exponentially stable.

Proof: From Proposition 17.1, we only need to prove $f(0) = 0$ and $yf(y) \geq 0$ for all $y \in R$. From (17.13) and Fig. 17.2 we have that $f(0) = \bar{y}^{N+1} = 0$. If $y < 0$, then from (17.13) and Fig. 17.2 we have

$$f(y) = \frac{\bar{y}^{l_1} \mu_{A^{l_1}}(y) + \bar{y}^{l_1+1} \mu_{A^{l_1+1}}(y)}{\mu_{A^{l_1}}(y) + \mu_{A^{l_1+1}}(y)} \tag{17.14}$$

or $f(y) = \bar{y}^{l_1}$ for some $l_1 \in \{1, 2, ..., N\}$. Since $\bar{y}^{l_1} \leq 0, \bar{y}^{l_1+1} \leq 0$ and the membership functions are non-negative, we have $f(y) \leq 0$; therefore, $yf(y) \geq 0$. Similarly, we can prove that $yf(y) \geq 0$ if $y > 0$. $\square$

From the three steps of the design procedure we see that in designing the fuzzy controller $f(y)$, we do not need to know the specific values of the process parameters A, b and c. Also, there is much freedom in choosing the parameters of the fuzzy controller. Indeed, we only require that the $\bar{y}^l$'s satisfy (17.12) and that the membership functions A^l are in the form shown in Fig.17.2. In Chapter 18, we will develop an approach to choosing the fuzzy controller parameters in an optimal fashion.

17.1.2 Input-Output Stability of Fuzzy Control Systems

The closed-loop fuzzy control system in Fig. 17.1 does not have an explicit input. In order to study input-output stability, we introduce an extra input and the system is shown in Fig. 17.3. We now cite a well-known result in control theory (its proof can be found in Vidyasagar [1993]).

Proposition 17.2. Consider the system in Fig.17.3 and suppose that the nonlinear controller $f(y)$ is globally Lipschitz continuous, that is,

$$|f(y_1) - f(y_2)| \leq \alpha |y_1 - y_2|, \quad \forall y_1, y_2 \in R \tag{17.15}$$

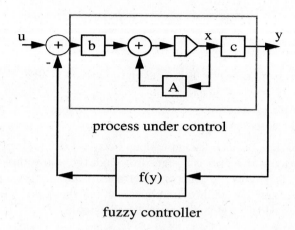

Figure 17.3. Closed-loop fuzzy control system with external input.

for some constant α. If the open-loop unforced system $\dot{x} = Ax$ is globally exponentially stable (or equivalently, the eigenvalues of A lie in the open left-half complex plane), then the forced closed-loop system in Fig. 17.3 is L_p-stable for all $p \in [1, \infty]$.

According to Proposition 17.2, if we can show that a designed fuzzy controller $f(y)$ satisfies the Lipschitz condition (17.15), then the closed-loop fuzzy control system in Fig.17.3 is L_p-stable, provided that the eigenvalues of A lie in the open left-half complex plane. It is interesting to see whether the fuzzy controller (17.13) designed through the three steps in Subsection 17.1.1 satisfies the Lipschitz condition (17.15). We first show that the fuzzy controller $f(y)$ of (17.13) is a continuous, bounded, and piece-wise linear function, from which we conclude that it satisfies the Lipschitz condition.

Lemma 17.1. The fuzzy controller $f(y)$ of (17.13) is continuous, bounded, and piece-wise linear.

Proof: Let $\bar{x}^l$ be the center of fuzzy set A^l as shown in Fig.17.2 ($l = 1, 2, ..., 2N+1$). Since the membership functions in Fig. 17.2 are continuous, the $f(y)$ of (17.13) is continuous. Since $f(y) = \bar{y}^1$ for $y \leq \bar{x}^1$, $f(y) = \bar{y}^{2N+1}$ for $y \geq \bar{x}^{2N+1}$, and $f(y)$ equals the weighted average of $\bar{y}^l$ and $\bar{y}^{l+1}$ for $y \in [\bar{x}^l, \bar{x}^{l+1}]$ ($l = 1, 2, ..., 2N$), we conclude that the $f(y)$ is bounded. To show that the $f(y)$ is a piece-wise linear function, we partition the real line R into $R = (-\infty, \bar{x}^1] \cup [\bar{x}^1, \bar{x}^2] \cup \cdots \cup [\bar{x}^{2N}, \bar{x}^{2N+1}] \cup [\bar{x}^{2N+1}, \infty)$. For $y \in (-\infty, \bar{x}^1]$ and $y \in [\bar{x}^{2N+1}, \infty)$, we have $f(y) = \bar{y}^1$ and $f(y) = \bar{y}^{2N+1}$, respectively, which are linear functions. For $y \in [\bar{x}^l, \bar{x}^{l+1}]$ with some $l \in$

$\{1, 2, ..., 2N\}$, we have from Fig. 17.2 that

$$f(y) = \frac{\bar{y}^l \mu_{A^l}(y) + \bar{y}^{l+1} \mu_{A^{l+1}}(y)}{\mu_{A^l}(y) + \mu_{A^{l+1}}(y)}$$

$$= \frac{\bar{y}^l \frac{\bar{x}^{l+1} - y}{\bar{x}^{l+1} - \bar{x}^l} + \bar{y}^{l+1} \frac{y - \bar{x}^l}{\bar{x}^{l+1} - \bar{x}^l}}{\frac{\bar{x}^{l+1} - y}{\bar{x}^{l+1} - \bar{x}^l} + \frac{y - \bar{x}^l}{\bar{x}^{l+1} - \bar{x}^l}}$$

$$= \frac{(\bar{y}^{l+1} - \bar{y}^l)y + \bar{y}^l \bar{x}^{l+1} - \bar{y}^{l+1} \bar{x}^l}{\bar{x}^{l+1} - \bar{x}^l} \qquad (17.16)$$

which is a linear function of y. Since $f(y)$ is continuous, it is a piece-wise linear function. $\square$

Combining Lemma 17.1 and Proposition 17.2, we obtain the following theorem.

Theorem 17.2. Consider the closed-loop fuzzy control system in Fig.17.3. Suppose that the fuzzy controller $f(y)$ is designed as in (17.13) and that all the eigenvalues of A lie in the open left-half complex plane. Then, the closed-loop fuzzy control system in Fig. 17.3 is L_p-stable for all $p \in [0, \infty]$.

Proof: Since a continuous, bounded, and piece-wise linear function satisfies the Lipschitz condition (17.15), this theorem follows from Proposition 17.2 and Lemma 17.1. $\square$

17.2 Stable Fuzzy Control of Multi-Input-Multi-Output Systems

17.2.1 Exponential Stability

Consider the multi-input-multi-output (MIMO) linear system

$$\dot{x}(t) = Ax(t) + Bu(t) \qquad (17.17)$$
$$y(t) = Cx(t) \qquad (17.18)$$

where the input $u(t) \in R^m$, the output $y(t) \in R^m$, and the state $x(t) \in R^n$. We assume that the number of input variables equals the number of output variables; this is called "squared" systems. In this case, the control $u(t) = (u_1(t), ..., u_m(t))^T$ consists of m fuzzy systems, that is,

$$u_j(t) = -f_j[y(t)] \qquad (17.19)$$

where $j = 1, 2, ..., m$, and $f_j[y(t)]$ are m-input-1-output fuzzy systems. The closed-loop fuzzy control system is still of the structure in Fig.17.1, except that the vector b is replaced by the matrix B, the vector c is replaced by the matrix C, and the scalar function f is replaced by the vector function $f = (f_1, ..., f_m)^T$. For the MIMO system (17.17)-(17.18), controllability and observability are still defined by (17.6) and (17.7) with b changed to B and c changed to C. For strictly positive real, let

$H(s) = C(sI - A)^{-1}B$ be the transfer matrix and $H(s)$ is said to be *strictly positive real* if

$$\inf_{w \in R} \lambda_{min}[H(jw) + H^*(jw)] > 0 \qquad (17.20)$$

where $*$ denotes conjugate transpose, and $\lambda_{min}[H(jw) + H^*(jw)]$ is the smallest eigenvalue of the matrix $H(jw) + H^*(jw)$.

As before, we first cite a result from control theory and then design a fuzzy controller that satisfies the conditions. The following proposition can be found in Vidyasagar [1993].

Proposition 17.3. Consider the closed-loop system (17.17)-(17.19), and suppose that: (a) all eigenvalues of A lie in the open left-half complex plane, (b) the system (17.17)-(17.18) is controllable and observable, and (c) the transfer matrix $H(s) = C(sI - A)^{-1}B$ is strictly positive real. If the control vector $f(y)$ satisfies $f(0) = 0$ and

$$y^T f(y) \geq 0, \ \forall y \in R^m \qquad (17.21)$$

then the equilibrium point $x = 0$ of the closed-loop system (17.17)-(17.19) is globally exponentially stable.

Note that in order to satisfy (17.21), the m fuzzy systems $\{f_1(y), ..., f_m(y)\}$ cannot be designed independently. We now design these m fuzzy systems in such a way that the resulting fuzzy controller $f(y) = (f_1(y), ..., f_m(y))^T$ satisfies (17.21).

Design of Stable Fuzzy Controller:

- **Step 1.** Suppose that the output $y_i(t)$ takes values in the interval $U_i = [\alpha_i, \beta_i] \subset R$, where $i = 1, 2, ..., m$. Define $2N_i + 1$ fuzzy sets $A_i^{l_i}$ in U_i which are normal, consistent, and complete with the triangular membership functions shown in Fig. 17.2 (with the subscribe i added to all the variables).

- **Step 2.** Consider m groups of fuzzy IF-THEN rules where the $j'th$ group ($j = 1, 2, ..., m$) consists of the following $\prod_{i=1}^{m}(2N_i + 1)$ rules:

$$IF \ y_1 \ is \ A_1^{l_1} \ and \ \cdots \ and \ y_m \ is \ A_m^{l_m}, \ THEN \ u \ is \ B_j^{l_1 \cdots l_m} \qquad (17.22)$$

where $l_i = 1, 2, ..., 2N_i + 1, i = 1, 2, ..., m$, and the center $\bar{y}_j^{l_1 \cdots l_m}$ of the fuzzy set $B_j^{l_1 \cdots l_m}$ are chosen such that

$$\bar{y}_j^{l_1 \cdots l_m} \begin{cases} \leq 0 & if \quad l_j = 1, 2, ..., N_j \\ = 0 & if \quad l_j = N_j + 1 \\ \geq 0 & if \quad l_j = N_j + 2, ..., 2N_j + 1 \end{cases} \qquad (17.23)$$

where l_i for $i = 1, 2, ..., m$ with $i \neq j$ can take any values from $\{1, 2, ..., 2N_i + 1\}$.

- **Step 3.** Design m fuzzy systems $f_j(y)$ each from the $\prod_{i=1}^{n}(2N_i + 1)$ rules in (17.22) using product inference engine, singleton fuzzifier, and center average defuzzifier; that is, the designed fuzzy controllers $f_j(y)$ are

$$u_j = -f_j(y) = -\frac{\sum_{l_1=1}^{2N_1+1} \cdots \sum_{l_m=1}^{2N_m+1} \bar{y}_j^{l_1 \cdots l_m} (\prod_{i=1}^{m} \mu_{A_i^{l_i}}(y_i))}{\sum_{l_1=1}^{2N_1+1} \cdots \sum_{l_m=1}^{2N_m+1} (\prod_{i=1}^{m} \mu_{A_i^{l_i}}(y_i))} \qquad (17.24)$$

where $j = 1, 2, ..., m$.

We see from these three steps that as in the SISO case, we do not need to know the process parameters A, B and C in order to design the fuzzy controller; we only require that the membership functions are of the form in Fig.17.2 and the parameters $\bar{y}_j^{l_1 \cdots l_m}$ satisfy (17.23). There is much freedom in choosing these parameters. We now show that the fuzzy controller $u = (u_1, ..., u_m)^T$ with u_j designed as in (17.24) guarantees a stable closed-loop system.

Theorem 17.3. Consider the closed-loop fuzzy control system (17.17)-(17.19). If the fuzzy controller $u = (u_1, ..., u_m)^T = (-f_1(y), ..., -f_m(y))^T$ is designed through the above three steps, (that is, $u_j = -f_j(y)$ is given by (17.24)) and the process under control satisfies conditions (a)-(c) in Proposition 17.3, then the equilibrium point $x = 0$ of the closed-loop system is globally exponentially stable.

Proof: If we can show that $f(0) = 0$ and $y^T f(y) \geq 0$ for all $y \in R^m$ where $f(y) = (f_1(y), ..., f_m(y))^T$, then this theorem follows from Proposition 17.3. From Fig.17.2, (17.23) and (17.24) we have that $f_j(0) = \bar{y}_j^{(N_1+1) \cdots (N_m+1)} = 0$ for $j = 1, 2, ..., m$. Since $y^T f(y) = y_1 f_1(y) + \cdots + y_m f_m(y)$, if we can show $y_j f_j(y) \geq 0$ for all $y_j \in R$ and all $j = 1, 2, ..., m$, then we have $y^T f(y) \geq 0$ for all $y \in R^m$. If $y_j < 0$, then from Fig. 17.2 we have that $\mu_{A_j^{l_j}}(y_j) = 0$ for $l_j = N_j + 2, ..., 2N_j + 1$. Hence, from (17.24) we have

$$f_j(y) = \frac{\sum_{l_1=1}^{2N_1+1} \cdots \sum_{l_{j-1}=1}^{2N_{j-1}+1} \sum_{l_j=1}^{N_j+1} \sum_{l_{j+1}=1}^{2N_{j+1}+1} \cdots \sum_{l_m=1}^{2N_m+1} \bar{y}_j^{l_1 \cdots l_m} (\prod_{i=1}^{m} \mu_{A_i^{l_i}}(y_i))}{\sum_{l_1=1}^{2N_1+1} \cdots \sum_{l_{j-1}=1}^{2N_{j-1}+1} \sum_{l_j=1}^{N_j+1} \sum_{l_{j+1}=1}^{2N_{j+1}+1} \cdots \sum_{l_m=1}^{2N_m+1} (\prod_{i=1}^{m} \mu_{A_i^{l_i}}(y_i))}$$

$$(17.25)$$

From (17.23) we have that $\bar{y}_j^{l_1 \cdots l_m} \leq 0$ for $l_j = 1, 2, ..., N_j + 1$, therefore $f_j(y) \leq 0$ and $y_j f_j(y) \geq 0$. Similarly, we can prove that $y_j f_j(y) \geq 0$ if $y_j > 0$. $\square$

17.2.2   Input-Output Stability

Consider again the closed-loop fuzzy control system (17.17)-(17-19). Similar to Proposition 17.2, we have the following result concerning the L_p-stability of the control system.

Proposition 17.4 (Vidgasagar [1993]). Consider the closed-loop control system (17.17)-(17.19) and suppose that the open-loop unforced system $\dot{x} = Ax$ is globally exponentially stable. If the nonlinear controller $f(y) = (f_1(y), ..., f_m(y))^T$

is globally Lipschitz continuous, that is,

$$||f(y_1) - f(y_2)|| \leq \alpha ||y_1 - y_2||, \; \forall y_1, y_2 \in R^m \tag{17.26}$$

where α is a constant, then the closed-loop system (17.17)-(17.19) is L_p-stable for all $p \in [1, \infty]$.

Using the same arguments as for Lemma 17.1, we can prove that the fuzzy systems (17.24) are continuous, bounded, and piece-wise linear functions. Since a vector of continuous, bounded, and piece-wise linear functions satisfies the Lipschitz condition (17.26), we obtain the following result according to Proposition 17.4.

Theorem 17.4. The fuzzy control system (17.17)-(17.19), with the fuzzy systems $f_j(y)$ given by (17.24), is L_p-stable for all $p \in [1, \infty]$, provided that the eigenvalues of A lie in the open left-half complex plane.

17.3   Summary and Further Readings

In this chapter we have demonstrated the following:

- The classical results of control systems with linear process and nonlinear controller (Propositions 17.1-17.4).

- Design of stable (exponentially or input-output stable) fuzzy controllers for SISO and MIMO linear systems and the proof of the properties of the designed fuzzy controllers.

Vidyasagar [1993] and Slotine and Li [1991] are excellent textbooks on nonlinear control; the first one presented a rigorous treatment to nonlinear control and the second one emphasized readability for practitioners. An early attempt to use the results in nonlinear control theory to design stable fuzzy controllers was Langari and Tomizuka [1990]. Chiu, Chand, Moor, and Chaudhary [1991] also gave sufficient conditions for stable fuzzy controllers for linear plants. The approaches in this chapter are quite preliminary and have not been reported in the literature.

17.4   Exercises

Exercise 17.1. Suppose $\Phi : R^m \to R^m$ and $a, b \in R$ with $a < b$. The Φ is said to *belong to the sector* $[a, b]$ if: (a) $\Phi(0) = 0$, and (b)

$$[\Phi(y) - ay]^T [by - \Phi(y)] \geq 0, \; \forall y \in R^m \tag{17.27}$$

Design a SISO fuzzy system with center average defuzzifier that belongs to the sector $[a, b]$.

Exercise 17.2. Design a SISO fuzzy system with maximum defuzzifier that belongs to the sector $[a, b]$.

Exercise 17.3. Design a 2-input-2-output fuzzy system (that is, two 2-input-1-output fuzzy systems) with center average defuzzifier that belongs to the sector $[a, b]$.

Exercise 17.4. Design a 2-input-2-output fuzzy system with maximum defuzzifier that belongs to the sector $[a, b]$.

Exercise 17.5. Show that the equilibrium point $x = 0$ of the system

$$\dot{x} = -x^2 \tag{17.28}$$

is asymptotically stable but not exponentially stable. How about the system

$$\dot{x} = -(1 + sin^2 x)x \tag{17.29}$$

Exercise 17.6. Suppose there exist constants $a, b, c, r > 0$, $p \geq 1$, and a continuous function $V : R^n \rightarrow R$ such that

$$a||x||^p \leq V(x) \leq b||x||^p \tag{17.30}$$
$$\dot{V}(x) \leq c||x||^p \tag{17.31}$$

Prove that the equilibrium 0 is exponentially stable.

Exercise 17.7. Simulate the fuzzy controller designed in Section 17.1 for the linear system with the transfer function

$$h(s) = \frac{1}{(s + 1)(s + 2)(s + 3)} \tag{17.32}$$

Make your own choice of the fuzzy controller parameters that satisfy the conditions in the design steps. Plot the closed-loop system outputs with different initial conditions.

Chapter 18

Fuzzy Control of Linear Systems II: Optimal and Robust Controllers

In Chapter 17 we gave conditions under which the closed-loop fuzzy control system is stable. Usually, we determine ranges for the fuzzy controller parameters such that stability is guaranteed if the parameters are in these ranges. We did not show how to choose the parameters within these ranges. In this chapter, we will first study how to determine the specific values of the fuzzy controller parameters such that certain performance criterion is optimized; that is, we first will consider the design of optimal fuzzy controllers for linear systems. This is a more difficult problem than designing stable-only fuzzy controllers. The approach in this chapter is very preliminary. In designing the optimal fuzzy controller, we must know the values of the plant parameters A, B and C, which are not required in designing the stable-only fuzzy controllers in Chapter 17.

The second topic in this chapter is robust fuzzy control of linear systems. This field is totally open and we will only show some very preliminary ideas of using the Small Gain Theorem to design robust fuzzy controllers.

18.1 Optimal Fuzzy Control

We first briefly review the Pontryagin Minimum Principle for solving the optimal control problem. Then, we constrain the controller to be a fuzzy system and develop a procedure for determining the fuzzy controller parameters such that the performance criterion is optimized.

18.1.1 The Pontryagin Minimum Principle

Consider the system

$$\dot{x}(t) = g[x(t), u(t)] \tag{18.1}$$

with initial condition $x(0) = x_0$, where $x \in R^n$ is the state, $u \in R^m$ is the control, and g is a linear or nonlinear function. The optimal control problem for the system (18.1) is as follows: determine the control $u(t)$ such that the performance criterion

$$J = S[x(T)] + \int_0^T L[x(t), u(t)]dt \tag{18.2}$$

is minimized, where S and L are some given functions, and the final time T may be given.

This optimal control problem can be solved by using the Pontryagin Minimum Principle, which is given as follows. First, define the Hamilton function

$$H(x, u, p) = L[x(t), u(t)] + p^T(t)g[x(t), u(t)] \tag{18.3}$$

and find $u = h(x, p)$ such that $H(x, u, p)$ is minimized with this u. Substituting $u = h(x, p)$ into (18.3) and define

$$H^*(x, p) = H[x, h(x, p), p] \tag{18.4}$$

Then, solve the $2n$ differential equations (with the two-point boundary condition)

$$\dot{x}(t) = \frac{\partial H^*}{\partial p}, \quad x(0) = x_0 \tag{18.5}$$

$$\dot{p}(t) = -\frac{\partial H^*}{\partial x}, \quad p(T) = \frac{\partial S}{\partial x}\Big|_{x(T)} \tag{18.6}$$

and let $x^*(t)$ and $p^*(t)$ be the solution of (18.5) and (18.6) (they are called the optimal trajectory). Finally, the optimal control is obtained as

$$u^*(t) = h[x^*(t), u^*(t)] \tag{18.7}$$

18.1.2 Design of Optimal Fuzzy Controller

Suppose that the system under control is the time-invariant linear system

$$\dot{x}(t) = Ax(t) + Bu(t), \quad x(0) = x_0 \tag{18.8}$$

where $x \in R^n$ and $u \in R^m$, and that the performance criterion is the quadratic function

$$J = x^T(T)Mx(T) + \int_0^T [x^T(t)Qx(t) + u^T(t)Ru(t)]dt \tag{18.9}$$

where the matrices $M \in R^{n \times n}, Q \in R^{n \times n}$ and $R \in R^{m \times m}$ are symmetric and positive definite.

Now, assume that the controller $u(t)$ is a fuzzy system in the form of (17.24) except that we change the system output y in (17.24) to the state x; that is, $u(t) = (u_1, ..., u_m)^T$ with

$$u_j = -f_j(x) = -\frac{\sum_{l_1=1}^{2N_1+1} \cdots \sum_{l_n=1}^{2N_n+1} \bar{y}_j^{l_1 \cdots l_n} (\prod_{i=1}^{n} \mu_{A_i^{l_i}}(x_i))}{\sum_{l_1=1}^{2N_1+1} \cdots \sum_{l_n=1}^{2N_n+1} (\prod_{i=1}^{n} \mu_{A_i^{l_i}}(x_i))} \tag{18.10}$$

We assume that the membership functions are fixed; they can be in the form shown in Fig.17.2 or other forms. Our task is to determine the parameters $\bar{y}_j^{l_1 \cdots l_n}$ such that J of (18.9) is minimized.

Define the fuzzy basis functions $b(x) = (b_1(x), ..., b_N(x))^T$ as

$$b_l(x) = \frac{\prod_{i=1}^{n} \mu_{A_i^{l_i}}(x_i)}{\sum_{l_1=1}^{2N_1+1} \cdots \sum_{l_n=1}^{2N_n+1} (\prod_{i=1}^{n} \mu_{A_i^{l_i}}(x_i))} \tag{18.11}$$

where $l_i = 1, 2, ..., 2N_i + 1, l = 1, 2, ..., N$ and $N = \prod_{i=1}^{n}(2N_i + 1)$. Define the parameter matrix $\Theta \in R^{m \times N}$ as

$$\Theta = \begin{bmatrix} -\Theta_1^T \\ \cdots \\ -\Theta_m^T \end{bmatrix} \tag{18.12}$$

where $\Theta_j^T \in R^{1 \times N}$ consists of the N parameters $\bar{y}_j^{l_1 \cdots l_n}$ for $l_i = 1, 2, ..., 2N_i + 1$ in the same ordering as $b_l(x)$ for $l = 1, 2, ..., N$. Using these notations, we can rewrite the fuzzy controller $u = (u_1, ..., u_m)^T = (-f_1(x), ..., -f_n(x))^T$ as

$$u = \Theta b(x) \tag{18.13}$$

To achieve maximum optimization, we assume that the parameter matrix Θ is time-vary, that is, $\Theta = \Theta(t)$.

Substituting (18.13) into (18.8) and (18.9), we obtain the closed-loop system

$$\dot{x}(t) = Ax(t) + B\Theta(t)b[x(t)] \tag{18.14}$$

and the performance criterion

$$J = x^T(T)Mx(T) + \int_0^T [x^T(t)Qx(t) + b^T(x(t))\Theta^T(t)R\Theta(t)b(x(t))]dt \tag{18.15}$$

Hence, the problem of designing the optimal fuzzy controller becomes the problem of determining the optimal $\Theta(t)$ such that J of (18.15) is minimized. Viewing the

$\Theta(t)$ as the control $u(t)$ in the Pontryagin Minimum Principle, we can determine the optimal $\Theta(t)$ from (18.3)-(18.7). Specifically, define the Hamilton function

$$H(x, \Theta, p) = x^T Q x + b^T(x)\Theta^T R\Theta b(x) + p^T[Ax + B\Theta b(x)] \qquad (18.16)$$

From $\frac{\partial H}{\partial \Theta} = 0$, that is,

$$\frac{\partial H}{\partial \Theta} = 2R\Theta b(x)b^T(x) + B^T p b^T(x) = 0 \qquad (18.17)$$

we obtain

$$\Theta = -\frac{1}{2}R^{-1}B^T p b^T(x)[b(x)b^T(x)]^{-1} \qquad (18.18)$$

Substituting (18.18) into (18.16), we obtain

$$H^*(x, p) = x^T Q x + p^T Ax + \frac{1}{4}b^T(x)[b(x)b^T(x)]^{-1}b(x)p^T BR^{-1}B^T p b^T(x)$$

$$[b(x)b^T(x)]^{-1}b(x) - \frac{1}{2}p^T BR^{-1}B^T p b^T(x)[b(x)b^T(x)]^{-1}b(x)$$

$$= x^T Q x + p^T Ax + [\alpha^2(x) - \alpha(x)]p^T BR^{-1}B^T p \qquad (18.19)$$

where $\alpha(x)$ is defined as

$$\alpha(x) = \frac{1}{2}b^T(x)[b(x)b^T(x)]^{-1}b(x) \qquad (18.20)$$

Using this H^* in (18.5) and (18.6), we have

$$\dot{x} = \frac{\partial H^*}{\partial p} = Ax + 2[\alpha^2(x) - \alpha(x)]BR^{-1}B^T p \qquad (18.21)$$

$$\dot{p} = -\frac{\partial H^*}{\partial x} = -2Qx - A^T p - [2\alpha(x) - 1]\frac{\partial \alpha(x)}{\partial x}p^T BR^{-1}B^T p \qquad (18.22)$$

with initial condition $x(0) = x_0$ and $p(T) = 2Mx(T)$. Let $x^*(t)$ and $p^*(t)$ ($t \in [0, T]$) be the solution of (18.21) and (18.22), then the optimal fuzzy controller parameters are

$$\Theta^*(t) = -\frac{1}{2}R^{-1}B^T p^*(t)b^T(x^*(t))[b(x^*(t))b^T(x^*(t))]^{-1} \qquad (18.23)$$

and the optimal fuzzy controller is

$$u^* = \Theta^*(t)b(x) \qquad (18.24)$$

Note that the optimal fuzzy controller (18.24) is a state feedback controller with time-varying coefficients. The design procedure of this optimal fuzzy controller is now summarized as follows.

Design of the Optimal Fuzzy Controller:

- **Step 1.** Specify the membership functions $\mu_{A_i^{l_i}}(x_i)$ to cover the state space, where $l_i = 1, 2, ..., 2N_i + 1$ and $i = 1, 2, ..., n$. We may not choose the membership functions as in Fig. 17.2 because the function $\alpha(x)$ with these membership functions is not differentiable (we need $\frac{\partial \alpha(x)}{\partial x}$ in (18.22)). We may choose $\mu_{A_i^{l_i}}(x_i)$ to be the Gaussian functions.

- **Step 2.** Compute the fuzzy basis functions $b_l(x)$ from (18.11) and the function $\alpha(x)$ from (18.20). Compute the derivative $\frac{\partial \alpha(x)}{\partial x}$.

- **Step 3.** Solve the two-point boundary differential equations (18.21) and (18.22) and let the solution be $x^*(t)$ and $p^*(t)$, $t \in [0, T]$. Compute $\Theta^*(t)$ for $t \in [0, T]$ according to (18.23).

- **Step 4.** The optimal fuzzy controller is obtained as (18.24)

Note that Steps 1-3 are off-line operations; that is, we first compute $\Theta^*(t)$ following Steps 1-3 and store the $\Theta^*(t)$ for $t \in [0, T]$ in the computer, then in on-line operation we simply substitute the stored $\Theta^*(t)$ into (18.24) to obtain the optimal fuzzy controller.

The most difficult part in designing this optimal fuzzy controller is to solve the two-point boundary differential equations (18.21) and (18.22). Since these differential equations are nonlinear, numerical integration is usually used to solve them.

18.1.3 Application to the Ball-and-Beam System

Consider the ball-and-beam system in Fig. 16.7, with the state-space model given by (16.4)-(16.5). Since the ball-and-beam system is nonlinear, to apply our optimal fuzzy controller we have to linearize it around the equilibrium point $x = 0$. The linearized system is in the form of (18.8) with

$$A = \begin{bmatrix} 0 & 1 & 0 & 0 \\ 0 & 0 & -\alpha\beta & 0 \\ 0 & 0 & 0 & 1 \\ 0 & 0 & 0 & 0 \end{bmatrix}, \quad B = \begin{bmatrix} 0 \\ 0 \\ 0 \\ 1 \end{bmatrix} \tag{18.25}$$

We now design an optimal fuzzy controller for the linearized ball-and-beam system, and apply the designed controller to the original nonlinear system (16.4)-(16.5). We choose $M = 0, Q = I, R = I, T = 30$, and $N_i = 2$ for $i = 1, 2, 3, 4$. The membership functions $\mu_{A_i^{l_i}}(x_i)$ are chosen as

$$\mu_{A_i^{l_i}}(x_i) = exp[-2(x_i - \bar{x}_i^{l_i})^2] \tag{18.26}$$

where $i = 1, 2, 3, 4, l_i = 1, 2, ..., 5$ and $\bar{x}_i^{l_i} = a_i + b_i(l_i - 1)$ with $a_1 = a_2 = -2, a_3 = a_4 = -1, b_1 = b_2 = 1$ and $b_3 = b_4 = 0.5$. Fig.18.1 shows the closed-loop output $y(t)$ for three initial conditions: $x(0) = (1, 0, 0, 0)^T, (2, 0, 0, 0)^T$ and $(3, 0, 0, 0)^T$. We see from Fig.18.1 that although our optimal fuzzy controller was designed based on the linearized model, it could smoothly regulate the ball to the origin from a number of initial positions.

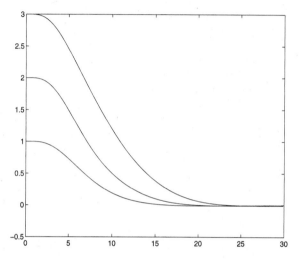

Figure 18.1. Closed-loop outputs $y(t)$ $(= r(t)$, the ball position) with the optimal fuzzy controller for three initial conditions.

18.2   Robust Fuzzy Control

For almost all practical systems, especially for industrial systems, the plant models can never be exactly obtained. Because the controllers are designed for the models, good performance in theory and in simulations (where models are used) do not imply good performance in real implementations (where controllers are directly applied to the real systems, not to the models). The objective of robust control is to design controllers that guarantee the stability of the closed-loop systems for a wide range of plants. The wider the range of the plants, the more robust the controller is. We now consider the design of robust fuzzy controllers for linear plants.

Consider the fuzzy control system in Fig. 17.1, where the process under control is now represented by its transfer function $G(s)$. For simplicity, we assume that

$G(s)$ is a SISO system. Define

$$\gamma(G) = \sup_{w \in R} |G(jw)| \tag{18.27}$$

and

$$\Omega_\alpha = \{G(s) : \gamma(G) \leq \alpha\} \tag{18.28}$$

Clearly, the larger the α, the more plants in Ω_α. The *objective* of robust fuzzy control is to design a fuzzy controller that stabilizes all the plants in Ω_α while allowing the Ω_α to be as large as possible. This is intuitively appealing because the larger the Ω_α, the more plants the fuzzy controller can stabilize, which means that the fuzzy controller is more robust.

To achieve the objective, the key mathematical tool is the famous *Small Gain Theorem* (Vidyasagar [1993]). To understand this theorem, we must first introduce the concept of the gain of a nonlinear mapping.

Definition 18.1. Let $f : R \to R$ satisfy $|f(x)| \leq \gamma|x|$ for some constant γ and all $x \in R$. Then the *gain of f* is defined as

$$\gamma(f) = \inf\{\gamma : |f(x)| \leq \gamma|x|, \ \forall x \in R\} \tag{18.29}$$

The Small Gain Theorem is now stated as follows.

Small Gain Theorem. Suppose that the linear plant $G(s)$ is stable and the nonlinear mapping $f(y)$ is bounded for bounded input. Then the closed-loop system in Fig. 17.1 (with the process under control represented by $G(s)$) is stable if

$$\gamma(G)\gamma(f) < 1 \tag{18.30}$$

From (18.30) we see that a smaller $\gamma(f)$ will permit a larger $\gamma(G)$, and according to (18.27)-(18.28) a larger $\gamma(G)$ means a larger set Ω_α of plants that the controller $f(y)$ can stabilize. Therefore, for the purpose of robust control, we should choose $\gamma(f)$ as small as possible. The extreme case is $f = 0$, which means no control at all; this is impractical because feedback control is required to improve performance. Since the most commonly used fuzzy controller is a weighted average of the $\bar{y}^l$'s (the centers of the THEN part membership functions), our recommendation for designing robust fuzzy controllers is the following: design the fuzzy controller using any method in Chapters 16-26, but try to use smaller $\bar{y}^l$'s in the design.

Robust fuzzy control is an open field and much work remains to be done.

18.3 Summary and Further Readings

In this chapter we have demonstrated the following:

- The Pontryagin Minimum Principle for solving the optimal control problem.

- How to use the Pontryagin Minimum Principle to design the optimal fuzzy controller for linear plants.

- The Small Gain Theorem and the basic idea of designing robust fuzzy controllers.

Good textbooks on optimal control are abundant in the literature, for example Anderson and Moore [1990] and Bryson and Ho [1975]. A good book on robust control is Green and Limebeer [1995].

18.4 Exercises

Exercise 18.1. A simplified model of the linear motion of an automobile is $\dot{x} = u$, where $x(t)$ is the vehicle velocity and $u(t)$ is the acceleration or deceleration. The car is initially moving at x_0 m/s. Using the Pontryagin Minimum Principle to design an optimal $u(t)$ which brings the velocity $x(t_f)$ to zero in *minimum time* t_f. Assume that acceleration and braking limitations require $|u(t)| \leq M$ for all t, where M is a constant.

Exercise 18.2. Use the Pontryagin Minimum Principle to solve the optimization problems:

(a) $\dot{x}_1 = x_2, \dot{x}_2 = u, x_1(0) = 1, x_2(0) = 1, x_1(2) = 0, x_2(2) = 0$, minimize $J = \frac{1}{2}\int_0^2 u^2(t)dt$.

(b) $\dot{x} = u, x(0) = 0, x(1) = 1$, minimize $J = \int_0^1 (x^2 + u^2)dt$.

Exercise 18.3. Derive the detailed formulas of the differential equations (18.21)-(18.22) for the example in Subsection 18.1.3. (You may or may not write out the details of $\frac{\partial \alpha(x)}{\partial x}$.)

Exercise 18.4. Design a fuzzy system $f(x)$ with center average defuzzifier on $U = [-1, 1]$ with at least five rules such that: (a) $f(0) = 0$, and (b) $\gamma(f) = 1$, where the R in (18.29) is changed to U.

Exercise 18.5. Design a fuzzy system $f(x)$ with maximum defuzzifier on $U = [-1, 1]$ with at least five rules such that: (a) $f(0) = 0$, and (b) $\gamma(f) = 1$, where the R in (18.29) is changed to U.

Chapter 19

Fuzzy Control of Nonlinear Systems I: Sliding Control

Sliding control is a powerful approach to controlling nonlinear and uncertain systems. It is a robust control method and can be applied in the presence of model uncertainties and parameter disturbances, provided that the bounds of these uncertainties and disturbances are known. A careful comparison of sliding control and fuzzy control shows that their operations are similar in many cases. In this chapter, we will explore the relationship between fuzzy control and sliding control, and design fuzzy controllers based on sliding control principles.

19.1 Fuzzy Control As Sliding Control: Analysis

19.1.1 Basic Principles of Sliding Control

Consider the SISO nonlinear system

$$x^{(n)} = f(\mathbf{x}) + u \tag{19.1}$$

where $u \in R$ is the control input, $x \in R$ is the output, and $\mathbf{x} = (x, \dot{x}, ..., x^{(n-1)})^T \in R^n$ is the state vector. In (19.1), the function $f(\mathbf{x})$ is not exactly known, but the uncertainty of $f(\mathbf{x})$ is bounded by a known function of $\mathbf{x}$; that is,

$$f(\mathbf{x}) = \hat{f}(\mathbf{x}) + \Delta f(\mathbf{x}) \tag{19.2}$$

and

$$|\Delta f(\mathbf{x})| \leq F(\mathbf{x}) \tag{19.3}$$

where $\Delta f(\mathbf{x})$ is unknown but $\hat{f}(\mathbf{x})$ and $F(\mathbf{x})$ are known. The control objective is to determine a feedback control $u = u(\mathbf{x})$ such that the state $\mathbf{x}$ of the closed-loop system will follow the desired state $\mathbf{x}_d = (x_d, \dot{x}_d, ..., x_d^{(n-1)})^T$; that is, the tracking error

$$\mathbf{e} = \mathbf{x} - \mathbf{x}_d = (e, \dot{e}, ..., e^{(n-1)})^T \tag{19.4}$$

238

should converge to zero, where $e = x - x_d$.

The basic idea of sliding control is as follows. Define a scalar function

$$s(\mathbf{x}, t) = (\frac{d}{dt} + \lambda)^{n-1} e$$
$$= e^{(n-1)} + C_{n-1}^1 \lambda e^{(n-2)} + C_{n-1}^2 \lambda^2 e^{(n-3)} + \cdots + \lambda^{n-1} e \qquad (19.5)$$

where λ is a positive constant. Then,

$$s(\mathbf{x}, t) = 0 \qquad (19.6)$$

defines a time-varying surface $S(t)$ in the state space R^n. For example, if $n = 2$ then the surface $S(t)$ is

$$s(\mathbf{x}, t) = \dot{e} + \lambda e = \dot{x} + \lambda x - \dot{x}_d - \lambda x_d = 0 \qquad (19.7)$$

which is a straight line in the $x - \dot{x}$ phase plane, as shown in Fig. 19.1. Since $\dot{x}_d$ and x_d are usually time-varying functions, the $S(t)$ is also time-varying. If the initial state $\mathbf{x}(0)$ equals the initial desired state $\mathbf{x}_d(0)$, that is, if $\mathbf{e}(0) = 0$, then from (19.5) and (19.6) we see that if the state vector $\mathbf{x}$ remains on the surface $S(t)$ for all $t \geq 0$, we will have $\mathbf{e}(t) = 0$ for all $t \geq 0$. Indeed, $s(\mathbf{x}, t) = 0$ represents a linear differential equation whose unique solution is $\mathbf{e}(t) = 0$ for the initial condition $\mathbf{e}(0) = 0$. Thus, *our tracking control problem is equivalent to keeping the scalar function $s(\mathbf{x}, t)$ at zero.* To achieve this goal, we can choose the control u such that

$$\frac{1}{2} \frac{d}{dt} s^2 \leq -\eta |s| \qquad (19.8)$$

if the state is outside of $S(t)$, where η is a positive constant. (19.8) is called the *sliding condition*; it guarantees that $|s(\mathbf{x}, t)|$ will decrease if $\mathbf{x}$ is not on the surface $S(t)$, that is, the state trajectory will move towards the surface $S(t)$, as illustrated in Fig. 19.1. The surface $S(t)$ is referred to as the *sliding surface*, the system on the surface is in the *sliding mode*, and the control that guarantees (19.8) is called *sliding mode control* or *sliding control*. To summarize the discussions above, we have the following lemma.

Lemma 19.1. Consider the nonlinear system (19.1) and let $s(\mathbf{x}, t)$ be defined as in (19.5). If we can design a controller u such that the sliding condition (19.8) is satisfied, then:

(a) The state will reach the sliding surface $S(t)$ within finite time.

(b) Once the state is on the sliding surface, it will remain there.

(c) If the state remains on the sliding surface, the tracking error $\mathbf{e}(t)$ will converge to zero.

Therefore, our goal is to design a controller u such that the closed-loop system satisfies the sliding condition (19.8).

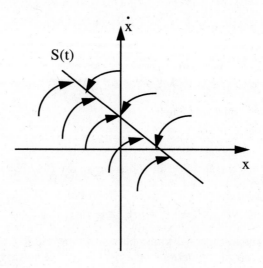

Figure 19.1. Sliding surface in two-dimensional phase plane.

We now derive the details of sliding control for a second-order system, that is, the system is (19.1) with $n = 2$. In this case, (19.8) becomes

$$s[f(\mathbf{x}) + u - \ddot{x}_d + \lambda\dot{e}] \leq -\eta|s| \tag{19.9}$$

where we used (19.5) and (19.1). If we choose

$$u = -\hat{f}(\mathbf{x}) + \ddot{x}_d - \lambda\dot{e} - K(x,\dot{x})sgn(s) \tag{19.10}$$

then (19.9) becomes

$$sgn(s)[f(\mathbf{x}) - \hat{f}(\mathbf{x}) - K(x,\dot{x})sgn(s)] \leq -\eta \tag{19.11}$$

where $sgn(s) = 1$ if $s > 0$, $sgn(s) = -1$ if $s < 0$, and $\hat{f}(\mathbf{x})$ is the estimate of $f(\mathbf{x})$ as in (19.2). Furthermore, (19.11) is equivalent to

$$K(x,\dot{x}) \geq \eta + sgn(s)[\Delta f(\mathbf{x})] \tag{19.12}$$

Therefore, if we choose

$$K(x,\dot{x}) = \eta + F(\mathbf{x}) \tag{19.13}$$

then from (19.3) we see that (19.12) is guaranteed, which in turn implies that the sliding condition (19.8) is satisfied. In conclusion, the sliding controller is given by (19.10) with $K(x,\dot{x})$ given by (19.13).

19.1.2 Analysis of Fuzzy Controllers Based on Sliding Control Principle

We now use the sliding control principle to analyze fuzzy controllers. For simplicity, we consider the system (19.1) with $n = 2$; but the approach can be generalized to high-order systems. Suppose that we choose the control u to be a fuzzy controller $u_{fuzz}(\mathbf{x})$, that is,

$$u = u_{fuzz}(\mathbf{x}) \tag{19.14}$$

The following theorem specifies conditions on the fuzzy controller $u_{fuzz}(\mathbf{x})$ such that the tracking error $\mathbf{e}$ converges to zero.

Theorem 19.1. Consider the nonlinear system (19.1) with $n = 2$ and assume that the control u is given by (19.14). If the fuzzy controller $u_{fuzz}(\mathbf{x})$ satisfies the following condition:

$$u_{fuzz}(\mathbf{x}) \leq -\eta - [f(\mathbf{x}) + \lambda \dot{e} - \ddot{x}_d], \; if \; sgn(s) > 0 \tag{19.15}$$
$$u_{fuzz}(\mathbf{x}) \geq \eta - [f(\mathbf{x}) + \lambda \dot{e} - \ddot{x}_d], \; if \; sgn(s) < 0 \tag{19.16}$$

where η and λ are positive constants and $s = \dot{e} + \lambda e$, then it is guaranteed that the tracking error $\mathbf{e} = (x - x_d, \dot{x} - \dot{x}_d)^T$ will converge to zero.

Proof: Substituting (19.14) into the sliding condition (19.9), we have

$$sgn(s)[f(\mathbf{x}) + u_{fuzz}(\mathbf{x}) - \ddot{x}_d + \lambda \dot{e}] \leq -\eta \tag{19.17}$$

Clearly, if $u_{fuzz}(\mathbf{x})$ satisfies (19.15) and (19.16), then (19.17) is true, which means that the sliding condition (19.8) is satisfied and therefore the tracking error will converge to zero according to Lemma 19.1. $\square$

From Theorem 19.1 we see that if the designed fuzzy controller satisfies (19.15) and (19.16), then the tracking error is guaranteed to converge to zero. However, since the fuzzy controller $u_{fuzz}(\mathbf{x})$ must change discontinuously across the sliding surface $s = 0$, it is difficult to use (19.15) and (19.16) as design constraints for the fuzzy controller. Therefore, Theorem 19.1 is more of analytical value rather than design value. Additionally, the discontinuous control (19.10) or (19.15)-(19.16) will cause chattering (see the next section) across the sliding surface, which is undesirable. In the next section, we will propose another condition for designing fuzzy controllers based on the sliding control principle.

19.2 Fuzzy Control As Sliding Control: Design

19.2.1 Continuous Approximation of Sliding Control Law

From the last section we see that the sliding control law (for example, (19.10)) has to be discontinuous across the sliding surface $S(t)$. Since the implementation of the control switchings could not be perfect and we have to sample the signals in digital control systems, this leads to *chattering*, as shown in Fig. 19.2. Chattering is undesirable because it involves high control activity and may excite high-frequency dynamics. A way to eliminate chattering is to introduce a thin *boundary layer* neighboring the sliding surface:

$$B(t) = \{\mathbf{x} : |s(\mathbf{x}, t) \leq \Phi\} \tag{19.18}$$

such that the control changes continuously within this boundary layer; see Fig. 19.2. Φ is called the *thickness* of the boundary layer and $\epsilon = \Phi/\lambda^{n-1}$ is called the *width* of the boundary layer. We now show that if the control law guarantees that the sliding condition (19.8) is satisfied outside the boundary layer $B(t)$, then the tracking error is guaranteed to be within the precision of ϵ.

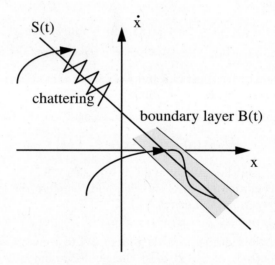

Figure 19.2. Chattering and boundary layer.

Lemma 19.2. If the sliding condition (19.8) is satisfied outside the boundary layer $B(t)$ of (19.18), then it is guaranteed that after finite time we will have

$$|e(t)| \leq \epsilon \tag{19.19}$$

Proof: Since the sliding condition (19.8) is satisfied outside $B(t)$, we have that the state vector $\mathbf{x}$ will enter the boundary layer $B(t)$ within finite time and will stay inside it afterwards. That is, no matter where the initial state $\mathbf{x}(0)$ is, after finite time we will have $|s(\mathbf{x}, t)| \leq \Phi$. Let

$$y_i(t) = (\frac{d}{dt} + \lambda)^{n-i-1} e \tag{19.20}$$

so we have

$$y_0(t) = s(\mathbf{x}, t) \tag{19.21}$$
$$y_{n-1}(t) = e(t) \tag{19.22}$$

and

$$y_i(t) = (\frac{d}{dt} + \lambda)y_{i+1}(t) \tag{19.23}$$

for $i = 0, 1, ..., n - 2$. Hence,

$$y_{i+1}(t) = \int_0^t e^{-\lambda(t-\tau)} y_i(\tau) d\tau \tag{19.24}$$

When $i = 0$, we have from (19.24), (19.21) and $|s(\mathbf{x}, t)| \leq \Phi$ that

$$|y_1(t)| \leq \int_0^t e^{-\lambda(t-\tau)}|s(\mathbf{x}, \tau)|d\tau$$
$$\leq \Phi \frac{1 - e^{-\lambda t}}{\lambda}$$
$$\leq \Phi/\lambda \tag{19.25}$$

When $i = 1$, we have

$$|y_2(t)| \leq \int_0^t e^{-\lambda(t-\tau)}|y_1(\tau)|d\tau$$
$$\leq (\Phi/\lambda)\frac{1 - e^{-\lambda t}}{\lambda}$$
$$\leq \Phi/\lambda^2 \tag{19.26}$$

Continuing this process until $i = n - 2$, we have

$$|y_{n-1}(t)| = |e(t)| \leq \Phi/\lambda^{n-1} = \epsilon \tag{19.27}$$

which is (19.19). $\square$

Lemma 19.2 shows that if we are willing to sacrifice precision, that is, from perfect tracking $e(t) = 0$ to tracking within precision $|e(t)| \leq \epsilon$, the requirement for control law is reduced from satisfying the sliding condition (19.8) all the time to satisfying the sliding condition only when $\mathbf{x}(t)$ is outside of the boundary layer $B(t)$. Consequently, we are able to design a smooth controller that does not need to switch discontinuously across the sliding surface. Specifically, for the second-order system, we change the control law (19.10) to

$$u = -\hat{f}(\mathbf{x}) + \ddot{x}_d - \lambda\dot{e} - K(x, \dot{x})sat(s/\Phi) \tag{19.28}$$

where the saturation function $sat(s/\Phi)$ is defined as

$$sat(s/\Phi) = \begin{cases} -1 & if \quad s/\Phi \leq -1 \\ s/\Phi & if \quad -1 < s/\Phi \leq 1 \\ 1 & if \quad s/\Phi > 1 \end{cases} \qquad (19.29)$$

Clearly, if the state is outside of the boundary layer, that is, if $|s/\Phi| > 1$, then $sat(s/\Phi) = sgn(s)$ and thus the control law (19.28) is equivalent to the control law (19.10). Therefore, the control law (19.28) with $K(x, \dot{x})$ given by (19.13) guarantees that the sliding condition (19.8) is satisfied outside the boundary layer $B(t)$. The control law (19.28) is a smooth control law and does not need to switch discontinuously across the sliding surface.

19.2.2 Design of Fuzzy Controller Based on the Smooth Sliding Control Law

From the last subsection we see that if we design a fuzzy controller according to (19.28), then the tracking error is guaranteed to satisfy (19.19) within finite time. For a given precision ϵ, we can choose Φ and λ such that $\epsilon = \Phi/\lambda^{n-1}$; that is, we can specify the design parameters Φ and λ such that the tracking error converges to any precision band $|e(t)| \leq \epsilon$. Since the control u of (19.28) is a smooth function of x and $\dot{x}$, we can design a fuzzy controller to approximate the u of (19.28). From the last subsection, we have the following theorem.

Theorem 19.2. Consider the nonlinear system (19.1) with $n = 2$ and assume that the control u is a fuzzy controller u_{fuzz}. If the fuzzy controller is designed as

$$u_{fuzz}(\mathbf{x}) = -\hat{f}(\mathbf{x}) + \ddot{x}_d - \lambda\dot{e} - [\eta + F(\mathbf{x})]sat(s/\Phi) \qquad (19.30)$$

then after finite time the tracking error $e(t) = x(t) - x_d(t)$ will satisfy (19.19).

Proof: Since the fuzzy controller (19.30) satisfies the sliding condition (19.8) when the state is outside of the boundary layer $B(t)$, this theorem follows from Lemma 19.2. □

Our task now is to design a fuzzy controller that approximates the right-hand side of (19.30). Since all the functions in the right-hand side of (19.30) are known, we can compute the values of the right-hand side of (19.30) at some regular points in the phase plane and design the fuzzy controller according to the methods in Chapters 10 and 11. Specifically, we have the following design method.

Design of the Fuzzy Controller:

- **Step 1.** Determine the domains of interest for e and $\dot{e}$; that is, determine the intervals $[\alpha_1, \beta_1]$ and $[\alpha_2, \beta_2]$ such that $\mathbf{e} = (e, \dot{e})^T \in U = [\alpha_1, \beta_1] \times [\alpha_2, \beta_2]$.

- **Step 2.** Let $g(e, \dot{e}) = -\hat{f}(\mathbf{x}) + \ddot{x}_d - \lambda\dot{e} - [\eta + F(\mathbf{x})]sat(s/\Phi)$ and view this $g(e, \dot{e})$ as the g in (10.9). Design the fuzzy controller through the three steps in Sections 10.2 or 11.1; that is, the designed fuzzy controller is the fuzzy system (10.10).

The approximation accuracy of this designed fuzzy controller to the ideal fuzzy controller $u_{fuzz}(\mathbf{x})$ of (19.30) is given by Theorems 10.1 or 11.1. As we see from these theorems that if we sufficiently sample the domains of interest, the approximation error can be as small as desired.

Example 19.1. Consider the first-order nonlinear system

$$\dot{x}(t) = \frac{1 - e^{-x(t)}}{1 + e^{-x(t)}} + u(t) \tag{19.31}$$

where the nonlinear function $f(x) = \frac{1-e^{-x(t)}}{1+e^{-x(t)}}$ is assumed to be unknown. Our task is to design a fuzzy controller based on the smooth sliding control law such that the $x(t)$ converges to zero.

Since

$$f(x) = 1 - \frac{-2e^{-x(t)}}{1 + e^{-x(t)}} \tag{19.32}$$

we choose $\hat{f}(x) = 1, \Delta f(x) = \frac{-2e^{-x(t)}}{1+e^{-x(t)}}$, and $F(x) = 2$, so that $|\Delta f(x)| \leq F(x)$. For this example, $x_d = 0, e = x - 0 = x, s = e = x$, and the sliding condition (19.8) becomes

$$x(f + u) \leq -\eta|x| \tag{19.33}$$

Hence, the sliding control law is

$$u(x) = -\hat{f}(x) - K(x)sgn(x) \tag{19.34}$$

where $K(x) = \eta + F(x) = \eta + 2$. Fig. 19.3 shows the closed-loop state $x(t)$ using this sliding controller for four initial conditions, where we chose $\eta = 0.1$ and the sampling rate equal to 0.02s. We see that chattering occurred.

The smooth sliding controller is

$$u_{sm}(x) = -\hat{f}(x) - K(x)sat(x/\Phi) \tag{19.35}$$

Viewing this $u_{sm}(x)$ as the $g(x)$ in (10.9), we designed a fuzzy controller following the three steps in Section 11.1 (this is a one-dimensional system, a special case of the system in Section 11.1), where we chose $\Phi = 0.2, U = [-2, 2], N = 9$, and the e^j's to be uniformly distributed over $[-2, 2]$. Fig.19.4 shows the closed-loop $x(t)$ with this fuzzy controller for the same initial conditions as in Fig. 19.3. Comparing Figs.19.4 with 19.3, we see that chattering disappeared, but a steady tracking error appeared; this is as expected: chattering was smoothed out with the sacrifice of precision. □

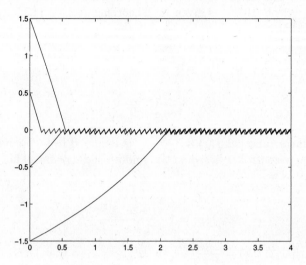

Figure 19.3. Closed-loop state $x(t)$ for the nonlinear system (19.31) with the sliding controller (19.34) for four different initial conditions.

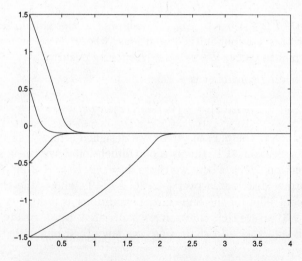

Figure 19.4. Closed-loop state $x(t)$ for the nonlinear system (19.31) with the fuzzy controller for four different initial conditions.

19.3   Summary and Further Readings

In this chapter we have demonstrated the following:

- How to design sliding controllers for nonlinear systems and what are the fundamental assumptions.

- What is chattering and how to smooth it (the balance between tracking precision and smooth control).

- How to design fuzzy controllers based on the smooth sliding control laws.

Sliding control was studied in books by Utkin [1978] and Slotine and Li [1991]. Applying sliding mode concept to fuzzy control was due to Palm [1992]. Books by Driankov, Hellendoorn, and Reinfrank [1993] and Yager and Filev [1994] also examined sliding mode fuzzy control.

19.4   Exercises

Exercise 19.1. Consider the second-order system

$$\ddot{x} + a(t)\dot{x}^2 cos3x = u \tag{19.36}$$

where $a(t)$ is unknown but verifies

$$1 \le a(t) \le 2 \tag{19.37}$$

Design a sliding controller u such that x converges to the desired trajectory x_d.

Exercise 19.2. Consider the nonlinear system

$$\ddot{x} = f(x, \dot{x}) + g(x, \dot{x})u \tag{19.38}$$

where f and g are unknown but $f(\mathbf{x}) = \hat{f}(\mathbf{x}) + \Delta f(\mathbf{x})$, $|\Delta f(\mathbf{x})| \le F(\mathbf{x})$, with $\hat{f}(\mathbf{x})$ and $F(\mathbf{x})$ known, and $0 < g_{min}(\mathbf{x}) \le g(\mathbf{x}) \le g_{max}(\mathbf{x})$, with $g_{min}(\mathbf{x})$ and $g_{max}(\mathbf{x})$ known. Let $\hat{g}(\mathbf{x}) = [g_{min}(\mathbf{x})g_{max}(\mathbf{x})]^{1/2}$ be the estimate of $g(\mathbf{x})$ and $\beta = [g_{max}(\mathbf{x})/g_{min}(\mathbf{x})]^{1/2}$. Show that the control law

$$u = \hat{g}^{-1}(\mathbf{x})[-\hat{f}(\mathbf{x}) + \ddot{x}_d - \lambda\dot{e} - K(\mathbf{x})sgn(s)] \tag{19.39}$$

with

$$K(\mathbf{x}) \ge \beta(F + \eta) + (\beta - 1)| - \hat{f}(\mathbf{x}) + \ddot{x}_d - \lambda\dot{e}| \tag{19.40}$$

satisfies the sliding condition (19.8).

Exercise 19.3. Simulate the sliding controller (19.34) in Example 19.1 using different sampling rate and observe the phenomenon that the larger the sampling rate, the stronger (in terms of the magnitude of the oscillation) the chattering.

Example 19.4. Consider the nonlinear system

$$\dot{x}_1 = sinx_2 + \sqrt{t+1}x_2 \qquad (19.41)$$
$$\dot{x}_2 = \alpha_1(t)x_1^4 cosx_2 + \alpha_2(t)u \qquad (19.42)$$

where $\alpha_1(t)$ and $\alpha_2(t)$ are unknown time-varying functions with the known bounds

$$|\alpha_1(t)| \leq 10, \ 1 \leq \alpha_2(t) \leq 2, \ \forall t \geq 0 \qquad (19.43)$$

Design a fuzzy controller based on the smooth sliding controller such that the closed-loop state $x_1(t)$ tracks a given desired trajectory $x_d(t)$.

Exercise 19.5. Design a sliding controller, using the approach in Subsection 19.1.1, and a fuzzy controller, using the trial-and-error approach in Chapter 16, for the inverted pendulum system. Compare the two controllers by plotting: (a) the control surfaces of the two controllers, and (b) the responses of the closed-loop systems with the two controllers. What are the conclusions of your comparison?

Chapter 20

Fuzzy Control of Nonlinear Systems II: Supervisory Control

20.1 Multi-level Control Involving Fuzzy Systems

The fuzzy controllers considered in Chapters 17-19 are single-loop (or single-level) controllers; that is, the whole control system consists of the process and the fuzzy controller connected in a single loop. For complex practical systems, the single-loop control systems may not effectively achieve the control objectives, and a multi-level control structure turns out to be very helpful. Usually, the lower-level controllers perform fast direct control and the higher-level controllers perform low-speed supervision. In this chapter, we consider two-level control structures where one of the levels is constructed from fuzzy systems. We have two possibilities: (i) the first-level controller is a fuzzy controller and the second level is a nonfuzzy supervisory controller (see Fig. 20.1), and (ii) the first level consists of a conventional controller (for example, a PID controller) and the second level comprises fuzzy systems performing supervisory operations (see Fig. 20.2).

The main advantage of two-level control is that different controllers can be designed to target different objectives, so that each controller is simpler and performance is improved. Specifically, for the two-level control system in Fig. 20.1, we can design the fuzzy controller without considering stability and use the supervisory controller to deal with stability related problems. In this way, we have much freedom in choosing the fuzzy controller parameters and consequently, the design of the fuzzy controller is simplified and performance is improved. We will show the details of this approach in the next section. For the two-level control system in Fig. 20.2, we will consider the special case where the first level is a PID controller and the second-level fuzzy system adjusts the PID parameters according to certain heuristic rules; the details will be given in Section 20.3.

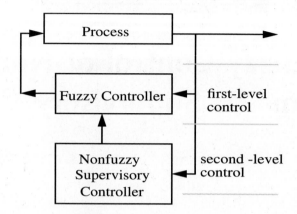

Figure 20.1. Architecture of a two-level fuzzy control system, where the fuzzy controller performs the main control action and the nonfuzzy supervisory controller monitors the operation and takes action when something undesirable happens.

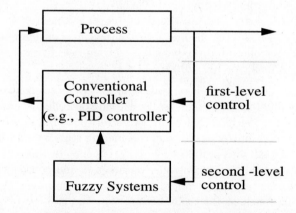

Figure 20.2. Architecture of a two-level fuzzy control system, where the first level is a conventional controller and the second level consists of fuzzy systems that supervise and modify the operations of the conventional controller.

20.2   Stable Fuzzy Control Using Nonfuzzy Supervisor

20.2.1   Design of the Supervisory Controller

Conceptually, there are at least two different approaches to guarantee the stability of a fuzzy control system. The first approach is to specify the structure and parameters of the fuzzy controller such that the closed-loop system with the fuzzy controller is stable (for example, the fuzzy controllers in Chapters 17 and 19). In the second approach, the fuzzy controller is designed first without any stability consideration, then another controller is appended to the fuzzy controller to take care of the stability requirement. Because there is much flexibility in designing the fuzzy controller in the second approach, the resulting fuzzy control system is expected to show better performance.

The key in the second approach is to design the appended second-level nonfuzzy controller to guarantee stability. Because we want the fuzzy controller to perform the main control action, the second-level controller would be better a safeguard rather than a main controller. Therefore, we choose the second-level controller to operate in the following supervisory fashion: if the fuzzy controller works well, the second-level controller is idle; if the pure fuzzy control system tends to be unstable, the second-level controller starts working to guarantee stability. Thus, we call the second-level controller a *supervisory controller*.

Consider the nonlinear system governed by the differential equation

$$x^{(n)} = f(x, \dot{x}, ..., x^{(n-1)}) + g(x, \dot{x}, ..., x^{(n-1)})u \qquad (20.1)$$

where $x \in R$ is the output of the system, $u \in R$ is the control, $\mathbf{x} = (x, \dot{x}, ..., x^{(n-1)})^T$ is the state vector that is assumed to be measurable or computable, and f and g are unknown nonlinear functions. We assume that $g > 0$. From nonlinear control theory (Isidori [1989]) we know that this system is in normal form and many general nonlinear systems can be transformed into this form. The main restriction is that the control u must appear linearly in the equation.

Now suppose that we have already designed a fuzzy controller

$$u = u_{fuzz}(\mathbf{x}) \qquad (20.2)$$

for the system. This can be done, for example, by the trial-and-error approach in Chapter 16. Our task is to guarantee the stability of the closed-loop system and, at the same time, without changing the existing design of the fuzzy controller $u_{fuzz}(\mathbf{x})$. More specifically, we are required to design a controller whose main control action is the fuzzy control $u_{fuzz}(\mathbf{x})$ and that the closed-loop system with this controller is globally stable in the sense that the state $\mathbf{x}$ is uniformly bounded, that is, $|\mathbf{x}(t)| \leq M_x, \forall t > 0$, where M_x is a constant given by the designer.

For this task, we append the fuzzy controller $u_{fuzz}(\mathbf{x})$ with a *supervisory controller* $u_s(\mathbf{x})$, which is nonzero only when the state $\mathbf{x}$ hits the boundary of the constraint set $\{\mathbf{x} : |\mathbf{x}| \le M_x\}$; that is, the control now is

$$u = u_{fuzz}(\mathbf{x}) + I^* u_s(\mathbf{x}) \tag{20.3}$$

where the *indicator function* $I^* = 1$ if $|\mathbf{x}| \ge M_x$ and $I^* = 0$ if $|\mathbf{x}| < M_x$. Therefore, the main control action is still the fuzzy controller $u_{fuzz}(\mathbf{x})$. Our task now is to design the u_s such that $|\mathbf{x}(t)| \le M_x$ for all $t > 0$.

Let us first examine whether it is possible to design such a supervisory controller without any additional assumption. Substituting (20.3) into (20.1) we have that the closed-loop system satisfies

$$x^{(n)} = f(\mathbf{x}) + g(\mathbf{x})u_{fuzz}(\mathbf{x}) + g(\mathbf{x})I^* u_s(\mathbf{x}) \tag{20.4}$$

Now suppose $|\mathbf{x}| = M_x$ and thus $I^* = 1$. Because we assume that $f(\mathbf{x})$ and $g(\mathbf{x})$ are totally unknown and can be arbitrary nonlinear functions, for any $u_s(\mathbf{x})$ we can always find $f(\mathbf{x})$ and $g(\mathbf{x})$ such that the right-hand side of (20.4) is positive, and therefore we will have $|\mathbf{x}| > M_x$. Thus, we must make additional assumptions for $f(\mathbf{x})$ and $g(\mathbf{x})$ in order for such u_s design to be possible. We need the following assumption.

Assumption 20.1: We can determine functions $f^U(\mathbf{x})$ and $g_L(\mathbf{x})$ such that $|f(\mathbf{x})| \le f^U(\mathbf{x})$ and $0 < g_L(\mathbf{x}) \le g(\mathbf{x})$, that is, we assume that we know the upper bound of $|f(\mathbf{x})|$ and the lower bound of $g(\mathbf{x})$.

In practice, the bounds $f^U(\mathbf{x})$ and $g_L(\mathbf{x})$ usually are not difficult to find because we only require to know the loose bounds, that is, $f^U(\mathbf{x})$ can be very large and $g_L(\mathbf{x})$ can be very small. Also, we require to have state-dependent bounds, which is weaker than requiring fixed bounds.

Before we design the supervisory controller u_s, we need to write the closed-loop system equation into a vector form. First, define

$$u^* = \frac{1}{g(\mathbf{x})}[-f(\mathbf{x}) - \mathbf{k}^T \mathbf{x}] \tag{20.5}$$

where $\mathbf{k} = (k_n, ..., k_1)^T \in R^n$ is such that all roots of the polynomial $s^n + k_1 s^{n-1} + ... + k_n$ are in the left-half complex plane. Using this u^*, we can rewrite (20.4) as

$$x^{(n)} = -\mathbf{k}^T \mathbf{x} + g[u_{fuzz} - u^* + I^* u_s] \tag{20.6}$$

Define

$$\Lambda = \begin{bmatrix} 0 & 1 & 0 & 0 & \cdots & 0 & 0 \\ 0 & 0 & 1 & 0 & \cdots & 0 & 0 \\ \cdots & \cdots & \cdots & \cdots & \cdots & \cdots & \cdots \\ 0 & 0 & 0 & 0 & \cdots & 0 & 1 \\ -k_n & -k_{n-1} & \cdots\cdots\cdots\cdots\cdots & -k_1 \end{bmatrix} \tag{20.7}$$

$$\mathbf{b} = \begin{bmatrix} 0 \\ \cdots \\ 0 \\ g \end{bmatrix} \tag{20.8}$$

then (20.6) can be written into the vector form

$$\dot{\mathbf{x}} = \Lambda \mathbf{x} + \mathbf{b}[u_{fuzz} - u^* + I^* u_s]. \tag{20.9}$$

We now design the supervisory controller u_s to guarantee $|\mathbf{x}| \leq M_x$. Define the Lyapunov function candidate

$$V = \frac{1}{2} \mathbf{x}^T P \mathbf{x} \tag{20.10}$$

where P is a symmetric positive definite matrix satisfying the Lyapunov equation

$$\Lambda^T P + P\Lambda = -Q \tag{20.11}$$

where $Q > 0$ is specified by the designer. Because Λ is stable, such P always exists. Using (20.9) and (20.11) and considering the case $|\mathbf{x}| \geq M_x$, we have

$$\dot{V} = -\frac{1}{2} \mathbf{x}^T Q \mathbf{x} + \mathbf{x}^T P \mathbf{b}[u_{fuzz} - u^* + u_s]$$
$$\leq |\mathbf{x}^T P \mathbf{b}|(|u_{fuzz}| + |u^*|) + \mathbf{x}^T P \mathbf{b} u_s \tag{20.12}$$

Our goal now is to design u_s such that $\dot{V} \leq 0$, that is, the right-hand side of (20.12) is non-positive. Observing (20.12) and (20.5), we choose the u_s as follows:

$$u_s = -sign(\mathbf{x}^T P \mathbf{b}) \left[\frac{1}{g_L} (f^U + |\mathbf{k}^T \mathbf{x}|) + |u_{fuzz}| \right] \tag{20.13}$$

Substituting (20.13) into (20.12) we have $\dot{V} \leq 0$. Therefore, the supervisory controller u_s of (20.13) guarantees that $|\mathbf{x}|$ is decreasing if $|\mathbf{x}| \geq M_x$. Consequently, if we choose the initial $|\mathbf{x}(0)| \leq M_x$, we will have $|\mathbf{x}(t)| \leq M_x$ for all $t \geq 0$. Because $g > 0$ and $\mathbf{x}$ and P are available, $sign(\mathbf{x}^T P \mathbf{b})$ in (20.13) can be computed. Also, all other terms in (20.13) are available, thus the u_s of (20.13) can be implemented on-line.

Because the I^* in (20.3) is a step function, the supervisory controller begins operation as soon as $\mathbf{x}$ hits the boundary $|\mathbf{x}| = M_x$ and is idle as soon as the $\mathbf{x}$ is back to the interior of the constraint set $|\mathbf{x}| \leq M_x$, hence the system may oscillate across the boundary $|\mathbf{x}| = M_x$. One way to overcome this "chattering" problem is to let I^* *continuously* change from 0 to 1. Specifically, we may choose the I^* as follows:

$$I^* = \begin{cases} 0 & |\mathbf{x}| < a \\ \frac{|\mathbf{x}| - a}{M_x - a} & a \leq |\mathbf{x}| < M_x \\ 1 & |\mathbf{x}| \geq M_x \end{cases} \tag{20.14}$$

where $a \in (0, M_x)$ is a parameter specified by the designer. With this I^* in (20.3), the supervisory controller u_s operates continuously from zero to full strength as $\mathbf{x}$ changes from a to M_x. Obviously, this I^* can also guarantee that $|\mathbf{x}| \leq M_x$ (Exercise 20.1).

20.2.2   Application to Inverted Pendulum Balancing

In this subsection, we apply a fuzzy controller together with the supervisory controller to the inverted pendulum balancing problem. The control objective is to balance the inverted pendulum and, at the same time, to guarantee that the state is bounded. The inverted pendulum system is illustrated in Fig. 1.9 and its dynamic equations are

$$\dot{x}_1 = x_2 \tag{20.15}$$

$$\dot{x}_2 = \frac{gsinx_1 - \frac{mlx_2^2cosx_1sinx_1}{m_c+m}}{l(\frac{4}{3} - \frac{mcos^2x_1}{m_c+m})} + \frac{\frac{cosx_1}{m_c+m}}{l(\frac{4}{3} - \frac{mcos^2x_1}{m_c+m})}u, \tag{20.16}$$

where $g = 9.8m/s^2$ is the acceleration due to gravity, m_c is the mass of cart, m is the mass of pole, l is the half length of pole, and u is the applied force (control). We chose $m_c = 1kg$, $m = 0.1kg$, and $l = 0.5m$ in the following simulations. Clearly, (20.15)-(20.16) is in the form of (20.1), thus our approach applies to this system.

Assume that the fuzzy controller u_{fuzz} is constructed from the following four fuzzy IF-THEN rules:

IF x_1 is positive and x_2 is positive, THEN u is negative big (20.17)

IF x_1 is positive and x_2 is negative, THEN u is zero   (20.18)

IF x_1 is negative and x_2 is positive, THEN u is zero   (20.19)

IF x_1 is negative and x_2 is negative, THEN u is positive big (20.20)

where the fuzzy sets "positive," "negative," "negative big," "zero," and "positive big" are characterized by the membership functions

$$\mu_{positive}(x) = \frac{1}{1 + e^{-30x}} \tag{20.21}$$

$$\mu_{negative}(x) = \frac{1}{1 + e^{30x}} \tag{20.22}$$

$$\mu_{negative\ big}(u) = e^{-(u+5)^2} \tag{20.23}$$

$$\mu_{zero}(u) = e^{-u^2} \tag{20.24}$$

$$\mu_{positive\ big}(u) = e^{-(u-5)^2} \tag{20.25}$$

respectively. Using the center average defuzzifier and the product inference engine, we obtain the fuzzy controller u_{fuzz} as follows:

$$u_{fuzz}(\mathbf{x}) = (5\frac{1}{1+e^{30x_1}}\frac{1}{1+e^{30x_2}} - 5\frac{1}{1+e^{-30x_1}}\frac{1}{1+e^{-30x_2}})/(\frac{1}{1+e^{30x_1}}\frac{1}{1+e^{30x_2}}$$

$$+ \frac{1}{1+e^{-30x_1}} \frac{1}{1+e^{30x_2}} + \frac{1}{1+e^{30x_1}} \frac{1}{1+e^{-30x_2}}$$

$$+ \frac{1}{1+e^{-30x_1}} \frac{1}{1+e^{-30x_2}}). \tag{20.26}$$

To design the supervisory controller, we first need to determine the bounds f^U and g_L. For this system, we have

$$|f(x_1, x_2)| = \left| \frac{gsinx_1 - \frac{mlx_2^2 cosx_1 sinx_1}{m_c + m}}{l(\frac{4}{3} - \frac{mcos^2 x_1}{m_c + m})} \right|$$

$$\leq \frac{9.8 + \frac{0.025}{1.1} x_2^2}{\frac{2}{3} - \frac{0.05}{1.1}}$$

$$= 15.78 + 0.0366 x_2^2 := f^U(x_1, x_2). \tag{20.27}$$

If we require that $|x_1| \leq \pi/9$ (we will specify the design parameters such that this requirement is satisfied), then

$$|g(x_1, x_2)| \geq \frac{cos\pi/9}{1.1(\frac{2}{3} + \frac{0.05}{1.1} cos^2 \pi/)} \dot{=} 1.1 := g_L(x_1, x_2). \tag{20.28}$$

Our control objective is to balance the inverted pendulum from arbitrary initial angles $x_1 \in [-\pi/9, \pi/9]$ and at the same time to guarantee $||(x_1, x_2)||_2 \leq \pi/9 \equiv M_x$.

The design parameters are specified as follows: $a = \pi/18$, $k_1 = 2$, $k_2 = 1$ (so that $s^2 + k_1 s + k_2$ is stable) and $Q = diag(10, 10)$. Then, we solve the Lyapunov equation (20.11) and obtain

$$P = \begin{bmatrix} 15 & 5 \\ 5 & 5 \end{bmatrix} \tag{20.29}$$

We simulated three cases: (i) without the supervisory controller, that is, only use the fuzzy controller (20.26), (ii) use the supervisory controller together with the fuzzy controller, and (iii) same as (ii) except that a white Gaussian noise with variance 3 was added to the control u, which represents wind-gusts disturbance. For each case, we simulated the closed-loop system for five initial conditions: $(x_1(0), x_2(0)) = (4^o, 0), (8^o, 0), (12^o, 0), (16^o, 0), (20^o, 0)$. The simulation results for cases (i), (ii) and (iii) are shown in Figs. 20.3, 20.4, and 20.5, respectively, where we show the angle $x_1(t)$ as a function of t for the five initial conditions. We see from these results that: (a) the pure fuzzy controller could balance the inverted pendulum for smaller initial angles $4^o, 8^o$ and 12°, but the system became unstable for larger initial angles 16° and 20°, (b) by appending the supervisory controller to the fuzzy controller, we successfully balanced the inverted pendulum for all the five initial angles and guaranteed that the angle is within $[-20^o, 20^o]$, and (c) the fuzzy controller was robust to random disturbance.

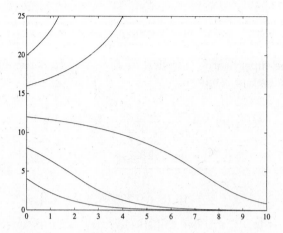

Figure 20.3. The closed-loop system state $x_1(t)$ for the five initial conditions using only the fuzzy controller.

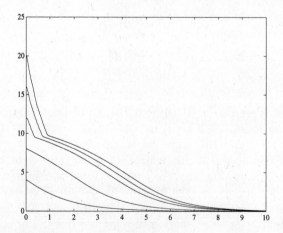

Figure 20.4. The closed-loop system state $x_1(t)$ for the five initial conditions using the fuzzy controller with the supervisory controller.

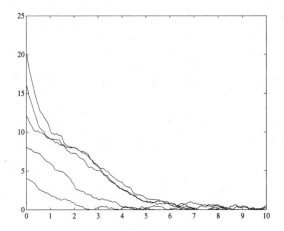

Figure 20.5. The same as Fig. 20.4 except that a white Gaussian noise with variance 3 was added to the control u.

20.3 Gain Scheduling of PID Controller Using Fuzzy Systems

20.3.1 The PID Controller

Due to their simple structure and robust performance, proportional-integral-derivative (PID) controllers are the most commonly used controllers in industrial process control. The transfer function of a PID controller has the following form:

$$G(s) = K_p + K_i/s + K_d s \qquad (20.30)$$

where K_p, K_i and K_d are called the propositional, integral, and derivative gains, respectively. Another equivalent form of the PID controller is

$$u(t) = K_p[e(t) + \frac{1}{T_i}\int_0^t e(\tau)d\tau + T_d \dot{e}(t)] \qquad (20.31)$$

where $T_i = K_p/K_i$ and $T_d = K_d/K_p$ are known as the integral and derivative time constants, respectively.

The success of the PID controller depends on an appropriate choice of the PID gains. Turning the PID gains to optimize performance is not a trivial task. In practice, the PID gains are usually turned by experienced human experts based on some "rule of thumb." In the next subsection, we will first determine a set of turning rules (fuzzy IF-THEN rules) for the PID gains by analyzing a typical response of the system, and then combine these rules into a fuzzy system that is used to adjust the PID gains on-line. We will follow the approach proposed by Zhao, Tomizuka, and Isaka [1993].

20.3.2   A Fuzzy System for Turning the PID Gains

Consider the two-level control system of Fig. 20.2, where the conventional controller is a PID controller in the form of (20.30) (or equivalently (20.31)) and the fuzzy system turns the PID gains in real time. The fuzzy system is constructed from a set of fuzzy IF-THEN rules that describe how to choose the PID gains under certain operation conditions. We first reformulate the problem and then derive the fuzzy IF-THEN rules.

Suppose that we can determine the ranges $[K_{pmin}, K_{pmax}] \subset R$ and $[K_{dmin}, K_{dmax}] \subset R$ such that the proportional gain $K_p \in [K_{pmin}, K_{pmax}]$ and the derivative gain $K_d \in [K_{dmin}, K_{dmax}]$. For convenience, K_p and K_d are normalized to the range between zero and one by the following linear transformation:

$$K'_p = \frac{K_p - K_{pmin}}{K_{pmax} - K_{pmin}} \tag{20.32}$$

$$K'_d = \frac{K_d - K_{dmin}}{K_{dmax} - K_{dmin}} \tag{20.33}$$

Assume that the integral time constant is determined with reference to the derivative time constant by

$$T_i = \alpha T_d \tag{20.34}$$

from which we obtain

$$K_i = K_p/(\alpha T_d) = K_p^2/(\alpha K_d) \tag{20.35}$$

Hence, the parameters to be turned by the fuzzy system are K'_p, K'_d and α. If we can determine these parameters, then the PID gains can be obtained from (20.32), (20.33) and (20.35). Assume that the inputs to the fuzzy system are $e(t)$ and $\dot{e}(t)$, so the fuzzy system turner consists of three two-input-one-output fuzzy systems, as shown in Fig. 20.6. We now derive the fuzzy IF-THEN rules that constitute these fuzzy systems.

Let the fuzzy IF-THEN rules be of the following form:

$$IF \ e(t) \ is \ A^l \ and \ \dot{e}(t) \ is \ B^l, \ THEN \ K'_p \ is \ C^l, \ K'_d \ is \ D^l, \ \alpha \ is \ E^l \tag{20.36}$$

where A^l, B^l, C^l, D^l and E^l are fuzzy sets, and $l = 1, 2, ..., M$. Suppose that the domains of interest of $e(t)$ and $\dot{e}(t)$ are $[e_M^-, e_M^+]$ and $[e_{Md}^-, e_{Md}^+]$, respectively, and we define 7 fuzzy sets, as shown in Fig. 20.7, to cover them. Thus, a complete fuzzy rule base consists of 49 rules. For simplicity, assume that C^l and D^l are either the fuzzy set *big* or the fuzzy set *small* whose membership functions are shown in Fig. 20.8. Finally, assume that E^l can be the four fuzzy sets shown in Fig. 20.9. We are now ready to derive the rules.

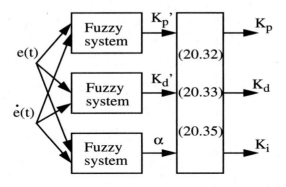

Figure 20.6. Fuzzy system turner for the PID gains.

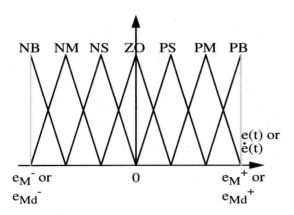

Figure 20.7. Membership functions for $e(t)$ and $\dot{e}(t)$.

Here we derive the rules experimentally based on the typical step response of the process. Fig. 20.10 shows an example of the typical time response. At the beginning, that is, around a_1, a big control signal is needed in order to achieve a

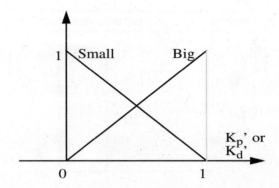

Figure 20.8. Membership functions for K'_p and K'_d.

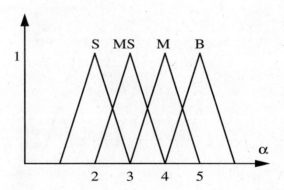

Figure 20.9. Membership functions for α.

fast rise time. To produce a big control signal, we need a large proportional gain K'_p, a small derivative gain K'_d, and a large integral gain. From (20.35) we see that for fixed K_p and K_d, the integral gain is inversely proportional to α, therefore a larger integral gain means a smaller α. Consequently, the rule around a_1 reads

$$IF\ e(t)\ is\ PB\ and\ \dot{e}(t)\ is\ ZO,\ THEN\ K'_p\ is\ Big,\ K'_d\ is\ Small,\ \alpha\ is\ S\ \ (20.37)$$

where the membership functions for the fuzzy sets *PB, ZO, Big, Small,* and *S* are shown in Figs. 20.7-20.9.

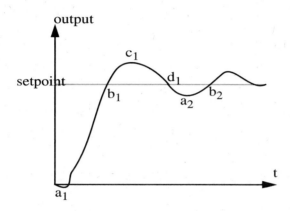

Figure 20.10. The typical step response of process.

Around point b_1 in Fig. 20.10, we expect a small control signal to avoid a large overshoot. So we need a small proportional gain, a large derivative gain, and a small integral gain. Thus, the following rule is taken:

$$IF\ e(t)\ is\ ZO\ and\ \dot{e}(t)\ is\ NB,\ THEN\ K'_p\ is\ Small,\ K'_d\ is\ Big,\ \alpha\ is\ B\ \ (20.38)$$

The control actions around points c_1 and d_1 are similar to those around points a_1 and b_1, respectively. Using this kind of idea, we can determine three sets of rules for K'_p, K'_d and α, and each set consists of 49 rules. These three sets of rules are shown in Figs. 20.11-20.13, respectively.

We combine the 49 rules in each set using product inference engine, singleton fuzzifier, and center average defuzzifier; that is, the parameters K'_p, K'_d and α are turned on-line according to

$$K'_p(t) = \frac{\sum_{l=1}^{49} \bar{y}_p^l \mu_{A^l}(e(t))\mu_{B^l}(\dot{e}(t))}{\sum_{l=1}^{49} \mu_{A^l}(e(t))\mu_{B^l}(\dot{e}(t))} \tag{20.39}$$

$$K'_d(t) = \frac{\sum_{l=1}^{49} \bar{y}_d^l \mu_{A^l}(e(t))\mu_{B^l}(\dot{e}(t))}{\sum_{l=1}^{49} \mu_{A^l}(e(t))\mu_{B^l}(\dot{e}(t))} \tag{20.40}$$

		$\dot{e}(t)$						
		NB	NM	NS	ZO	PS	PM	PB
$e(t)$	NB	B	B	B	B	B	B	B
	NM	S	B	B	B	B	B	S
	NS	S	S	B	B	B	S	S
	ZO	S	S	S	B	S	S	S
	PS	S	S	B	B	B	S	S
	PM	S	B	B	B	B	B	S
	PB	B	B	B	B	B	B	B

Figure 20.11. Fuzzy turning rules for K_p'.

		$\dot{e}(t)$						
		NB	NM	NS	ZO	PS	PM	PB
$e(t)$	NB	S	S	S	S	S	S	S
	NM	B	B	S	S	S	B	B
	NS	B	B	B	S	B	B	B
	ZO	B	B	B	B	B	B	B
	PS	B	B	B	S	B	B	B
	PM	B	B	S	S	S	B	B
	PB	S	S	S	S	S	S	S

Figure 20.12. Fuzzy turning rules for K_d'.

$$\alpha(t) = \frac{\sum_{l=1}^{49} \bar{y}_\alpha^l \mu_{A^l}(e(t)) \mu_{B^l}(\dot{e}(t))}{\sum_{l=1}^{49} \mu_{A^l}(e(t)) \mu_{B^l}(\dot{e}(t))} \qquad (20.41)$$

		$\dot{e}(t)$						
		NB	NM	NS	ZO	PS	PM	PB
	NB	2	2	2	2	2	2	2
	NM	3	3	2	2	2	3	3
	NS	4	3	3	2	3	3	4
$e(t)$	ZO	5	4	3	3	3	4	5
	PS	4	3	3	2	3	3	4
	PM	3	3	2	2	2	3	3
	PB	2	2	2	2	2	2	2

Figure 20.13. Fuzzy turning rules for α.

where A^l and B^l are shown in Fig. 20.7, and $\bar{y}_p^l, \bar{y}_d^l$ and $\bar{y}_\alpha^l$ are the centers of the corresponding fuzzy sets in Figs. 20.8 and 20.9 (according to the rules in Figs. 20.11-20.13).

Simulation results and comparison with classical methods can be found in Zhao, Tomizuka, and Isaka [1993].

20.4 Summary and Further Readings

In this chapter we have demonstrated the following:

- The structure and working principles of two-level fuzzy control.

- How to design the nonfuzzy supervisory controller to guarantee the global stability of the fuzzy control system.

- How to derive fuzzy IF-THEN rules for turning the PID gains based on heuristic analysis of the typical step response.

Multi-level control has been studied in the field of intelligent robotics, see Valavanis and Saridis [1992] and Tzafestas [1991]. The supervisory control idea in this chapter was proposed in Wang [1994b]. Using fuzzy systems to turn the PID parameters was studied by a number of researchers and the approach in this chapter was due to Zhao, Tomizuka, and Isaka [1993].

20.5   Exercises

Exercise 20.1. Show that if we use the continuous indicator function I^* of (20.14) in the controller (20.3) with the u_s given by (20.13), we can still guarantee that $|\mathbf{x}(t)| \leq M_x$ for all $t \geq 0$ if $|\mathbf{x}(0)| \leq M_x$. Furthermore, show that in this case there exists $M_x^* < M_x$ such that $|\mathbf{x}(t)| \leq M_x^*$ for all $t \geq 0$ if $|\mathbf{x}(0)| \leq M_x^*$.

Exercise 20.2. Discuss the similarities and differences between the supervisory controller in this chapter and the sliding controller in Chapter 19.

Exercise 20.3. Repeat the simulations in Subsection 20.2.2 with slightly different fuzzy control rules and different initial conditions.

Exercise 20.4. In the PID controller (20.30), $K_p, K_i/s$, and $K_d s$ are called proportional, integral, and derivative modes, respectively. Use examples to show that:

(a) The proportional mode provides a rapid adjustment of the manipulated variable, does not provide zero steady-atate offset although it reduces the error, speeds up dynamic response, and can cause instability if tuned improperly.

(b) The integral mode achieves zero steady-state offset, adjusts the manipulated variable in a slower manner than the proportional mode, and can cause instability if tuned improperly.

(c) The derivative mode does not influence the final steady-state value of error, provides rapid correction based on the rate of change of the controlled variable, and can cause undesirable high-frequency variation in the manipulated variable.

Exercise 20.5. Consider the two-level fuzzy control system in Fig. 20.2, where the process is given by

$$G(s) = \frac{27}{(s+1)(s+3)^3} \tag{20.42}$$

the conventional controller is the PID controller, and the fuzzy systems are given by (20.39)-(20.41). Simulate this system and plot the process output. You may choose any reasonable values for $K_{pmin}, K_{pmax}, K_{dmin}, K_{dmax}, e_M^-, e_M^+, e_{Md}^-$ and e_{Md}^+.

Chapter 21

Fuzzy Control of Fuzzy System Model

In Chapters 17-20, we studied the fuzzy control systems where the processes under control are represented by ordinary linear or nonlinear dynamic system models. In many practical problems, human experts may provide linguistic descriptions (in terms of fuzzy IF-THEN rules) about the process that can be combined into a model of the process; this model is called a *fuzzy system model*. Therefore, it is interesting to study the fuzzy control system in which the process is modeled by fuzzy systems and the feedback controller is a fuzzy controller; this is the topic of this chapter.

We will first introduce the Takagi-Sugeno-Kang (TSK) fuzzy system and derive the detailed formula of the closed-loop system in which the process and the controller are represented by the TSK fuzzy system. Then, we will analyze the stability of the closed-loop system. Finally, a design procedure for stable fuzzy controllers will be introduced.

21.1   The Takagi-Sugeno-Kang Fuzzy System

The Takagi-Sugeno-Kang (TSK) fuzzy system was proposed as an alternative to the fuzzy systems we have been using in most parts of this book. The TSK fuzzy system is constructed from the following rules:

$$IF \ x_1 \ is \ C_1^l \ and \ \cdots \ and \ x_n \ is \ C_n^l, \ THEN \ y^l = c_0^l + c_1^l x_1 + \cdots + c_n^l x_n \quad (21.1)$$

where C_i^l are fuzzy sets, c_i^l are constants, and $l = 1, 2, ..., M$. That is, the IF parts of the rules are the same as in the ordinary fuzzy IF-THEN rules, but the THEN parts are linear combinations of the input variables. Given an input $x = (x_1, ..., x_n)^T \in U \subset R^n$, the output $f(x) \in V \subset R$ of the TSK fuzzy system is computed as the weighted average of the y^l's in (21.1), that is,

$$f(x) = \frac{\sum_{l=1}^{M} y^l w^l}{\sum_{l=1}^{M} w^l} \quad (21.2)$$

where the weights w^l are computed as

$$w^l = \prod_{i=1}^{n} \mu_{C_i^l}(x_i) \tag{21.3}$$

We see that the TSK fuzzy system is still a mapping from $U \subset R^n$ to $V \subset R$. The physical meaning of the rule (21.1) is that when x is constrained to the fuzzy range characterized by the IF part of the rule, the output is a linear function of the input variables. Therefore, the TSK fuzzy system can be viewed as a somewhat piece-wise linear function, where the change from one piece to the other is smooth rather than abrupt. If $c_i^l = 0$ for $i = 1, 2, ..., n$ and c_0^l equals the center $\bar{y}^l$ of the fuzzy set B^l in the ordinary fuzzy IF-THEN rule (7.1), then the TSK fuzzy system is identical to the fuzzy system with product inference engine, singleton fuzzifier, and center average defuzzifier (comparing (21.2)-(21.3) with (9.1)).

If the output of a TSK fuzzy system appears as one of its inputs, we obtain the so-called dynamic TSK fuzzy system. Specifically, a *dynamic TSK fuzzy system* is constructed from the following rules:

$$IF\ x(k)\ is\ A_1^p\ and\ \cdots\ and\ x(k-n+1)\ is\ A_n^p\ and\ u(k)\ is\ B^p$$
$$THEN\ x^p(k+1) = a_1^p x(k) + \cdots + a_n^p x(k-n+1) + b^p u(k) \tag{21.4}$$

where A_i^p and B^p are fuzzy sets, a_i^p and b^p are constants, $p = 1, 2, ..., N$, $u(k)$ is the input to the system, and $\mathbf{x}(k) = (x(k), x(k-1), ..., x(k-n+1))^T \in R^n$ is the state vector of the system. The output of the dynamic TSK fuzzy system is computed as

$$x(k+1) = \frac{\sum_{p=1}^{N} x^p(k+1) v^p}{\sum_{p=1}^{N} v^p} \tag{21.5}$$

where $x^p(k+1)$ is given in (21.4) and

$$v^p = \prod_{i=1}^{n} \mu_{A_i^p}[x(k-i+1)] \mu_{B^p}[u(k)] \tag{21.6}$$

We will use this dynamic TSK fuzzy system to model the process under control.

21.2 Closed-Loop Dynamics of Fuzzy Model with Fuzzy Controller

Consider the feedback control system in Fig. 21.1, where the process under control is modeled by the dynamic TSK fuzzy model (21.5), and the controller is the TSK fuzzy system (21.2) with $c_0^l = 0$ and $x_i = x(k-i+1)$ for $i = 1, 2, ..., n$. The following theorem gives the closed-loop dynamics of the control system in Fig. 21.1.

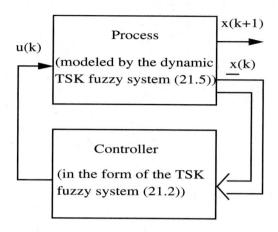

Figure 21.1. Fuzzy control of fuzzy system model.

Theorem 21.1. The closed-loop fuzzy control system in Fig. 21.1 is equivalent to the dynamic TSK fuzzy system constructed from the following rules:

$IF\ x(k)\ is\ (C_1^l\ and\ A_1^p)\ and\ \cdots\ and\ x(k-n+1)\ is\ (C_n^l\ and\ A_n^p)\ and\ u(k)\ is\ B^p,$

$$THEN\ x^{lp}(k+1) = \sum_{i=1}^{n}(a_i^p + b^p c_i^l)x(k-i+1) \tag{21.7}$$

where $u(k)$ is the output of the controller, $l = 1, 2, ..., M$, $p = 1, 2, ..., N$, and the fuzzy sets $(C_i^l\ and\ A_i^p)$ are characterized by the membership functions $\mu_{C_i^l}(x(k-i+1))\mu_{A_i^p}(x(k-i+1))$. The output of this dynamic TSK fuzzy system is computed as

$$x(k+1) = \frac{\sum_{l=1}^{M}\sum_{p=1}^{N}x^{lp}(k+1)w^l v^p}{\sum_{l=1}^{M}\sum_{p=1}^{N}w^l v^p} \tag{21.8}$$

where

$$w^l = \prod_{i=1}^{n}\mu_{C_i^l}(x(k-i+1)) \tag{21.9}$$

$$v^p = \prod_{i=1}^{n}\mu_{A_i^p}(x(k-i+1))\mu_{B^p}(u(k)) \tag{21.10}$$

Proof: From Fig. 21.1 we see that the $u(k)$ in (21.4) equals the $f(x)$ of (21.2). Hence, $x^p(k+1)$ in (21.4) becomes

$$x^p(k+1) = a_1^p x(k) + \cdots + a_n^p x(k-n+1) + b^p \frac{\sum_{l=1}^{M} [c_1^l x(k) + \cdots + c_n^l x(k-n+1)]w^l}{\sum_{l=1}^{M} w^l}$$

$$= \frac{\sum_{l=1}^{M} [\sum_{i=1}^{n} (a_i^p + b^p c_i^l)x(k-i+1)]w^l}{\sum_{l=1}^{M} w^l} \tag{21.11}$$

Substituting (21.11) into (21.5), we obtain the output of the closed-loop system

$$x(k+1) = \frac{\sum_{l=1}^{M} \sum_{p=1}^{N} [\sum_{i=1}^{n} (a_i^p + b^p c_i^l)x(k-i+1)]w^l v^p}{\sum_{l=1}^{M} \sum_{p=1}^{N} w^l v^p} \tag{21.12}$$

which is (21.8). $\square$

Example 21.1. Suppose that the process in Fig. 21.1 is modeled by a second-order dynamic TSK fuzzy system that is constructed from the following two rules:

$L^1:$ IF $x(k)$ is A_1^1 and $x(k-1)$ is A_2^1 and $u(k)$ is B^1,
          $THEN$ $x^1(k+1) = 1.5x(k) + 2.1x(k-1) - u(k)$ $\qquad$ (21.13)

$L^2:$ IF $x(k)$ is A_1^2 and $x(k-1)$ is A_2^2 and $u(k)$ is B^2,
          $THEN$ $x^2(k+1) = 0.3x(k) - 3.4x(k-1) + 0.5u(k)$ $\qquad$ (21.14)

and that the controller in Fig. 21.1 is a TSK fuzzy system constructed from the following two rules:

$R^1:$ IF $x(k)$ is C_1^1 and $x(k-1)$ is C_2^1,
          $THEN$ $u^1(k) = k_1^1 x(k) + k_2^1 x(k-1)$ $\qquad$ (21.15)

$R^2:$ IF $x(k)$ is C_1^2 and $x(k-1)$ is C_2^2,
          $THEN$ $u^2(k) = k_1^2 x(k) + k_2^2 x(k-1)$ $\qquad$ (21.16)

Then from Theorem 21.1 we have that the closed-loop system is a dynamic TSK fuzzy system constructed from the following four rules:

$S^{11}:$ IF $x(k)$ is $(A_1^1$ and $C_1^1)$ and $x(k-1)$ is $(A_2^1$ and $C_2^1)$ and $u(k)$ is B^1,
            $THEN$ $x^{11}(k+1) = (1.5 - k_1^1)x(k) + (2.1 - k_2^1)x(k-1)$ $\qquad$ (21.17)

$S^{12}:$ IF $x(k)$ is $(A_1^1$ and $C_1^2)$ and $x(k-1)$ is $(A_2^1$ and $C_2^2)$ and $u(k)$ is B^1,
            $THEN$ $x^{12}(k+1) = (1.5 - k_1^2)x(k) + (2.1 - k_2^2)x(k-1)$ $\qquad$ (21.18)

S^{21} :   IF $x(k)$ is $(A_1^2$ and $C_1^1)$ and $x(k-1)$ is $(A_2^2$ and $C_2^1)$ and $u(k)$ is B^2,

THEN $x^{21}(k+1) = (0.3 + 0.5k_1^1)x(k) + (-3.4 + 0.5k_2^1)x(k-1)$ (21.19)

S^{22} :   IF $x(k)$ is $(A_1^2$ and $C_1^2)$ and $x(k-1)$ is $(A_2^2$ and $C_2^2)$ and $u(k)$ is B^2,

THEN $x^{22}(k+1) = (0.3 + 0.5k_1^2)x(k) + (-3.4 + 0.5k_2^2)x(k-1)$ (21.20)

The dynamic equation can be obtained according to (21.8)-(21.10). □

Since the closed-loop fuzzy control system in Fig. 21.1 is equivalent to a dynamic TSK fuzzy system, it is therefore important to study the stability of the dynamic TSK fuzzy system; this is the topic of the next section.

21.3   Stability Analysis of the Dynamic TSK Fuzzy System

Consider the dynamic TSK fuzzy system (21.5) with b^p in (21.4) equal zero. We assume $b^p = 0$ because comparing (21.4) with (21.7) we see that there is no $b^p u(k)$ term in (21.7). Define the state vector $\mathbf{x}(k) = (x(k), x(k-1), ..., x(k-n+1))^T$ and

$$A_p = \begin{bmatrix} a_1^p & a_2^p & \cdots & a_{n-1}^p & a_n^p \\ 1 & 0 & \cdots & 0 & 0 \\ 0 & 1 & \cdots & 0 & 0 \\ & & \cdots & & \\ 0 & 0 & \cdots & 1 & 0 \end{bmatrix} \qquad (21.21)$$

Then the dynamic TSK fuzzy system (21.5) can be rewritten as

$$\mathbf{x}(k+1) = \frac{\sum_{p=1}^N A_p \mathbf{x}(k) v^p}{\sum_{p=1}^N v^p} \qquad (21.22)$$

where v^p is defined in (21.6). Since the right-hand side of (21.22) equals zero when $\mathbf{x}(k) = 0$, the origin in R^n is an equilibrium point of the dynamic system (21.22). We now use the following well-known Lyapunov stability theorem to study the stability of the dynamic system (21.22)

Lyapunov Stability Theorem: Consider the discrete-time system described by

$$\mathbf{x}(k+1) = f[\mathbf{x}(k)] \qquad (21.23)$$

where $\mathbf{x}(k) \in R^n$ and $f(0) = 0$. Suppose that there exists a scalar function $V[\mathbf{x}(k)]$ such that: (a) $V(0) = 0$, (b) $V[\mathbf{x}(k)] > 0$ for $\mathbf{x}(k) \neq 0$, (c) $V[\mathbf{x}(k)] \to \infty$ as $||\mathbf{x}(k)|| \to \infty$, and (d) $\Delta V[\mathbf{x}(k)] = V[\mathbf{x}(k+1)] - V[\mathbf{x}(k)] < 0$ for $\mathbf{x}(k) \neq 0$, then the equilibrium point 0 of the system (21.23) is globally asymptotically stable.

In order to apply this theorem to the system (21.22), we need the following lemma.

Matrix Inequality Lemma: If P is a positive definite matrix such that

$$A^T PA - P < 0 \ and \ B^T PB - P < 0 \tag{21.24}$$

where $A, B, P \in R^{n \times n}$, then

$$A^T PB + B^T PA - 2P < 0 \tag{21.25}$$

Proof of this lemma is left as an exercise. Using the Lyapunov Stability Theorem and the Matrix Inequality Lemma, we obtain the following theorem on the stability of the dynamic TSK fuzzy system (21.22).

Theorem 21.2. The equilibrium point 0 of the dynamic TSK fuzzy system (21.22) is globally asymptotically stable if there exists a common positive definite matrix P such that

$$A_p^T PA_p - P < 0 \tag{21.26}$$

for all $p = 1, 2, ..., N$.

Proof: Consider the Lyapunov function candidate

$$V[\mathbf{x}(k)] = \mathbf{x}^T(k)P\mathbf{x}(k) \tag{21.27}$$

where P is a positive definite matrix. This $V[\mathbf{x}(k)]$ satisfies conditions (a)-(c) in the Lyapunov Stability Theorem; we now show that it also satisfies condition (d). Using (21.22), we have

$$\Delta V[\mathbf{x}(k)] = \mathbf{x}^T(k+1)P\mathbf{x}(k+1) - \mathbf{x}^T(k)P\mathbf{x}(k)$$

$$= \mathbf{x}^T(k) \left[(\frac{\sum_{p=1}^{N} A_p^T v^p}{\sum_{p=1}^{N} v^p})P(\frac{\sum_{p=1}^{N} A_p v^p}{\sum_{p=1}^{N} v^p}) - P \right] \mathbf{x}(k)$$

$$= \frac{\sum_{p=1}^{N} \sum_{q=1}^{N} v^p v^q \mathbf{x}^T(k)(A_p^T PA_q - P)\mathbf{x}(k)}{\sum_{p=1}^{N} \sum_{q=1}^{N} v^p v^q} \tag{21.28}$$

Rearrange the summations, (21.28) becomes

$$\Delta V[\mathbf{x}(k)] = [\sum_{p=1}^{N} (v^p)^2 \mathbf{x}^T(k)(A_p^T PA_p - P)\mathbf{x}(k) + \sum_{p<q}^{N} v^p v^q \mathbf{x}^T(k)(A_p^T PA_p$$

$$+ A_q^T PA_p - 2P)\mathbf{x}(k)]/[\sum_{p=1}^{N} \sum_{q=1}^{N} v^p v^q] \tag{21.29}$$

From (21.26), the Matrix Inequality Lemma, and the fact that $v^p \geq 0$, we conclude that $\Delta V[\mathbf{x}(k)] < 0$. Hence, this theorem follows from the Lyapunov Stability Theorem. □

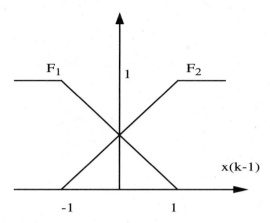

Figure 21.2. Membership functions for the fuzzy sets in the rules (21.30)-(21.31).

Theorem 21.2 gives a sufficient condition for ensuring the stability of the dynamic TSK fuzzy system (21.22). We may intuitively guess that the nonlinear system (21.22) is stable if all locally approximate linear systems A_p ($p = 1, 2, ..., N$) are stable (the linear system $\mathbf{x}(k + 1) = A_p\mathbf{x}(k)$ is stable if all the eigenvalues of A_p are within the unit circle). However, this is not true in general, as we notice that even if all the $A'_p s$ are stable, there may not exist a *common* positive definite matrix P such that (21.26) is true. The following example shows that two locally stable linear systems result in an unstable nonlinear system.

Example 21.2. Consider a dynamic TSK fuzzy system constructed from the following two rules:

$$IF\ x(k-1)\ is\ F_1,\ THEN\ x^1(k+1) = x(k) - 0.5x(k-1) \quad (21.30)$$
$$IF\ x(k-1)\ is\ F_2,\ THEN\ x^2(k+1) = -x(k) - 0.5x(k-1) \quad (21.31)$$

where the membership functions of the fuzzy sets F_1 and F_2 are shown in Fig. 21.2. For this system, we have

$$A_1 = \begin{bmatrix} 1 & -0.5 \\ 1 & 0 \end{bmatrix}, A_2 = \begin{bmatrix} -1 & -0.5 \\ 1 & 0 \end{bmatrix} \quad (21.32)$$

The eigenvalues of A_1 and A_2 are $\frac{1\pm j}{2}$ and $\frac{-1\pm j}{2}$, respectively, which are all within the unit circle, so the two locally approximate linear systems $\mathbf{x}(k+1) = A_1\mathbf{x}(k)$ and $\mathbf{x}(k+1) = A_2\mathbf{x}(k)$ are stable. However, the dynamic TSK fuzzy system constructed from the two rules (21.30) and (21.31)

$$\mathbf{x}(k+1) = \frac{A_1\mathbf{x}(k)\mu_{F_1}(x(k-1)) + A_2\mathbf{x}(k)\mu_{F_2}(x(k-1))}{\mu_{F_1}(x(k-1)) + \mu_{F_2}(x(k-1))} \tag{21.33}$$

is unstable, as illustrated in Fig. 21.3, which shows the $x(k)$ resulting from (21.33) with initial condition $\mathbf{x}(1) = (x(1), x(0))^T = (-1.7, 1.9)^T$. $\square$

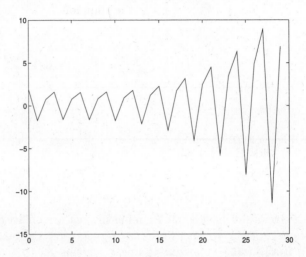

Figure 21.3. Response of the dynamic TSK fuzzy system (21.33) $(x(k))$ with initial condition $\mathbf{x}(1) = (x(1), x(0))^T = (-1.7, 1.9)^T$.

Obviously, in Example 21.2 there does not exist a common P such that (21.26) is true, since the final dynamic TSK fuzzy system is unstable. We now give a necessary condition for ensuring the existence of the common P.

Lemma 21.1. Assume that A_p $(p = 1, 2, ..., N)$ are stable and nonsingular matrices. If there exists a common positive matrix P such that $A_p^T P A_p - P < 0$ for $p = 1, 2, ..., N$, then $A_p A_q$ are stable matrices for all $p, q = 1, 2, ..., N$.

Proof of this lemma is left as an exercise. Lemma 21.1 shows that if any $A_p A_q$ is an unstable matrix, then the common P does not exist and therefore it is possible that the dynamic TSK fuzzy system is unstable. For Example 21.2, we have

$$A_1 A_2 = \begin{bmatrix} -1.5 & -0.5 \\ -1 & -0.5 \end{bmatrix} \tag{21.34}$$

whose eigenvalues are $\frac{-2\pm\sqrt{3}}{2}$ and one of which is outside of the unit circle.

There is no general procedure that guarantees to find such common P. Usually, a trial-and-error approach has to be taken, as we will show in the next section.

21.4   Design of Stable Fuzzy Controllers for the Fuzzy Model

In Theorem 21.2, a sufficient condition was given to ensure the stability of the dynamic TSK fuzzy system. In Theorem 21.1, it was proven that the closed-loop fuzzy control system in Fig.21.1 can be represented as a dynamic TSK fuzzy system. Therefore, we can use Theorem 21.2 to design stable fuzzy controllers for the fuzzy system model; this is the topic of this section.

Since there is no systematic way to find the common P in Theorem 21.2, our design procedure, which is given below, has to be trial and error in nature.

Design of Stable Fuzzy Controller for Fuzzy System Model:

- **Step 1**. Use Theorem 20.1 to represent the closed-loop fuzzy control system as a dynamic TSK fuzzy system. The parameters a_i^p and b^p and the membership functions $\mu_{A_i^p}$ for the process are known, and those for the controller (that is, c_i^l and $\mu_{C_i^l}$) are to be designed. Usually, fix the $\mu_{C_i^l}$ and design the c_i^l according to Theorem 21.2.

- **Step 2**. Choose the parameters c_i^l such that all the locally approximate linear systems are stable, where, according to (21.7), the locally approximate linear systems are $\mathbf{x}(k+1) = A_{lp}\mathbf{x}(k)$ with

$$
A_{lp} = \begin{bmatrix}
a_1^p + b^p c_1^l & a_2^p + b^p c_2^l & \cdots & a_{n-1}^p + b^p c_{n-1}^l & a_n^p + b^p c_n^l \\
1 & 0 & \cdots & 0 & 0 \\
0 & 1 & \cdots & 0 & 0 \\
& & \cdots & & \\
0 & 0 & \cdots & 1 & 0
\end{bmatrix} \tag{21.35}
$$

where $l = 1, 2, ..., M$, and $p = 1, 2, ..., N$.

- **Step 3**. Find positive definite matrices P_{lp} such that

$$
A_{lp}^T P_{lp} A_{lp} - P_{lp} < 0 \tag{21.36}
$$

for $l = 1, 2, ..., M$ and $p = 1, 2, ..., N$. If there exists $P_{l^*p^*}$ for some fixed $l^* \in \{1, 2, ..., M\}$ and $p^* \in \{1, 2, ..., N\}$ such that

$$
A_{lp}^T P_{l^*p^*} A_{lp} - P_{l^*p^*} < 0 \tag{21.37}
$$

for all $l = 1, 2, ..., M$ and $p = 1, 2, ..., N$, then select this $P_{l^*p^*}$ as the common P; otherwise, go to Step 2 to redesign the parameters c_i^l until the common $P = P_{l^*p^*}$ is found.

Since there is much freedom in choosing the parameters c_i^l in Step 2, it is possible that we can find the common P after several iterations. Of course, we cannot guarantee that such common P can be found. We now test the design procedure through an example.

Example 21.3. Suppose that the process is modeled by a dynamic TSK fuzzy system with the following two rules:

$$IF\ x(k)\ is\ G_1,\ THEN\ x^1(k+1) = 2.18x(k) - 0.59x(k-1)$$
$$-0.603u(k) \tag{21.38}$$
$$IF\ x(k)\ is\ G_2,\ THEN\ x^2(k+1) = 2.26x(k) - 0.36x(k-1)$$
$$-1.120u(k) \tag{21.39}$$

and the controller is a TSK fuzzy system constructed from the two rules:

$$IF\ x(k)\ is\ G_1,\ THEN\ u^1(k) = c_1^1 x(k) + c_2^1 x(k-1) \tag{21.40}$$
$$IF\ x(k)\ is\ G_2,\ THEN\ u^2(k) = c_1^2 x(k) + c_2^2 x(k-1) \tag{21.41}$$

where the membership functions for G_1 and G_2 are shown in Fig. 21.4, and the controller parameters c_1^1, c_2^1, c_1^2 and c_2^2 are to be designed such that the closed-loop system is stable.

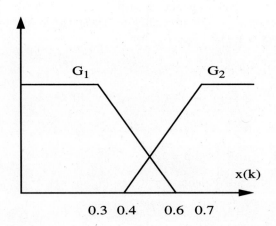

Figure 21.4. Membership functions for the fuzzy sets in Example 21.3.

Step 1. From Theorem 21.1 we obtain that the closed-loop system is a dynamic TSK fuzzy system constructed from the following four rules:

$$IF\ x(k)\ is\ (G_1\ and\ G_2),\ THEN\ x^{11}(k+1) = (2.18 - 0.603c_1^1)x(k)$$
$$+(-0.59 - 0.603c_2^1)x(k-1) \tag{21.42}$$
$$IF\ x(k)\ is\ (G_1\ and\ G_2),\ THEN\ x^{12}(k+1) = (2.18 - 0.603c_1^2)x(k)$$
$$+(-0.59 - 0.603c_2^2)x(k-1) \tag{21.43}$$

$$IF\ x(k)\ is\ (G_1\ and\ G_2),\ THEN\ x^{21}(k+1) = (2.26 - 1.120c_1^1)x(k)$$
$$+(-0.36 - 1.120c_2^1)x(k-1) \qquad (21.44)$$
$$IF\ x(k)\ is\ (G_1\ and\ G_2),\ THEN\ x^{22}(k+1) = (2.26 - 1.120c_1^2)x(k)$$
$$+(-0.36 - 1.120c_2^2)x(k-1) \qquad (21.45)$$

Step 2. The four matrices A_{lp} for the four linear subsystems (21.42)-(21.45) are

$$A_{11} = \begin{bmatrix} 2.18 - 0.603c_1^1 & -0.59 - 0.603c_2^1 \\ 1 & 0 \end{bmatrix} \qquad (21.46)$$

$$A_{12} = \begin{bmatrix} 2.18 - 0.603c_1^2 & -0.59 - 0.603c_2^2 \\ 1 & 0 \end{bmatrix} \qquad (21.47)$$

$$A_{21} = \begin{bmatrix} 2.26 - 1.120c_1^1 & -0.36 - 1.120c_2^1 \\ 1 & 0 \end{bmatrix} \qquad (21.48)$$

$$A_{22} = \begin{bmatrix} 2.26 - 1.120c_1^2 & -0.36 - 1.120c_2^2 \\ 1 & 0 \end{bmatrix} \qquad (21.49)$$

After much trial and error, we found that if we choose

$$c_1^1 = 1.564,\ c_2^1 = -0.223,\ c_1^2 = 0.912,\ c_2^2 = 0.079 \qquad (21.50)$$

then all the four linear subsystems are stable.

Step 3. We found that a common P can be found for the parameters in (21.50). Therefore, our stable fuzzy controller is the TSK fuzzy system constructed from the two rules (21.40)-(21.41) with the parameters given by (21.50). $\square$

21.5   Summary and Further Readings

In this chapter we have demonstrated the following:

- The static and dynamic TSK fuzzy systems.

- The dynamic equation of the closed-loop system in which the process is modeled by a dynamic TSK fuzzy system and the controller is a static TSK fuzzy system.

- Conditions that ensure the stability of the above closed-loop system.

- How to design the fuzzy controller such that the closed-loop system above is stable.

Using the TSK fuzzy systems to model nonlinear systems was studied in Takagi and Sugeno [1985] and Sugeno and Kang [1988]. The stability analysis in this chapter was due to Tanaka and Sugeno [1992] where more examples can be found.

21.6   Exercises

Exercise 21.1. Prove the Matrix Inequality Lemma.

Exercise 21.2. If we change condition (d) in the Lyapunov Stability Theorem to $\Delta V[x(k)] = V[x(k+1)] - V[x(k)] < 0$ for $x(k)$ in a neighborhood of 0, then the conclusion becomes that the system is asymptotically stable. Prove that the system

$$x_1(k+1) = x_1(k)[x_1^2(k) + x_2^2(k)] \tag{21.51}$$
$$x_2(k+1) = x_2(k)[x_1^2(k) + x_2^2(k)] \tag{21.52}$$

is asymptotically stable.

Exercise 21.3. Prove Lemma 21.1.

Exercise 21.4. Consider the dynamic TSK fuzzy system (21.22) with $N = 2$. Suppose that A_1 and A_2 are stable and $A_1A_2 = A_2A_1$. Given $P_0 > 0$, determine the matrices P_1 and P_2 from

$$A_1^T P_1 A_1 - P_1 = -P_0 \tag{21.53}$$
$$A_2^T P_2 A_2 - P_2 = -P_1 \tag{21.54}$$

Prove that P_2 is the common P in Theorem 21.2.

Exercise 21.5. Generalize the procedure of finding the common P in Exercise 21.4 to arbitrary N.

Exercise 21.6. Suppose that the process under control is modeled by a dynamic TSK fuzzy system constructed from the rules (21.38) and (21.39). Design a linear controller $u(k) = Kx(k)$ (that is, determine the constant K) such that the closed-loop system is stable.

Exercise 21.7. Repeat Exercise 21.6, with the fuzzy sets G_1 and G_2 replaced by the fuzzy sets F_1 and F_2 in Fig. 21.2, respectively.

Chapter 22

Qualitative Analysis of Fuzzy Control and Hierarchical Fuzzy Systems

22.1 Phase Plane Analysis of Fuzzy Control Systems

Phase plane analysis is a graphical method for studying second-order systems. The basic idea is to generate motion trajectories of the system in the state space corresponding to various initial conditions, and then to examine the qualitative features of the trajectories. In this way, information concerning stability, robustness and other properties of the system can be obtained. The state space for second-order systems is called the *phase plane*.

The main advantage of phase plane analysis is its graphical nature, which allows us to visualize what goes on in a nonlinear system starting from different initial conditions, without having to solve the nonlinear equations analytically. The main disadvantage of the method is that it is restricted to second-order systems and is difficult to generalize to higher-order systems.

We now use the phase plane method to study second-order fuzzy control systems. Let $x = (x_1, x_2)^T$ be the state and consider the fuzzy control system

$$\dot{x} = f(x) + bu \tag{22.1}$$

$$u = \Phi(x) \tag{22.2}$$

where $f(x)$ is a nonlinear vector function (vector field) representing the plant dynamic, b is a two-dimensional vector, u is a scalar control variable, and $\Phi(x)$ is a two-input-one-output fuzzy system. To study this closed-loop fuzzy control system, it is helpful to identify which rules are firing along a certain trajectory. Suppose that we define N_1 and N_2 fuzzy sets to cover the domains of x_1 and x_2, respectively, and that the fuzzy rule base of $\Phi(x)$ consists of $N_1 \times N_2$ rules. Let the $l'th$ rule in the fuzzy rule base be

$$IF\ x_1\ is\ A_1^l\ and\ x_2\ is\ A_2^l,\ THEN\ y\ is\ B^l \tag{22.3}$$

Then we say that the point (x_1, x_2) in the phase plane *belongs to rule l** if it holds that

$$\mu_{A_1^{l*}}(x_1) \star \mu_{A_2^{l*}}(x_2) \geq \mu_{A_1^l}(x_1) \star \mu_{A_2^l}(x_2) \tag{22.4}$$

for all $l \neq l*$, where $\star$ represents t-norm.

Consider the fuzzy rule base in Fig.22.1, where $N_1 = N_2 = 5$ and the fuzzy rule base contains 25 rules. As shown in Fig.22.1, the state space of interest is partitioned into 25 regions, with each region belonging to a rule. For a given initial condition, the closed-loop system trajectory of (22.1)-(22.2) can be mapped onto the partitioned state space of Fig. 22.1. A trajectory corresponds to a sequence of rules that fire along the trajectory; this sequence of rules is called the *linguistic trajectory*, corresponding to the state trajectory. For the trajectory in Fig. 22.1, the corresponding linguistic trajectory is

$$Linguistic\ trajectory = (rule16,\ rule21,\ rule22,\ rule23,\ rule24,\ rule19,\ rule14,$$
$$rule9,\ rule4,\ rule3,\ rule2,\ rule7,\ rule12,\ rule17,$$
$$rule18,\ rule13) \tag{22.5}$$

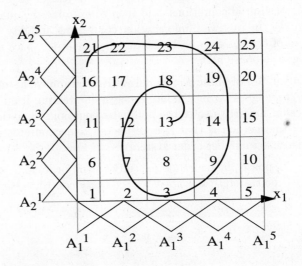

Figure 22.1. An example of linguistic trajectory.

By modifying the rules in the linguistic trajectory, we can change the corresponding state trajectory. For example, if we feel that the state trajectory in Fig.

22.1 converges too slowly in the early stage, we may modify, for example, rule 22 and rule 23 such that the convergence becomes faster. This can be done by noticing that the tangent vector of the state trajectory equals the summation of vector field $f(x)$ and vector field $b\Phi(x)$, as shown in Fig. 22.2. If the point x in Fig. 22.2 belongs to rule 23, then by increasing the center value of the THEN-part fuzzy set of rule 23, we can increase the value of $b\Phi(x)$ at this point and therefore speed up convergence. The advantage of the phase plane analysis is that it helps us to identify the rules that influence the behavior of the trajectory, so that we only need to modify these rules to achieve the desired behavior.

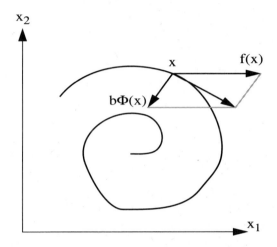

Figure 22.2. The state trajectory moves along the direction of the vector field $f(x) + b\Phi(x)$.

For the closed-loop fuzzy control system (22.1)-(22.2), we should pay special attention to the subspace $\Phi(x) = 0$, which is a line in the phase plane. This line separates the phase plane into positive and negative control regions. It also is called the switching line because when the state trajectory goes across this line, the control changes from positive to negative and vice versa. When the state vector is far away from the switching line, the control vector $b\Phi(x)$ usually has greater influence on the closed-loop system than the plant component $f(x)$. When the state vector gets closer to the switching line, $b\Phi(x)$ becomes smaller so that $f(x)$ has more influence on the closed-loop system. The relationship between $b\Phi(x)$ and $f(x)$ determines the behavior of the closed-loop system. Two situations are of most interests to us:

- *Stable closed-loop systems.* This is often the case where the open-loop system $\dot{x} = f(x)$ is stable and the control $u = \Phi(x)$ tries to lead the system trajectory towards the switching line $\Phi(x) = 0$. When the trajectory approaches the switching line, the plant component $f(x)$ has a greater influence, which makes the trajectory converge to the equilibrium point. See Fig. 22.3 for an example.

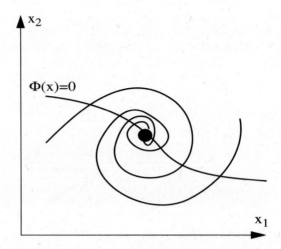

Figure 22.3. An example of stable closed-loop system.

- *Limit cycles.* This may be the case when the open-loop system $\dot{x} = f(x)$ is unstable and the control $u = \Phi(x)$ tries to stabilize the system. When the state is far away from the switch line $\Phi(x) = 0$, the control $\Phi(x)$ has a greater influence so that the state trajectory converges towards the switching line. When the state trajectory moves near the switching line, the unstable plant component $f(x)$ has a greater influence, which makes the state trajectory diverge away from the equilibrium point. This interaction between the control and the plant components makes the state oscillate around the equilibrium point and a *limit cycle* is thus formed. Fig. 22.4 shows an example.

Although the phase plane analysis is qualitative, it is helpful to characterize the dynamic behavior of a fuzzy control system. It also can serve as the basis for adequate selection or modification of the rules.

22.2   Robustness Indices for Stability

We know that a linear control system is stable if the eigenvalues of the closed-loop system are in the open left-half complex plane. If the eigenvalues are close to the imaginary axis, then small changes of the parameters may move some eigenvalues

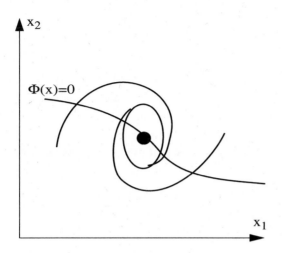

Figure 22.4. An example of limit cycle.

to the right-half plane and cause instability. Therefore, the further away the eigenvalues are from the imaginary axis, the more robust the stability is. So the distance of the smallest eigenvalue to the imaginary axis can be used as a measure of the robustness of the system's stability. In this section, we generalize this kind of analysis to fuzzy control systems and establish some robustness indices for stability. We will consider the one-dimensional case first to introduce the basic concepts, and then extend them to the n-dimensional case.

22.2.1 The One-Dimensional Case

Consider the fuzzy control system (22.1)-(22.2) with $x \in R$ be a scalar and $b = 1$. Suppose that the $f(x)$ is a monotone and increasing function with $f(0) = 0$, and that the fuzzy controller $\Phi(x)$ satisfies $\Phi(0) = 0$. The equilibrium point of the closed-loop system is determined by

$$\dot{x} = f(x) + \Phi(x) = 0 \tag{22.6}$$

Since $f(0) = \Phi(0) = 0$, the origin is an equilibrium point. For this equilibrium point to be stable, a sufficient condition is

$$\frac{d}{dx}[f(x) + \Phi(x)]|_{x=0} = f'(0) + \Phi'(0) < 0 \tag{22.7}$$

This can be proven by using the Lyapunov Linearization Theorem (Vidyasagar [1993]). Consequently, the closed-loop system (22.1)-(22.2) is globally stable if the following two conditions are satisfied:

$$Condition\ 1:\ f'(0) + \Phi'(0) < 0 \tag{22.8}$$

$$Condition\ 2:\ |\Phi(x)| < |f(x)|,\ \forall x \neq 0 \tag{22.9}$$

Condition 1 ensures that the origin is a stable equilibrium point, while Condition 2 guarantees no intersection of curves $\Phi(x)$ and $-f(x)$ and therefore prevents the appearance of other equilibrium points.

If stability is lost, then either the equilibrium at the origin becomes unstable, or new equilibrium points are produced due to the intersection of $\Phi(x)$ and $-f(x)$. Referring to (22.8) we see that the number

$$I_1 = -(f'(0) + \Phi'(0)) \tag{22.10}$$

can be used to measure the robustness of stability of the equilibrium at the origin. The larger the value of I_1, the more robust the stability of the origin. I_1 of (22.10) is our first *robustness index for stability* of the fuzzy control system.

Similarly, a measure can be associated with Condition 2. This measure should correspond to the minimum distance between $\Phi(x)$ and $-f(x)$, that is, $min|\Phi(x) + f(x)|$. However, since $min|\Phi(x) + f(x)|$ equals zero when $x = 0$, the min should be taken over the range that excludes a neighborhood of the origin. This gives our second *robustness index for stability*

$$I_2 = \min_{x \in R-(-\alpha,\alpha)} |\Phi(x) + f(x)| \tag{22.11}$$

where α is a positive constant. The larger the value of I_2, the more robust the stability of the fuzzy control system.

22.2.2 The n-Dimensional Case

Let us start with $n = 2$ and then generalize to arbitrary n. Assume that $f(0) = 0$ and $\Phi(0) = 0$, so the origin is an equilibrium point of the closed-loop system (22.1)-(22.2). From the Lyapunov Linearization Theorem (Vidyasagar [1993]) we know that the origin is stable if the two eigenvalues of the linearized system around the origin have negative real parts. Generally, there are two ways in which the system can become unstable:

- A real eigenvalue crosses the imaginary axis and acquires a positive sign. This is called *static bifurcation*.

- A pair of complex eigenvalues cross the imaginary axis and both of them take positive real parts. This is called *Hopf bifurcation*.

We now examine under what conditions these two bifurcations may occur.

Let

$$J = \frac{\partial(f(x) + b\Phi(x))}{\partial x}|_{x=0} = \begin{bmatrix} a_{11} & a_{12} \\ a_{21} & a_{22} \end{bmatrix} \tag{22.12}$$

be the Jacobian matrix of $f(x) + \Phi(x)$ at the origin. Then, the eigenvalues of the linearized system at the origin are the solution of the characteristic equation

$$|sI - J| = s^2 - (a_{11} + a_{22})s + a_{11}a_{22} - a_{12}a_{21} = 0 \tag{22.13}$$

A static bifurcation is produced when a real eigenvalue crosses the imaginary axis, that is, when one of the roots of (22.13) is zero. This will happen, as we can see from (22.13), only when $a_{11}a_{22} - a_{12}a_{21} = 0$. The larger the value of $|a_{11}a_{22} - a_{12}a_{21}|$, the further away the system is from static bifurcation. Therefore, we define

$$I_1 = |a_{11}a_{22} - a_{12}a_{21}| = |det(J)| \tag{22.14}$$

as a robustness index for stability for the second-order fuzzy control system. Similarly, a Hopf bifurcation may occur when the real parts of the two complex eigenvalues equal zero. From (22.13) we see that this will happen only when $a_{11} + a_{22} = 0$. The larger the value of $|a_{11}+a_{22}|$, the more unlikely the Hopf bifurcation. Therefore, we define

$$I_1' = |a_{11} + a_{22}| = |tr(J)| \tag{22.15}$$

as another robustness index. In summary, the larger the values of I_1 and I_1', the more robust the stability of the fuzzy control system is.

Similar to the one-dimensional case, stability may be lost when the vector field of the fuzzy controller $b\Phi(x)$ compensates exactly the vector field of the plant $f(x)$. Let $b = (b_1, b_2)$ and $f(x) = (f_1(x), f_2(x))$, then the compensation of the vector fields of the plant and the controller can occur only in the region of the state space where the plant component has the direction (b_1, b_2), that is, in the region defined by

$$C = \left\{ x \in R^2 | \frac{f_1(x)}{b_1} = \frac{f_2(x)}{b_2} \right\} \tag{22.16}$$

Similar to the I_2 of (22.11) for the one-dimensional case, we define the robustness index as

$$I_2 = \min_{x \in C - B} |f(x) + b\Phi(x)| \tag{22.17}$$

where $B = \{x \in R^2 ||x|^2 \le \alpha\}$ is a ball around the origin.

For the general $n > 2$ cases, let

$$J = \frac{\partial}{\partial x}[f(x) + b\Phi(x)]|_{x=0} \tag{22.18}$$

be the Jacobian matrix of the closed-loop system around the origin and

$$P(s) = |sI - J| = s^n + a_1 s^{n-1} + \cdots + a_{n-1}s + a_n \tag{22.19}$$

be its characteristic polynomial. Similar to the $n = 2$ case, $a_n = 0$ may cause static bifurcation. Therefore, we define

$$I_1 = a_n = |det(J)| \tag{22.20}$$

as a robustness index. Similar to I_1' of (22.15), we can define a robustness index for Hopf bifurcation. Finally, the generalization of I_2 (22.17) to the $n > 2$ case can be done by noticing that the subspace C of (22.16) now becomes

$$C = \left\{ x \in R^n | \frac{f_1(x)}{b_1} = \frac{f_2(x)}{b_2} = \cdots = \frac{f_n(x)}{b_n} \right\} \tag{22.21}$$

Example 22.1. Consider the fuzzy control system

$$\dot{x}_1 = x_2 \tag{22.22}$$
$$\dot{x}_2 = -12.74x_1 - 2.22x_2 + 2.22x_1^2 x_2 + 12.74\Phi(x_1, x_2) \tag{22.23}$$

The Jacobian matrix of this system at the origin is

$$J = \begin{pmatrix} 0 & 1 \\ -12.74 + 12.74\frac{\partial\Phi(x_1,x_2)}{\partial x_1}\big|_{x_1=x_2=0} & -2.22 + 12.74\frac{\partial\Phi(x_1,x_2)}{\partial x_2}\big|_{x_1=x_2=0} \end{pmatrix} \tag{22.24}$$

If there is no control, that is, if $\Phi(x_1, x_2) = 0$, then the robustness indices I_1 and I_1' are

$$I_1 = 12.74, \quad I_1' = 2.22 \tag{22.25}$$

With fuzzy controller $\Phi(x_1, x_2)$, these indices become

$$I_1 = \left| -12.74 + 12.74\frac{\partial\Phi(x_1,x_2)}{\partial x_1}\big|_{x_1=x_2=0} \right|, \quad I_1' = \left| -2.22 + 12,74\frac{\partial\Phi(x_1,x_2)}{\partial x_2}\big|_{x_1=x_2=0} \right| \tag{22.26}$$

Therefore, if we design the fuzzy controller such that $\frac{\partial\Phi(x_1,x_2)}{\partial x_1}\big|_{x_1=x_2=0} < 0$ and $\frac{\partial\Phi(x_1,x_2)}{\partial x_2}\big|_{x_1=x_2=0} < 0$, then the closed-loop fuzzy control system will be more robust than the uncontrolled open-loop system. □

22.3   Hierarchical Fuzzy Control

22.3.1   The Curse of Dimensionality

In Chapters 10 and 11 we saw that in order to design a fuzzy system with the required accuracy, the number of rules has to increase exponentially with the number of input variables to the fuzzy system. Specifically, suppose there are n input variables and m fuzzy sets are defined for each input variable, then the number of rules in the fuzzy system is m^n. For large n, m^n is a huge number. In practice, it is not uncommon to have, say, five input variables. With $n = 5$ and $m = 3$ (usually, at least three fuzzy sets should be defined for each variable), $m^n = 243$; if $m = 5$, which is more likely than $m = 3$, we have $m^n = 3120$. It is impractical to implement a fuzzy system with thousands of rules. A serious problem facing fuzzy system applications is how to deal with this rule explosion problem.

In fact, it is a common phenomenon that the complexity of a problem increases exponentially with the number of variables involved; this is not unique to fuzzy systems. This phenomenon was identified by Bellman as "the curse of dimensionality." Some approaches have been proposed to deal with this difficulty; using the *hierarchical fuzzy system* is one approach. We will see that the hierarchical fuzzy system has the nice property that the number of rules needed to the construct the fuzzy system increases only *linearly* with the number of variables.

22.3.2   Construction of the Hierarchical Fuzzy System

The idea of the hierarchical fuzzy system is to put the input variables into a collection of low-dimensional fuzzy systems, instead of a single high-dimensional fuzzy system as is the usual case. Each low-dimensional fuzzy system constitutes a level in the hierarchical fuzzy system. Suppose that there are n input variables $x_1, ..., x_n$, then the hierarchical fuzzy system is constructed as follows:

- The first level is a fuzzy system with n_1 input variables $x_1, ..., x_{n_1}$ which is constructed from the rules

$$IF\ x_1\ is\ A_1^l\ and\ \cdots\ and\ x_{n_1}\ is\ A_{n_1}^l,\ THEN\ y_1\ is\ B_1^l \qquad (22.27)$$

  where $2 \leq n_1 < n$, and $l = 1, 2, ..., M_1$.

- The $i'th$ level $(i > 1)$ is a fuzzy system with $n_i + 1$ $(n_i \geq 1)$ input variables, which is constructed from the rules

$$IF\ x_{N_i+1}\ is\ A_{N_i+1}^l\ and\ \cdots\ and\ x_{N_i+n_i}\ is\ A_{N_i+n_i}^l\ and\ y_{i-1}\ is\ C_{i-1}^l,$$
$$THEN\ y_i\ is\ B_i^l \qquad (22.28)$$

  where $N_i = \sum_{j=1}^{i-1} n_j$, and $l = 1, 2, ..., M_i$.

- The construction continues until $i = L$ such that $\sum_{j=1}^{L} n_j = n$, that is, until all the input variables are used in one of the levels.

We see that the first level converts n_1 variables $x_1, ..., x_{n_1}$ into one variable y_1, which is then sent to the second level. In the second level, other n_2 variables $x_{n_1+1}, ..., x_{n_1+n_2}$ and the variable y_1 are combined into another variable y_2, which is sent to the third level. This process continues until all the variables $x_1, ..., x_n$ are used.

A special case of the hierarchical fuzzy system is to choose $n_1 = 2$ and $n_i = 1$ for $i = 2, 3, ..., L$. In this case, all the fuzzy systems in the hierarchy have two inputs, and there are $L = n - 1$ levels. This special hierarchical fuzzy system is illustrated in Fig. 22.5.

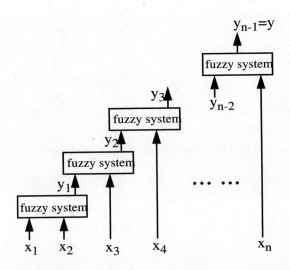

Figure 22.5. An example of hierarchical fuzzy system with $n_1 = 2$ and $n_i = 1$ for $i = 2, 3, ..., n - 1$. This hierarchical fuzzy system consists of $n - 1$ two-input fuzzy systems.

Next, we study the properties of the hierarchical fuzzy system.

22.3.3 Properties of the Hierarchical Fuzzy System

We now show that the number of rules in the hierarchical fuzzy system is a linear function of the number of input variables (Theorem 22.1) and that the rule number reaches its minimum in the special case of Fig. 22.5 (Theorem 22.2).

Theorem 22.1. Let n, n_i and L be the same as in the design of the hierarchical fuzzy system. Suppose that m fuzzy sets are defined for each variable, including the input variables $x_1, ..., x_n$ and the intermediate variables $y_1, ..., y_L$, and that the fuzzy system in the $i'th$ level $(i = 2, ..., L)$ is constructed from m^{n_i+1} rules (a complete fuzzy rule base) and for the first level the rule number is m^{n_1}. If $n_1 = n_i + 1 = c$ (constant) for $i = 2, 3, ..., L$, then the total number of rules in the whole hierarchical fuzzy system is

$$M = \frac{m^c}{c - 1}(n - 1) \tag{22.29}$$

Proof: Obviously, we have

$$M = m^{n_1} + \sum_{i=2}^{L} m^{n_i+1}$$

$$= Lm^c \qquad (22.30)$$

Since $n = \sum_{i=1}^{L} n_i = c + \sum_{i=2}^{L}(c-1) = Lc - L + 1$, we have

$$L = \frac{n-1}{c-1} \qquad (22.31)$$

Substituting (22.31) into (22.30), we obtain (22.29). $\square$

Since $\frac{m^c}{c-1}$ is a constant, we see from (22.29) that the number of rules in the hierarchical fuzzy system increases linearly with the number of input variables. For the case of $m = 3, c = 2$ and $n = 5$, we have $M = 3^2 4 = 36$; if we use the conventional fuzzy system, the number of rules is $m^n = 3^5 = 243$. The reduction of rules is even greater for larger m and n.

Finally, we show that the rule number M reaches its minimum when $c = 2$.

Theorem 22.2. Let the assumptions in Theorem 22.1 be true. If $m \geq 2$, then the total number of rules M of (22.29) is minimized when $c = 2$, that is, when the fuzzy systems in all the levels have two inputs as shown in Fig. 22.5.

The proof of this theorem is left as an exercise.

22.4 Summary and Further Readings

In this chapter we have demonstrated the following:

- Phase plane analysis of second-order fuzzy control systems, including the concept of linguistic trajectory and how to use it to modify the rules.

- The definitions and meanings of various robustness indices for stability.

- The construction and basic properties of the hierarchical fuzzy system.

The concept of linguistic analysis for fuzzy control systems was proposed by Braae and Rutherford [1979]. Stability indices for fuzzy control systems were studied in Aracil, Ollero, and Garcia-Cerezo [1989]. The book Driankov, Hellendoorn and Reinfrank [1993] also contains qualitative analysis of fuzzy control systems and a discussion on the robustness indices. Hierarchical fuzzy systems were proposed by Raju, Zhou, and Kisner [1991]. It was proven in Wang [1996] that hierarchical fuzzy systems also are universal approximators.

22.5   Exercises

Exercise 22.1. Show that the nonlinear system

$$\dot{x}_1 = x_2 - x_1(x_1^2 + x_2^2 - 1) \tag{22.32}$$
$$\dot{x}_2 = -x_1 - x_2(x_1^2 + x_2^2 - 1) \tag{22.33}$$

converges to a limit cycle no matter where the initial state is.

Exercise 22.2. Design a 25-rule fuzzy controller for the inverted pendulum system in Fig. 1.9 and determine the linguistic trajectories for initial conditions $(\theta(0), \dot{\theta}(0)) = (10^\circ, 0), (5^\circ, 0)$ and $(-8^\circ, 0)$.

Exercise 22.3. Design a 25-rule fuzzy controller for the ball-and-beam system in Fig. 16.7 and determine the linguistic trajectories for initial conditions $(r(0), \theta(0)) = (1, 0^\circ), (1, 10^\circ)$ and $(-1, 5^\circ)$.

Exercise 22.4. Design the fuzzy system $\Phi(x_1, x_2)$ in Example 22.1 such that the closed-loop system is more robust than the open-loop system.

Exercise 22.5. Consider the hierarchical fuzzy system in Fig. 22.5 with $n = 3$. Let $g(x_1, x_2, x_3)$ be an unknown function, but we know its values at $(x_1, x_2, x_3) = (e_i, e_j, e_k)$ for $i, j, k = 1, 2, ..., 6$, where $e_i = 0.2(i - 1)$. Let $f_1(x_1, x_2)$ and $f_2(y_1, x_3)$ be the two fuzzy systems and $f(x_1, x_2, x_3) = f_2[f_1(x_1, x_2), x_3]$ be the hierarchical fuzzy system. Design f_1 and f_2 such that $f(e_i, e_j, e_k) = g(e_i, e_j, e_k)$ for $i, j, k = 1, 2, ..., 6$.

Exercise 22.6. Use the ideas in Exercise 22.5 and Chapter 10 to prove that hierarchical fuzzy systems are universal approximators.

Exercise 22.7. Prove Theorem 22.2.

Part V

Adaptive Fuzzy Control

Fuzzy controllers are supposed to work in situations where there is a large uncertainty or unknown variation in plant parameters and structures. Generally, the basic objective of adaptive control is to maintain consistent performance of a system in the presence of these uncertainties. Therefore, advanced fuzzy control should be adaptive.

The basic configuration of an adaptive fuzzy control system is shown in Fig. 23.1. The *reference model* is used to specify the ideal response that the fuzzy control system should follow. The *plant* is assumed to contain unknown components. The *fuzzy controller* is constructed from fuzzy systems whose parameters θ are adjustable. The *adaptation law* adjusts the parameters θ online such that the plant output $y(t)$ tracks the reference model output $y_m(t)$.

From Fig. 23.1 we see that the main differences between adaptive fuzzy control systems and nonadaptive fuzzy control systems are: (i) the fuzzy controller in the adaptive fuzzy control system is changing during real-time operation, whereas the fuzzy controller in the nonadaptive fuzzy control system is fixed before real-time operation, and (ii) an additional component, the adaptation law, is introduced to the adaptive fuzzy control system to adjust the fuzzy controller parameters.

The main advantages of adaptive fuzzy control over nonadaptive fuzzy control are: (i) better performance is usually achieved because the adaptive fuzzy controller can adjust itself to the changing environment, and (ii) less information about the plant is required because the adaptation law can help to learn the dynamics of the plant during real-time operation.

The main disadvantages of adaptive fuzzy control over nonadaptive fuzzy control are: (i) the resulting control system is more difficult to analyze because it is not only nonlinear but also time varying, and (ii) implementation is more costly.

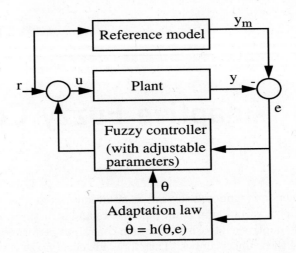

Figure 23.1. The basic configuration of adaptive fuzzy control systems.

In this part (Chapters 23-26), we will develop a number of adaptive fuzzy controllers for unknown or partially unknown nonlinear systems. In Chapter 23, we will classify the adaptive fuzzy controllers into three categories: indirect, direct, and combined indirect/direct schemes, and develop a basic indirect adaptive fuzzy controller. In Chapter 24, the basic elements in the direct and combined indirect/direct adaptive fuzzy controllers will be developed. Although the adaptive fuzzy controllers in Chapters 23 and 24 have nice convergence properties, it cannot be guaranteed that the resulting closed-loop nonlinear time-varying systems are globally stable. In Chapter 25, we will show how to use supervisory control and parameter projection to guarantee the uniform boundedness of all the variables. Finally, Chapter 26 will use the concept of input-output linearization to design adaptive fuzzy controllers for general nonlinear systems.

Chapter 23

Basic Adaptive Fuzzy Controllers I

23.1 Classification of Adaptive Fuzzy Controllers

Adaptive fuzzy control and conventional adaptive control have similarities and differences. They are similar in: (i) the basic configuration and principles are more or less the same, and (ii) the mathematical tools used in the analysis and design are very similar. The main differences are: (i) the fuzzy controller has a special nonlinear structure that is universal for different plants, whereas the structure of a conventional adaptive controller changes from plant to plant, and (ii) human knowledge about the plant dynamics and control strategies can be incorporated into adaptive fuzzy controllers, whereas such knowledge is not considered in conventional adaptive control systems. This second difference identifies the main advantage of adaptive fuzzy control over conventional adaptive control.

In order to develop methods to incorporate human knowledge, we must first consider what types of human knowledge we are going to use. From a high-level conceptual point of view, any control system consists at least a plant and a controller. Therefore, human knowledge about a control system can be classified into two categories: plant knowledge and control knowledge. In our fuzzy control framework, these two types of human knowledge are specified as:

- **Plant knowledge**: Fuzzy IF-THEN rules that describe the behavior of the unknown plant (for example, we can describe the behavior of a car using the fuzzy IF-THEN rule: "IF you apply more force to the accelerator, THEN the speed of the car will increase," where "more" and "increase" are characterized by fuzzy sets).

- **Control knowledge**: Fuzzy control rules that state in which situations what control actions should be taken (for example, we often use the following fuzzy IF-THEN rule to drive a car: "IF the speed is low, THEN apply more force to the accelerator," where "low" and "more" are characterized by fuzzy sets).

Depending upon the human knowledge used and the structure of the fuzzy controller, adaptive fuzzy control is classified into the following three categories:

- **Indirect adaptive fuzzy control**: The fuzzy controller comprises a number of fuzzy systems constructed (initially) from the plant knowledge.

- **Direct adaptive fuzzy control**: The fuzzy controller is a single fuzzy system constructed (initially) from the control knowledge.

- **Combined indirect/direct fuzzy control**: The fuzzy controller is a weighted average of the indirect and direct adaptive fuzzy controllers (therefore, both plant and control knowledge is used).

In the next section, we will develop the basics of indirect adaptive fuzzy control schemes; direct and combined indirect/direct fuzzy control will be studied in Chapter 24.

23.2   Design of the Indirect Adaptive Fuzzy Controller

23.2.1   Problem Specification

Suppose that the plant is a $n'th$ order nonlinear system described by the differential equation

$$x^{(n)} = f(x, \dot{x}, ..., x^{(n-1)}) + g(x, \dot{x}, ..., x^{(n-1)})u \qquad (23.1)$$
$$y = x \qquad\qquad\qquad\qquad\qquad\qquad\qquad\qquad\quad (23.2)$$

where f and g are *unknown* functions, $u \in R$ and $y \in R$ are the input and output of the plant, respectively, and $\mathbf{x} = (x_1, x_2, ..., x_n)^T = (x, \dot{x}, ..., x^{(n-1)})^T \in R^n$ is the state vector of the system that is assumed to be available for measurement. In order for (23.1) to be controllable, we require that $g(\mathbf{x}) \neq 0$. Without loss of generality we assume that $g(\mathbf{x}) > 0$. In the spirit of the nonlinear control literature (Isidori [1989]), these systems are in normal form and have the relative degree equal to n.

The *control objective* is to design a feedback controller $u = u(\mathbf{x}|\theta)$ based on fuzzy systems and an adaptation law for adjusting the parameter vector θ, such that the plant output y follows the ideal output y_m which and its time derivatives are known and bounded.

Since the functions $f(\mathbf{x})$ and $g(\mathbf{x})$ in the plant are nonlinear and are assumed to be unknown, we are dealing with a quite general single-input-single-output nonlinear control problem. Therefore, in the control objective we did not insist that the plant output y should converge to the ideal output y_m asymptotically; we only require that y follows y_m as close as possible. In Chapters 25-26, we will introduce more advanced algorithms that guarantee the stability and convergence of the adaptive fuzzy control systems. The goal of this chapter is to show the basic ideas of adaptive fuzzy control.

Since we design an indirect adaptive fuzzy controller in this section, some plant knowledge is available. Specifically, we assume that a collection of fuzzy IF-THEN rules are available that describe the input-output behavior of $f(\mathbf{x})$ and $g(\mathbf{x})$; these rules are given as follows:

$$IF \ x_1 \ is \ F_1^r \ and \ \cdots \ and \ x_n \ is \ F_n^r, \ THEN \ f(\mathbf{x}) \ is \ C^r \qquad (23.3)$$

which describe $f(\mathbf{x})$, and

$$IF \ x_1 \ is \ G_1^s \ and \ \cdots \ and \ x_n \ is \ G_n^s, \ THEN \ g(\mathbf{x}) \ is \ D^s \qquad (23.4)$$

which describe $g(\mathbf{x})$, where F_i^r, C^r, G_i^s and D^s are fuzzy sets, $r = 1, 2, ..., L_f$ and $s = 1, 2, ..., L_g$.

23.2.2 Design of the Fuzzy Controller

If the nonlinear functions $f(\mathbf{x})$ and $g(\mathbf{x})$ are known, then we can choose the control u to cancel the nonlinearity and design the controller based on linear control theory (for example, pole placement). Specifically, let $e = y_m - y = y_m - x$, $\mathbf{e} = (e, \dot{e}, ..., e^{(n-1)})^T$ and $\mathbf{k} = (k_n, ..., k_1)^T$ be such that all roots of the polynormal $s^n + k_1 s^{n-1} + \cdots + k_n$ are in the open left-half complex plane, and choose the control law as

$$u^* = \frac{1}{g(\mathbf{x})}[-f(\mathbf{x}) + y_m^{(n)} + \mathbf{k}^T \mathbf{e}] \qquad (23.5)$$

Substituting (23.5) into (23.1), we obtain the closed-loop system governed by

$$e^{(n)} + k_1 e^{(n-1)} + \cdots + k_n e = 0 \qquad (23.6)$$

Because of the choice of $\mathbf{k}$, we have $e(t) \to 0$ as $t \to \infty$, that is, the plant output y converges to the ideal output y_m asymptotically.

Since $f(\mathbf{x})$ and $g(\mathbf{x})$ are unknown, the ideal controller (23.5) cannot be implemented. However, we have the fuzzy IF-THEN rules (23.3)-(23.4) that describe the input-output behavior of $f(\mathbf{x})$ and $g(\mathbf{x})$. Therefore, a reasonable idea is to replace the $f(\mathbf{x})$ and $g(\mathbf{x})$ in (23.5) by fuzzy systems $\hat{f}(\mathbf{x})$ and $\hat{g}(\mathbf{x})$, which are constructed from the rules (23.3) and (23.4), respectively. Since the rules (23.3)-(23.4) provide only rough information about $f(\mathbf{x})$ and $g(\mathbf{x})$, the constructed fuzzy systems $\hat{f}(\mathbf{x})$ and $\hat{g}(\mathbf{x})$ may not approximate $f(\mathbf{x})$ and $g(\mathbf{x})$ well enough. To improve the accuracy of $\hat{f}(\mathbf{x})$ and $\hat{g}(\mathbf{x})$, one idea is to leave some parameters in $\hat{f}(\mathbf{x})$ and $\hat{g}(\mathbf{x})$ free to change during online operation so that the approximation accuracy improves over time. Let $\theta_f \in R^{M_f}$ and $\theta_g \in R^{M_g}$ be the free parameters in $\hat{f}(\mathbf{x})$ and $\hat{g}(\mathbf{x})$, respectively, so we denote $\hat{f}(\mathbf{x}) = \hat{f}(\mathbf{x}|\theta_f)$ and $\hat{g}(\mathbf{x}) = \hat{g}(\mathbf{x}|\theta_g)$. Replacing the $f(\mathbf{x})$

and $g(\mathbf{x})$ in (23.5) by the fuzzy systems $\hat{f}(\mathbf{x}|\theta_f)$ and $\hat{g}(\mathbf{x}|\theta_g)$, respectively, we obtain the fuzzy controller

$$u = u_I = \frac{1}{\hat{g}(\mathbf{x}|\theta_g)}[-\hat{f}(\mathbf{x}|\theta_f) + y_m^{(n)} + \mathbf{k}^T \mathbf{e}] \tag{23.7}$$

This fuzzy controller is called the *certainty equivalent controller*, because if the $\hat{f}$ and $\hat{g}$ equal the corresponding f and g (which means that there is no uncertainty about f and g), then the controller u_I becomes the ideal controller u^* of (23.5).

To implement the controller (23.7), we must specify the detailed formulas of $\hat{f}(\mathbf{x}|\theta_f)$ and $\hat{g}(\mathbf{x}|\theta_g)$. Since the number of rules in (23.3) and (23.4) may be small, it is generally not sufficient to construct $\hat{f}(\mathbf{x}|\theta_f)$ and $\hat{g}(\mathbf{x}|\theta_g)$ based only on the L_f rules in (23.3) and L_g rules in (23.4). We should construct $\hat{f}(\mathbf{x}|\theta_f)$ and $\hat{g}(\mathbf{x}|\theta_g)$ based on complete sets of rules that include the rules in (23.3) and (23.4) as special cases. Specifically, $\hat{f}(\mathbf{x}|\theta_f)$ and $\hat{g}(\mathbf{x}|\theta_g)$ are constructed from the following two steps:

- **Step 1.** For variable x_i ($i = 1, 2, ..., n$), define p_i fuzzy sets $A_i^{l_i}$ ($l_i = 1, 2, ..., p_i$), which include the F_i^r ($r = 1, 2, ..., L_f$) in (23.3) as special cases, and define q_i fuzzy sets $B_i^{l_i}$ ($l_i = 1, 2, ..., q_i$), which include the G_i^s ($s = 1, 2, ..., L_g$) in (23.4) as special cases.

- **Step 2.** Construct the fuzzy system $\hat{f}(\mathbf{x}|\theta_f)$ from the $\prod_{i=1}^n p_i$ rules:

$$IF\ x_1\ is\ A_1^{l_1}\ and\ \cdots\ and\ x_n\ is\ A_n^{l_n},\ THEN\ \hat{f}\ is\ E^{l_1\cdots l_n} \tag{23.8}$$

where $l_i = 1, 2, ..., p_i, i = 1, 2, ..., n$, and $E^{l_1\cdots l_n}$ equals C^r if the IF part of (23.8) agrees with the IF part of (23.3) and equals some arbitrary fuzzy set otherwise. Similarly, construct the fuzzy system $\hat{g}(\mathbf{x}|\theta_g)$ from the $\prod_{i=1}^n q_i$ rules:

$$IF\ x_1\ is\ B_1^{l_1}\ and\ \cdots\ and\ x_n\ is\ B_n^{l_n},\ THEN\ \hat{g}\ is\ H^{l_1\cdots l_n} \tag{23.9}$$

where $l_i = 1, 2, ..., q_i, i = 1, 2, ..., n$, and $H^{l_1\cdots l_n}$ equals D^s if the IF part of (23.9) agrees with the IF part of (23.4) and equals some arbitrary fuzzy set otherwise. Specifically, using the product inference engine, singleton fuzzifier and center average defuzzifier, we obtain

$$\hat{f}(\mathbf{x}|\theta_f) = \frac{\sum_{l_1=1}^{p_1} \cdots \sum_{l_n=1}^{p_n} \bar{y}_f^{l_1\cdots l_n}(\prod_{i=1}^n \mu_{A_i^{l_i}}(x_i))}{\sum_{l_1=1}^{p_1} \cdots \sum_{l_n=1}^{p_n}(\prod_{i=1}^n \mu_{A_i^{l_i}}(x_i))} \tag{23.10}$$

$$\hat{g}(\mathbf{x}|\theta_g) = \frac{\sum_{l_1=1}^{q_1} \cdots \sum_{l_n=1}^{q_n} \bar{y}_g^{l_1\cdots l_n}(\prod_{i=1}^n \mu_{B_i^{l_i}}(x_i))}{\sum_{l_1=1}^{q_1} \cdots \sum_{l_n=1}^{q_n}(\prod_{i=1}^n \mu_{B_i^{l_i}}(x_i))} \tag{23.11}$$

Let $\bar{y}_f^{l_1\cdots l_n}$ and $\bar{y}_g^{l_1\cdots l_n}$ be the free parameters that are collected into $\theta_f \in R\prod_{i=1}^{n} p_i$ and $\theta_g \in R\prod_{i=1}^{n} q_i$, respectively, so we can rewrite (23.10) and (23.11) as

$$\hat{f}(\mathbf{x}|\theta_f) = \theta_f^T \xi(\mathbf{x}) \tag{23.12}$$
$$\hat{g}(\mathbf{x}|\theta_g) = \theta_g^T \eta(\mathbf{x}) \tag{23.13}$$

where $\xi(\mathbf{x})$ is a $\prod_{i=1}^{n} p_i$-dimensional vector with its $l_1 \cdots l'_n th$ element

$$\xi_{l_1\cdots l_n}(\mathbf{x}) = \frac{\prod_{i=1}^{n} \mu_{A_i^{l_i}}(x_i)}{\sum_{l_1=1}^{p_1} \cdots \sum_{l_n=1}^{p_n} (\prod_{i=1}^{n} \mu_{A_i^{l_i}}(x_i))} \tag{23.14}$$

and $\eta(\mathbf{x})$ is a $\prod_{i=1}^{n} q_i$-dimensional vector with its $l_1 \cdots l'_n th$ element

$$\eta_{l_1\cdots l_n}(\mathbf{x}) = \frac{\prod_{i=1}^{n} \mu_{B_i^{l_i}}(x_i)}{\sum_{l_1=1}^{q_1} \cdots \sum_{l_n=1}^{q_n} (\prod_{i=1}^{n} \mu_{B_i^{l_i}}(x_i))} \tag{23.15}$$

From Step 2 we see that some parameters in θ_f and θ_g are chosen according to the rules (23.3)-(23.4), and the remaining parameters in θ_f and θ_g are chosen randomly (or assuming some structure). Since the parameters θ_f and θ_g will change during online operation, this parameter setting gives the *initial parameters*. Our next task is to design an adaptation law for θ_f and θ_g, such that the tracking error e is minimized.

23.2.3   Design of Adaptation Law

Substituting (23.7) into (23.1) and after some manipulation, we obtain the closed-loop dynamics of the fuzzy control system as

$$e^{(n)} = -\mathbf{k}^T \mathbf{e} + [\hat{f}(\mathbf{x}|\theta_f) - f(\mathbf{x})] + [\hat{g}(\mathbf{x}|\theta_g) - g(\mathbf{x})]u_I \tag{23.16}$$

Let

$$\Lambda = \begin{bmatrix} 0 & 1 & 0 & 0 & \cdots & 0 & 0 \\ 0 & 0 & 1 & 0 & \cdots & 0 & 0 \\ \cdots & \cdots & \cdots & \cdots & \cdots & \cdots & \cdots \\ 0 & 0 & 0 & 0 & \cdots & 0 & 1 \\ -k_n & -k_{n-1} & \cdots & \cdots & \cdots & \cdots & -k_1 \end{bmatrix}, \mathbf{b} = \begin{bmatrix} 0 \\ \cdots \\ 0 \\ 1 \end{bmatrix} \tag{23.17}$$

then the dynamic equation (23.16) can be rewritten into the vector form

$$\dot{\mathbf{e}} = \Lambda\mathbf{e} + \mathbf{b}\{[\hat{f}(\mathbf{x}|\theta_f) - f(\mathbf{x})] + [\hat{g}(\mathbf{x}|\theta_g) - g(\mathbf{x})]u_I\} \tag{23.18}$$

Define the optimal parameters as

$$\theta_f^* = arg \min_{\theta_f \in R\prod_{i=1}^{n} p_i} \left[\sup_{\mathbf{x} \in R^n} |\hat{f}(\mathbf{x}|\theta_f) - f(\mathbf{x})| \right] \tag{23.19}$$

$$\theta_g^* = arg \min_{\theta_g \in R\prod_{i=1}^{n} q_i} \left[\sup_{\mathbf{x} \in R^n} |\hat{g}(\mathbf{x}|\theta_g) - g(\mathbf{x})| \right] \tag{23.20}$$

thus $\hat{f}(\mathbf{x}|\theta_f^*)$ and $\hat{g}(\mathbf{x}|\theta_g^*)$ are the best (min-max) approximators of $f(\mathbf{x})$ and $g(\mathbf{x})$, respectively, among all the fuzzy systems in the form of (23.10) and (23.11). Define the *minimum approximation error*

$$w = [\hat{f}(\mathbf{x}|\theta_f^*) - f(\mathbf{x})] + [\hat{g}(\mathbf{x}|\theta_g^*) - g(\mathbf{x})]u_I \tag{23.21}$$

Using this w, we can rewrite (23.18) as

$$\dot{\mathbf{e}} = \Lambda\mathbf{e} + \mathbf{b}\{[\hat{f}(\mathbf{x}|\theta_f) - \hat{f}(\mathbf{x}|\theta_f^*)] + [\hat{g}(\mathbf{x}|\theta_g) - g(\mathbf{x}|\theta_g^*)]u_I + w\} \tag{23.22}$$

Substituting (23.12) and (23.13) into (23.22), we obtain the following closed-loop dynamic equation that specifies explicitly the relationship between the tracking error $\mathbf{e}$ and the controller parameters θ_f and θ_g:

$$\dot{\mathbf{e}} = \Lambda\mathbf{e} + \mathbf{b}[(\theta_f - \theta_f^*)^T \xi(\mathbf{x}) + (\theta_g - \theta_g^*)^T \eta(\mathbf{x})u_I + w] \tag{23.23}$$

The task of an adaptation law is to determine an adjusting mechanism for θ_f and θ_g, such that the tracking error $\mathbf{e}$ and the parameter errors $\theta_f - \theta_f^*$ and $\theta_g - \theta_g^*$ are minimized.

To complete this task, consider the Lyapunov function candidate

$$V = \frac{1}{2}\mathbf{e}^T P\mathbf{e} + \frac{1}{2\gamma_1}(\theta_f - \theta_f^*)^T(\theta_f - \theta_f^*) + \frac{1}{2\gamma_2}(\theta_g - \theta_g^*)^T(\theta_g - \theta_g^*) \tag{23.24}$$

where γ_1 and γ_2 are positive constants, and P is a positive definite matrix satisfying the Lyapunov equation

$$\Lambda^T P + P\Lambda = -Q \tag{23.25}$$

where Q is an arbitrary $n \times n$ positive definite matrix, and Λ is given by (23.17). The time derivative of V along the closed-loop system trajectory (23.23) is

$$\dot{V} = -\frac{1}{2}\mathbf{e}^T P\mathbf{e} + \mathbf{e}^T P\mathbf{b}w + \frac{1}{\gamma_1}(\theta_f - \theta_f^*)^T[\dot{\theta}_f + \gamma_1 \mathbf{e}^T P\mathbf{b}\xi(\mathbf{x})]$$

$$+ \frac{1}{\gamma_2}(\theta_g - \theta_g^*)^T[\dot{\theta}_g + \gamma_2 \mathbf{e}^T P\mathbf{b}\eta(\mathbf{x})u_I] \tag{23.26}$$

To minimize the tracking error $\mathbf{e}$ and the parameter errors $\theta_f - \theta_f^*$ and $\theta_g - \theta_g^*$, or equivalently, to minimize V, we should choose the adaptation law such that $\dot{V}$

is negative. Since $-\frac{1}{2}\mathbf{e}^T P\mathbf{e}$ is negative and we can choose the fuzzy systems such that the minimum approximation error w is small, a good strategy is to choose the adaptation law such that the last two terms in (23.26) are zero, that is, our adaptation law is

$$\dot{\theta}_f = -\gamma_1 \mathbf{e}^T P\mathbf{b}\xi(\mathbf{x}) \qquad (23.27)$$

$$\dot{\theta}_g = -\gamma_2 \mathbf{e}^T P\mathbf{b}\eta(\mathbf{x})u_I \qquad (23.28)$$

This approach to designing the adaptation law is called the *Lyapunov synthesis approach*, because the goal is to minimize the Lyapunov function V.

In summary, the whole indirect adaptive fuzzy control system is shown in Fig. 23.2. It should be remembered that the plant knowledge (fuzzy IF-THEN rules (23.3)-(23.4)) is incorporated through the initial parameters $\theta_f(0)$ and $\theta_g(0)$, as shown in Step 2 of the design procedure for $\hat{f}(\mathbf{x}|\theta_f)$ and $\hat{g}(\mathbf{x}|\theta_g)$.

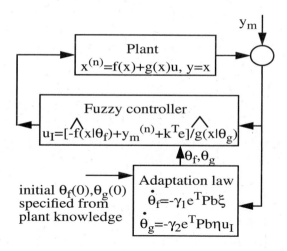

Figure 23.2. The indirect adaptive fuzzy control system.

23.3 Application to Inverted Pendulum Tracking Control

We now apply the indirect adaptive fuzzy controller to the inverted pendulum system shown in Fig. 1.9 whose dynamics are characterized by (20.15)-(20.16). We choose $k_1 = 2$ and $k_2 = 1$ (so that $s^2 + k_1 s + k_2$ is stable), $Q = diag(10, 10)$, and solve the Lyapunov equation (23.25) to obtain the P as in (20.29). We now consider two examples: one without any plant knowledge (Example 23.1) and the other with some plant knowledge (Example 23.2).

Example 23.1. In this example, we assume that there are no linguistic rules (23.3) and (23.4). We choose $p_1 = p_2 = q_1 = q_2 = 5$, and $A_1^l = A_2^l = B_1^l = B_2^l$ with $\mu_{A_1^1}(x_1) = exp[-(\frac{x_1+\pi/6}{\pi/24})^2]$, $\mu_{A_1^2}(x_1) = exp[-(\frac{x_1+\pi/12}{\pi/24})^2]$, $\mu_{A_1^3}(x_1) = exp[-(\frac{x_1}{\pi/24})^2]$, $\mu_{A_1^4}(x_1) = exp[-(\frac{x_1-\pi/12}{\pi/24})^2]$, and $\mu_{A_1^5}(x_i) = exp[-(\frac{x_1-\pi/6}{\pi/24})^2]$, which cover the interval $[-\pi/6, \pi/6]$. Since the range of $f(x_1, x_2)$ is much larger than that of $g(x_1, x_2)$, we choose $\gamma_1 = 50$ and $\gamma_2 = 1$. Figs. 23.3 and 23.4 show the closed-loop $x_1(t)$ together with the ideal output $y_m(t)$ for the initial conditions $\mathbf{x}(0) = (-\frac{\pi}{60}, 0)^T$ and $\mathbf{x}(0) = (\frac{\pi}{60}, 0)^T$, respectively. $\square$

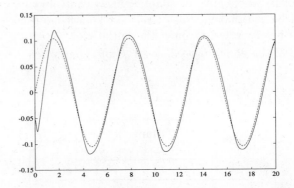

Figure 23.3. The state $x_1(t)$ (solid line) and its desired value $y_m(t) = \frac{\pi}{30} sin(t)$ (dashed line) for the initial condition $\mathbf{x}(0) = (-\frac{\pi}{60}, 0)^T$ in Example 23.1.

Example 23.2. Here we consider the same situation as in Example 23.1 except that there are some linguistic rules about $f(x_1, x_2)$ and $g(x_1, x_2)$ based on the following physical intuition. First, suppose that there is no control; that is, $u = 0$. In this case the acceleration of the angle $\theta = x_1$ equals $f(x_1, x_2)$. Based on physical intuition we have the following observation:

$$\textit{The bigger the } x_1, \textit{ the larger the } f(x_1, x_2) \qquad (23.29)$$

Our task now is to transform this observation into fuzzy IF-THEN rules about $f(x_1, x_2)$. Since $(x_1, x_2) = (0, 0)$ is an (unstable) equilibrium point of the system, we have the first rule:

$$R_f^{(1)}: \quad IF \ x_1 \ is \ F_1^3 \ and \ x_2 \ is \ F_2^3, \ THEN \ f(x_1, x_2) \ is \ near \ zero \qquad (23.30)$$

where F_i^j $(i = 1, 2, j = 1, 2, ..., 5)$ equal the A_i^j defined in Example 23.1, and "near zero" is a fuzzy set with center at zero. From Fig. 1.9 we see that the acceleration of x_1 is proportional to the gravity $mgsin(x_1)$; that is, we have approximately that

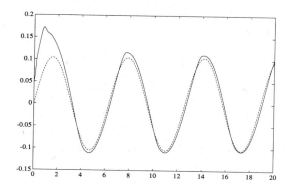

Figure 23.4. The state $x_1(t)$ (solid line) and its desired value $y_m(t) = \frac{\pi}{30}sin(t)$ (dashed line) for the initial condition $\mathbf{x}(0) = (\frac{\pi}{60}, 0)^T$ in Example 23.1.

$f(x_1, x_2) = \alpha sin(x_1)$, where α is a constant. Clearly, $f(x_1, x_2)$ achieves its maximum at $x_1 = \pi/2$; thus, from (20.16) we approximately have $\alpha = 16$. Therefore, we obtain the following fuzzy IF-THEN rules for $f(x_1, x_2)$:

$R_f^{(2)}$:   *IF x_1 is F_1^1 and x_2 is F_2^3, THEN $f(x_1, x_2)$ is near -8,*(23.31)

$R_f^{(3)}$:   *IF x_1 is F_1^2 and x_2 is F_2^3, THEN $f(x_1, x_2)$ is near -4,*(23.32)

$R_f^{(4)}$:   *IF x_1 is F_1^4 and x_2 is F_2^3, THEN $f(x_1, x_2)$ is near 4,*   (23.33)

$R_f^{(5)}$:   *IF x_1 is F_1^5 and x_2 is F_2^3, THEN $f(x_1, x_2)$ is near 8,*   (23.34)

where F_i^j ($i = 1, 2, j = 1, 2, ..., 5$) are identical to the corresponding A_i^j in Example 23.1, and the values after *near* are determined according to $16sin(\pi/6) = 8$ and $8sin(\pi/12) \doteq 4$. Also based on physical intuition we have that $f(x_1, x_2)$ is more sensitive to x_1 than to x_2, we therefore extend the rules (23.30)-(23.34) to the rules where x_2 is any F_2^j for $j = 1, 2, ..., 5$. In summary, the final rules characterizing $f(x_1, x_2)$ are shown in Fig.23.5. These rules are used to determine the initial parameters $\theta_f(0)$.

Next, we determine fuzzy IF-THEN rules for $g(x_1, x_2)$ based on physical intuition. Since $g(x_1, x_2)$ determines the strength of the control u on the system and clearly this strength is maximized at $x_1 = 0$, we have the following observation:

The smaller the x_1, the larger the $g(x_1, x_2)$.   (23.35)

Similar to the way of constructing the rules for $f(x_1, x_2)$, we transfer the observation (23.35) into 25 fuzzy rules for $g(x_1, x_2)$, which are shown in Fig. 23.6.

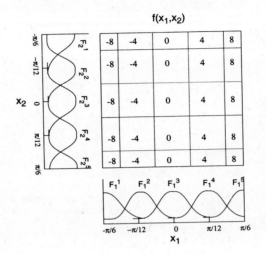

Figure 23.5. Linguistic fuzzy IF-THEN rules for $f(x_1, x_2)$.

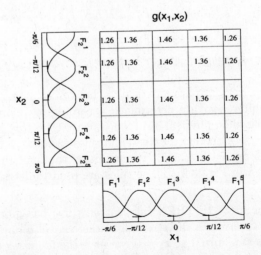

Figure 23.6. Linguistic fuzzy IF-THEN rules for $g(x_1, x_2)$.

Figs.23.7 and 23.8 show the closed-loop $x_1(t)$ together with the ideal output $y_m(t)$ for the two initial conditions $\mathbf{x}(0) = (-\frac{\pi}{60}, 0)^T$ and $\mathbf{x}(0) = (\frac{\pi}{60}, 0)^T$, respectively, after the fuzzy IF-THEN rules in Figs. 23.5 and 23.6 are incorporated. Comparing these results with those in Example 23.1, we see that the initial parts of control are apparently improved after incorporating these rules. $\square$

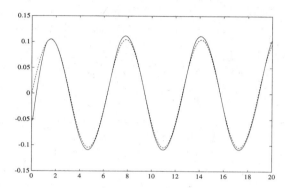

Figure 23.7. The state $x_1(t)$ (solid line) and its desired value $y_m(t) = \frac{\pi}{30}sin(t)$ (dashed line) for the initial condition $\mathbf{x}(0) = (-\frac{\pi}{60}, 0)^T$ in Example 23.2.

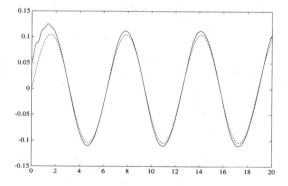

Figure 23.8. The state $x_1(t)$ (solid line) and its desired value $y_m(t) = \frac{\pi}{30}sin(t)$ (dashed line) for the initial condition $\mathbf{x}(0) = (\frac{\pi}{60}, 0)^T$ in Example 23.2.

23.4 Summary and Further Readings

In this chapter we have demonstrated the following:

- The motivation and classification of adaptive fuzzy controllers.

- Design of the indirect adaptive fuzzy controller using the certainty equivalent principle and the Lyapunov synthesis approach.

- How to apply the indirect adaptive fuzzy controller to the inverted pendulum tracking control problem.

There are many good textbooks on adaptive control, for example, Narendra and Annaswamy [1989] and Åström and Wittenmark [1995]. Adaptive control of nonlinear systems is currently an active research field and some earlier papers were Sastry and Isidori [1989], Narendra and Parthasarathy [1990], and Sanner and Slotine [1991]. The approach in this chapter is taken from Wang [1994a].

23.5 Exercises

Exercise 23.1. Consider the control of a mass on a frictionless surface by a motor force u, with the plant dynamics being

$$m\ddot{x} = u \tag{23.36}$$

The objective is to design a controller u such that $x(t)$ will converge to the reference signal $y_m(t)$, which is governed by

$$\ddot{y}_m + 2\dot{y}_m + y_m = r(t) \tag{23.37}$$

Consider the following two cases:

(a) m is known, and design a nonadaptive controller to achieve the objective.

(b) m is unknown, and design an adaptive controller to achieve the objective. Draw the block diagram of this adaptive control system, and discuss the intuitive reasonableness of the adaptation law.

Exercise 23.2. Consider the first-order nonlinear system

$$\dot{x} = f(x) + u \tag{23.38}$$

where $f(x)$ is unknown. Design an adaptive fuzzy controller u such that $x(t)$ follows the desired state $x_d(t) = sin(t)$ as t tends to infinity. Under what conditions can you guarantee that the tracking error $x_d(t) - x(t)$ converges to zero as t goes to infinity?

Exercise 23.3. Suppose that the plant knowledge provides very good estimates of f and g and we do not want to change them during online adaptation. Modify the indirect adaptive fuzzy control scheme in this chapter to achieve this objective.

Exercise 23.4. Consider the adaptation law (23.27) and (23.28) and discuss intuitively why parameter estimation can be easier on unstable systems than on stable systems.

Exercise 23.5. Consider the system

$$\dot{e} = -e + \theta w(t) \tag{23.39}$$
$$\dot{\theta} = -ew(t) \tag{23.40}$$

where $e, \theta \in R$, and $w(t)$ is bounded. Show that $e(t) \to 0$ as $t \to \infty$. Is the equilibrium at origin asymptotically stable?

Exercise 23.6. Repeat the simulations in Section 23.3 and simulate more cases.

Chapter 24

Basic Adaptive Fuzzy Controllers II

24.1 Design of the Direct Adaptive Fuzzy Controller

24.1.1 Problem Specification

Suppose that the plant is represented by

$$x^{(n)} = f(x, \dot{x}, ..., x^{(n-1)}) + bu \tag{24.1}$$

$$y = x \tag{24.2}$$

where f is an unknown function and b is an unknown positive constant. The control objective remains the same as in the indirect adaptive fuzzy control; that is, design a feedback controller $u = u(\mathbf{x}|\theta)$ based on fuzzy systems and an adaptation law for adjusting the parameter vector θ, such that the plant output y follows the ideal output y_m as close as possible. The main difference lies in the assumption about the available human knowledge. Specifically, instead of knowing the plant knowledge (23.3) and (23.4), here we are provided with some control knowledge; that is, the following fuzzy IF-THEN rules that describe human control actions:

$$IF \ x_1 \ is \ P_1^r \ and \ \cdots \ and \ x_n \ is \ P_n^r, \ THEN \ u \ is \ Q^r \tag{24.3}$$

where P_i^r and Q^r are fuzzy sets in R, and $r = 1, 2, ..., L_u$. The fuzzy controller should be designed in such a way that the rules (24.3) can be naturally incorporated.

24.1.2 Design of the Fuzzy Controller

To incorporate the rules (24.3), a natural choice is to use a single fuzzy system as the controller, that is, the fuzzy controller in this case is

$$u = u_D(\mathbf{x}|\theta) \tag{24.4}$$

where u_D is a fuzzy system and θ is the collection of adjustable parameters. Specifically, the fuzzy system $u_D(\mathbf{x}|\theta)$ is constructed from the following two steps:

- **Step 1.** For each variable x_i $(i = 1, 2, ..., n)$, define m_i fuzzy sets $A_i^{l_i}$ $(l_i = 1, 2, ..., m_i)$, which include the P_i^r $(r = 1, 2, ..., L_u)$ in (24.3) as special cases.

- **Step 2.** Construct the fuzzy system $u_D(\mathbf{x}|\theta)$ from the following $\prod_{i=1}^{n} m_i$ rules:

$$IF \ x_1 \ is \ A_1^{l_1} \ and \ \cdots \ and \ x_n \ is \ A_n^{l_n}, \ THEN \ u_D \ is \ S^{l_1 \cdots l_n} \qquad (24.5)$$

where $l_i = 1, 2, ..., m_i, i = 1, 2, ..., n$, and $S^{l_1 \cdots l_n}$ equals the Q^r in (24.3) if the IF part of (24.5) agrees with the IF part of (24.3) and equals some arbitrary fuzzy set otherwise. Specifically, using product inference engine, singleton fuzzifier and center average defuzzifier, we obtain

$$u_D(\mathbf{x}|\theta) = \frac{\sum_{l_1=1}^{m_1} \cdots \sum_{l_n=1}^{m_n} \bar{y}_u^{l_1 \cdots l_n} [\prod_{i=1}^{n} \mu_{A_i^{l_i}}(x_i)]}{\sum_{l_1=1}^{m_1} \cdots \sum_{l_n=1}^{m_n} [\prod_{i=1}^{n} \mu_{A_i^{l_i}}(x_i)]} \qquad (24.6)$$

Choose $\bar{y}_u^{l_1 \cdots l_n}$ as adjustable parameters and collect them into the vector $\theta \in R^{\prod_{i=1}^{n} m_i}$, the fuzzy controller becomes

$$u_D(\mathbf{x}|\theta) = \theta^T \xi(\mathbf{x}) \qquad (24.7)$$

where $\xi(\mathbf{x})$ is the same as in (23.14) except that the p_i there is now replaced by m_i.

From Step 2 we see that the initial values of some parameters in θ are chosen according to the rules (24.3), and the remainders are chosen randomly (or according to some strategy). Therefore, the control knowledge (24.3) is incorporated into the the fuzzy controller through the setting of its initial parameters.

24.1.3 Design of Adaptation Law

Let u^* be the same ideal control (23.5) as in Section 23.2, with $g(\mathbf{x}) = b$. Substituting (24.4) into (24.1) and by rearrangement, we obtain

$$e^{(n)} = -\mathbf{k}^T \mathbf{e} + b[u^* - u_D(\mathbf{x}|\theta)] \qquad (24.8)$$

Let Λ be as defined in (23.17) and $\mathbf{b} = (0, ..., 0, b)^T$, the closed-loop dynamics can be written into the vector form

$$\dot{\mathbf{e}} = \Lambda \mathbf{e} + \mathbf{b}[u^* - u_D(\mathbf{x}|\theta)] \qquad (24.9)$$

Define the optimal parameters

$$\theta^* = arg \min_{\theta \in R^{\prod_{i=1}^{n} m_i}} \left[\sup_{\mathbf{x} \in R^n} |u_D(\mathbf{x}|\theta) - u^*| \right] \qquad (24.10)$$

and the *minimum approximation error*

$$w = u_D(\mathbf{x}|\theta^*) - u^* \tag{24.11}$$

Using (24.11) and (24.7), we can rewrite the error equation (24.9) as

$$\dot{\mathbf{e}} = \Lambda\mathbf{e} + \mathbf{b}(\theta^* - \theta)^T\xi(\mathbf{x}) - \mathbf{b}w \tag{24.12}$$

Consider the Lyapunov function candidate

$$V = \frac{1}{2}\mathbf{e}^T P\mathbf{e} + \frac{b}{2\gamma}(\theta^* - \theta)^T(\theta^* - \theta) \tag{24.13}$$

where P is a positive definite matrix satisfying the Lyapunov equation (23.25), and γ is a positive constant (recall that $b > 0$ by assumption, so V is positive). Using (24.12) and (23.25), we have

$$\dot{V} = -\frac{1}{2}\mathbf{e}^T Q\mathbf{e} + \mathbf{e}^T P\mathbf{b}[(\theta^* - \theta)^T\xi(\mathbf{x}) - w] - \frac{b}{\gamma}(\theta^* - \theta)^T\dot{\theta} \tag{24.14}$$

Let p_n be the last column of P, then from $\mathbf{b} = (0, ..., 0, b)^T$ we have $\mathbf{e}^T P\mathbf{b} = \mathbf{e}^T p_n b$. So (24.14) can be rewritten as

$$\dot{V} = -\frac{1}{2}\mathbf{e}^T Q\mathbf{e} + \frac{b}{\gamma}(\theta^* - \theta)^T[\gamma\mathbf{e}^T p_n\xi(\mathbf{x}) - \dot{\theta}] - \mathbf{e}^T p_n bw \tag{24.15}$$

If we choose the adaptation law

$$\dot{\theta} = \gamma\mathbf{e}^T p_n\xi(\mathbf{x}) \tag{24.16}$$

then

$$\dot{V} = -\frac{1}{2}\mathbf{e}^T Q\mathbf{e} - \mathbf{e}^T p_n bw \tag{24.17}$$

Since $Q > 0$ and w is the minimum approximation error, we can hope that by designing the fuzzy system $u_D(\mathbf{x}|\theta)$ with a sufficiently large number of rules, the w would be small enough such that $|\mathbf{e}^T p_n bw| < \frac{1}{2}\mathbf{e}^T Q\mathbf{e}$, which results in $\dot{V} < 0$.

In summary, the whole direct adaptive fuzzy control system is shown in Fig. 24.1. The fuzzy control rules (24.3) (control knowledge) are incorporated by choosing the initial parameters of the fuzzy controller according them, as shown in Step 2 of the design procedure for the fuzzy controller.

24.1.4 Simulations

Example 24.1. Consider the first-order nonlinear system (19.31). Our objective is to use the direct adaptive fuzzy controller to regulate the state $x(t)$ to zero; that is, we have $y_m = 0$. It is clear that the plant (19.31) is unstable if the control equals

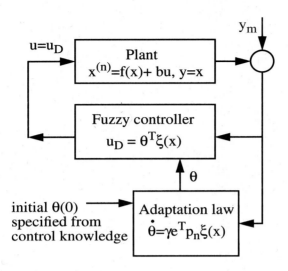

Figure 24.1. The direct adaptive fuzzy control system.

zero. Indeed, if $u(t) \equiv 0$, then $\dot{x} = \frac{1-e^{-x}}{1+e^{-x}} > 0$ for $x > 0$, and $\dot{x} = \frac{1-e^{-x}}{1+e^{-x}} < 0$ for $x < 0$. We choose $\gamma = 1$ and define six fuzzy sets $N3, N2, N1, P1, P2$, and $P3$ over the interval $[-3, 3]$ with membership functions $\mu_{N3}(x) = 1/(1+exp(5(x+2)))$, $\mu_{N2}(x) = exp(-(x+1.5)^2)$, $\mu_{N1}(x) = exp(-(x+0.5)^2)$, $\mu_{P1}(x) = exp(-(x-0.5)^2)$, $\mu_{P2}(x) = exp(-(x-1.5)^2)$, and $\mu_{P3}(x) = 1/(1 + exp(-5(x - 2)))$, which are shown in Fig. 24.2. We consider two cases: (i) there are no fuzzy control rules, and the initial $\theta_i(0)$'s are chosen randomly in the interval $[-2, 2]$, and (ii) there are two fuzzy control rules:

$$IF \ x \ is \ N2, \ THEN \ u(x) \ is \ PB \qquad (24.18)$$

$$IF \ x \ is \ P2, \ THEN \ u(x) \ is \ NB \qquad (24.19)$$

where $\mu_{PB}(u) = exp(-(u - 2)^2)$, and $\mu_{NB}(u) = exp(-(u + 2)^2)$. These two rules are obtained by considering the fact that our problem is to control $x(t)$ to zero; therefore, if x is negative, then the control $u(x)$ should be *positive big (PB)* so that it may happen that $\dot{x} > 0$ (see (19.31)). On the other hand, if x is positive, then the control $u(x)$ should be *negative big (NB)* so that it may happen that $\dot{x} < 0$. Figs.24.3 and 24.4 show the $x(t)$ for the cases without and with the linguistic control rules (24.18) and (24.19), respectively, for the initial condition $x(0) = 1$. We see from Figs. 24.3 and 24.4 that: (a) the direct adaptive fuzzy controller could regulate the plant to the origin without using the fuzzy control rules (24.18) and (24.19), and (b) by using the fuzzy control rules, the speed of convergence became much faster. $\square$

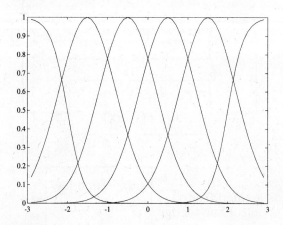

Figure 24.2. Membership functions defined over the state space for Example 24.1.

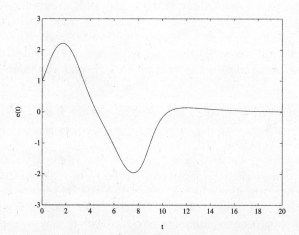

Figure 24.3. Closed-loop system state $x(t)$ using the direct adaptive fuzzy controller for the plant (19.31) without incorporating the fuzzy control rules.

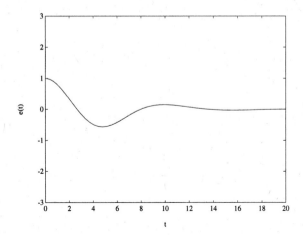

Figure 24.4. Closed-loop system state $x(t)$ using the direct adaptive fuzzy controller for the plant (19.31) after incorporating the fuzzy control rules (24.18) and (24.19).

Example 24.2. In this example, we consider the Duffing forced-oscillation system:

$$\dot{x}_1 = x_2 \qquad (24.20)$$
$$\dot{x}_2 = -0.1x_2 - x_1^3 + 12cos(t) + u(t) \qquad (24.21)$$

If the control $u(t)$ equals zero, the system is chaotic. The trajectory of the system with $u(t) \equiv 0$ is shown in the (x_1, x_2) phase plane in Fig.24.5 for the initial condition $x_1(0) = x_2(0) = 2$ and time period $t_0 = 0$ to $t_f = 60$. We now use the direct adaptive fuzzy controller to control the state x_1 to track the reference trajectory $y_m(t) = sin(t)$. In the phase plane, this reference trajectory is the unit circle: $y_m^2 + \dot{y}_m^2 = 1$. We choose $k_1 = 2, k_2 = 1, \gamma = 2$, and $Q = diag(10, 10)$. We use the six fuzzy sets shown in Fig. 24.2 for x_1 and x_2, and assume that there are no fuzzy control rules. The closed-loop trajectory is shown in Fig. 24.6 for initial condition $x_1(0) = x_2(0) = 2$ and time period from $t_0 = 0$ to $t_f = 60$. $\square$

24.2 Design of the Combined Direct/Indirect Adaptive Fuzzy Controller

The indirect adaptive fuzzy controller can make use of linguistic descriptions about the plant (plant knowledge), whereas the direct adaptive fuzzy controller can utilize linguistic control rules (control knowledge). In this section, we design the combined

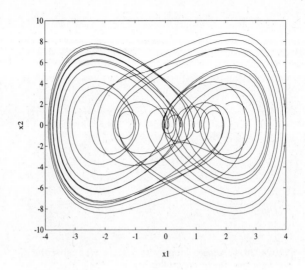

Figure 24.5. Trajectory of the chaotic system (24.20)
and (24.21) in the (x_1, x_2) phase plane with $u(t) \equiv 0$ and
$x_1(0) = x_2(0) = 2$.

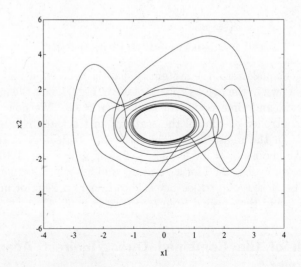

Figure 24.6. Closed-loop system trajectory $(x_1(t), x_2(t))$
using the direct adaptive fuzzy controller for the chaotic
system (24.20) and (24.21).

indirect/direct adaptive fuzzy controller that can incorporate both types of linguistic information.

24.2.1 Problem Specification

Consider the plant (24.1)-(24.2). For simplicity, assume that $b = 1$. Suppose that the following three pieces of information are available:

- **Information 1**: We know an approximate model of the plant; that is, we are given a function $\hat{f}$, which is an estimate of the f in (24.1).

- **Information 2**: We are given a set of fuzzy IF-THEN rules describing the difference $f - \hat{f}$ under various conditions; that is, we have

$$IF\ x_1\ is\ S_1^j\ and\ \cdots\ and\ x_n\ is\ S_n^j,\ THEN\ f - \hat{f}\ is\ E^j \qquad (24.22)$$

  where S_i^j and E^j are fuzzy sets in R, and $j = 1, 2, ..., L_e$.

- **Information 3**: We are given L_u fuzzy IF-THEN rules in the form of (24.3), which describe recommended control actions under various conditions.

Information 1 represents mathematical knowledge about the plant (obtained according to system configuration and physical laws), and Informations 2 and 3 are linguistic knowledge about the approximate mathematical model and the control actions, respectively. Our objective is to combine these three pieces of information into a controller and to design an adaptation law for the adjustable parameters in the controller, such that the closed-loop output $y(t)$ follows the ideal output $y_m(t)$.

24.2.2 Design of the Fuzzy Controller

Let $\mathbf{k}$ and $\mathbf{e}$ be as defined in Subsection 23.2.2. If $f(\mathbf{x})$ is known, then we know from Subsection 23.2.2 that the optimal control

$$u^* = -f(\mathbf{x}) + y_m^{(n)} + \mathbf{k}^T\mathbf{e} \qquad (24.23)$$

applied to (24.1) (with $b = 1$) guarantees $y(t) \to y_m(t)$. Let $\tilde{f}(\mathbf{x}|\theta_I)$ be a fuzzy system constructed from the rules including (24.22), then the best estimate of $f(\mathbf{x})$, based on Informations 1 and 2, is

$$\hat{f}(\mathbf{x}) + \tilde{f}(\mathbf{x}|\theta_I) \qquad (24.24)$$

Hence, to utilize Informations 1 and 2, we should use the controller

$$u_{12} = -\hat{f}(\mathbf{x}) - \tilde{f}(\mathbf{x}|\theta_I) + y_m^{(n)} + \mathbf{k}^T\mathbf{e} \qquad (24.25)$$

Since Information 3 consists of a set of fuzzy control rules, to use it we should consider the controller

$$u_3 = u_D(\mathbf{x}|\theta_D) \qquad (24.26)$$

where u_D is a fuzzy system constructed from the rules including (24.3). Therefore, a good choice of the final controller is a weighted average of u_{12} and u_3, that is, the final controller is

$$u = \alpha u_{12} + (1 - \alpha)u_3 \qquad (24.27)$$

where $\alpha \in [0,1]$ is a weighting factor. If the plant knowledge, that is, Informations 1 and 2, is more important and reliable than the control knowledge Information 3, we should choose a larger α; otherwise, a smaller α should be chosen.

The fuzzy systems $\tilde{f}(\mathbf{x}|\theta_I)$ and $u_D(\mathbf{x}|\theta_D)$ are designed following the same steps as in Sections 23.2 and 24.1; we omit the details. In terms of the adjustable parameters θ_I and θ_D, we represent $\tilde{f}(\mathbf{x}|\theta_I)$ and $u_D(\mathbf{x}|\theta_D)$ as

$$\tilde{f}(\mathbf{x}|\theta_I) = \theta_I^T \xi(\mathbf{x}) \qquad (24.28)$$
$$u_D(\mathbf{x}|\theta_D) = \theta_D^T \eta(\mathbf{x}) \qquad (24.29)$$

where $\xi(\mathbf{x})$ and $\eta(\mathbf{x})$ are vectors of fuzzy basis functions. Our next task is, as usual, to design an adaptation law for θ_I and θ_D, such that the tracking error $\mathbf{e}$ would be as small as possible.

24.2.3 Design of Adaptation Law

Substituting (24.27) into (24.1) (with $b = 1$) and after some straightforward manipulation, we obtain the following error equation governing the closed-loop system

$$e^{(n)} = -\mathbf{k}^T\mathbf{e} + \alpha(\hat{f} + \tilde{f} - f) + (1 - \alpha)(u^* - u_D) \qquad (24.30)$$

where u^* is given in (24.23), $e = y_m - y = y_m - x$, and $\mathbf{e} = (e, \dot{e}, ..., e^{(n-1)})^T$. Let Λ and $\mathbf{b}$ be defined in (23.17), then (24.30) can be rewritten as

$$\dot{\mathbf{e}} = \Lambda\mathbf{e} + \mathbf{b}[\alpha(\hat{f} + \tilde{f} - f) + (1 - \alpha)(u^* - u_D)] \qquad (24.31)$$

Suppose that θ_I and θ_D are M- and N-dimensional vectors, respectively, and define their optimal values as

$$\theta_I^* = arg \min_{\theta_I \in R^M} \left[\sup_{\mathbf{x} \in R^n} |\hat{f}(\mathbf{x}) + \tilde{f}(\mathbf{x}|\theta_I) - f(\mathbf{x})| \right] \qquad (24.32)$$

$$\theta_D^* = arg \min_{\theta_D \in R^N} \left[\sup_{\mathbf{x} \in R^n} |u^*(\mathbf{x}) - u_D(\mathbf{x}|\theta_D)| \right] \qquad (24.33)$$

Let w be the *minimum approximation error* defined by

$$w = \alpha[\hat{f}(\mathbf{x}) + \tilde{f}(\mathbf{x}|\theta_I^*) - f(\mathbf{x})] + (1 - \alpha)[u^*(\mathbf{x}) - u_D(\mathbf{x}|\theta_D^*)] \qquad (24.34)$$

Using w and (24.28)-(29.29), we can rewrite the error equation (24.31) as

$$\dot{\mathbf{e}} = \Lambda\mathbf{e} + \mathbf{b}[\alpha(\theta_I - \theta_I^*)^T\xi(\mathbf{x}) + (1 - \alpha)(\theta_D - \theta_D^*)^T\eta(\mathbf{x}) + w] \qquad (24.35)$$

Consider the Lyapunov function candidate

$$V = \frac{1}{2}\mathbf{e}^T P \mathbf{e} + \frac{\alpha}{2\gamma_1}(\theta_I - \theta_I^*)^T(\theta_I - \theta_I^*) + \frac{1-\alpha}{2\gamma_2}(\theta_D - \theta_D^*)^T(\theta_D - \theta_D^*) \quad (24.36)$$

where $P > 0$ satisfies the Lyapunov equation (23.25), and γ_1, γ_2 are positive constants. Using (24.35) and (23.25), we have

$$\dot{V} = -\frac{1}{2}\mathbf{e}^T Q \mathbf{e} + \frac{\alpha}{\gamma_1}(\theta_I - \theta_I^*)^T[\dot{\theta}_I + \gamma_1 \mathbf{e}^T P \mathbf{b}\xi(\mathbf{x})]$$

$$+ \frac{1-\alpha}{\gamma_2}(\theta_D - \theta_D^*)^T[\dot{\theta}_D + \gamma_2 \mathbf{e}^T P \mathbf{b}\eta(\mathbf{x})] + \mathbf{e}^T P \mathbf{b} w \quad (24.37)$$

Therefore, if we choose the adaptation law

$$\dot{\theta}_I = -\gamma_1 \mathbf{e}^T P \mathbf{b}\xi(\mathbf{x}) \quad (24.38)$$
$$\dot{\theta}_D = -\gamma_2 \mathbf{e}^T P \mathbf{b}\eta(\mathbf{x}) \quad (24.39)$$

then we have

$$\dot{V} = -\frac{1}{2}\mathbf{e}^T Q \mathbf{e} + \mathbf{e}^T P \mathbf{b} w \quad (24.40)$$

Since w is the minimum approximation error, (24.40) is the best we can get.

In summary, the combined direct/indirect adaptive fuzzy control system is shown in Fig.24.7. We see from the figure that Information 1 is directly used in the controller, and Informations 2 and 3 are incorporated throught the initial parameters $\theta_I(0)$ and $\theta_D(0)$, respectively.

24.2.4   Convergence Analysis

Up to this point, we have developed the basic schemes of indirect, direct, and combined indirect/direct adaptive fuzzy controllers. We used the Lyapunov synthesis approach to design these adaptive fuzzy controllers, so that they are intuitively appealing. However, we did not analyze the performance of the resulting adaptive fuzzy control systems in detail. There are two fundamental issues in performance analysis: stability and convergence, where by stability we mean the boundedness of the variables involved (states, parameters, etc.), and by convergence we mean the convergence of the tracking error to zero. We now show in the following theorem that if the fuzzy control system in Fig. 24.7 is stable, then under mild conditions the tracking error will converge to zero. In the next chapter, we will develop approaches to achieve stability.

Theorem 24.1. Consider the combined indirect/direct adaptive fuzzy control system in Fig. 24.7. If the state $\mathbf{x}$, the parameters θ_I and θ_D, and the minimum approximation error w are bounded, then:

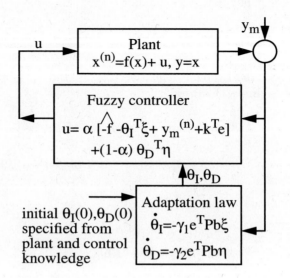

Figure 24.7. The combined indirect/direct adaptive fuzzy
control system.

(a) The tracking error satisfies

$$\int_0^t |\mathbf{e}(\tau)|^2 d\tau \le a + b \int_0^t |w(\tau)|^2 d\tau \qquad (24.41)$$

for all $t \ge 0$, where a and b are constants, and w is the minimum approximation
error defined by (24.34).

(b) If w is squared integrable, that is, if $\int_0^\infty |w(t)|^2 dt < \infty$, then $lim_{t\to\infty}|\mathbf{e}(t)| = 0$.

Proof: (a) Let λ_{Qmin} be the minimum eigenvalue of Q. Then from (24.40) we
have

$$\dot{V} \le -\frac{\lambda_{Qmin} - 1}{2}|\mathbf{e}|^2 - \frac{1}{2}(|\mathbf{e}|^2 - 2\mathbf{e}^T Pbw + |Pbw|^2) + \frac{1}{2}|Pbw|^2$$

$$\le -\frac{\lambda_{Qmin} - 1}{2}|\mathbf{e}|^2 + \frac{1}{2}|Pbw|^2 \qquad (24.42)$$

Integrating both sides of (24.42) and assuming that $\lambda_{Qmin} > 1$ (since Q is determined by the designer, we can choose such Q), we have

$$\int_0^t |\mathbf{e}(\tau)|^2 d\tau \le \frac{2}{\lambda_{Qmin} - 1}(|V(0)| + |V(t)|) + \frac{1}{\lambda_{Qmin} - 1}|Pb|^2 \int_0^t |w(\tau)|^2 d\tau$$

$$(24.43)$$

Define $a = \frac{2}{\lambda_{Qmin}-1}(|V(0)| + \sup_{t\geq 0}|V(t)|)$ and $b = \frac{1}{\lambda_{Qmin}-1}|Pb|^2$, (24.43) becomes (24.41). (Note that $\sup_{t\geq 0}|V(t)|$ is finite since e, $\theta_I - \theta_I^*$ and $\theta_D - \theta_D^*$ are all bounded by assumption.)

(b) If w is squared integrable, then from (24.41) we conclude that e is also squared integrable. Since all the variables in the right-hand side of (24.35) are bounded by assumption, we have that $\dot{e}$ is also bounded. Using the Barbalat's Lemma (Sastry and Bodson [1989]) (If e is squared integrable and bounded and $\dot{e}$ is bounded, then $lim_{t\to\infty}|e(t)| = 0$), we have $lim_{t\to\infty}|e(t)| = 0$. $\square$

Using the same ideas as in the proof of Theorem 24.1, we can prove that if all the variables in the indirect and direct adaptive fuzzy control systems in Figs. 23.2 and 24.1 are bounded and the minimum approximation errors are squared integrable, then the tracking errors will converge to zero.

24.3   Summary and Further Readings

In this chapter we have demonstrated the following:

- How to design the direct adaptive fuzzy controller.

- How to combine the approximate mathematical model, linguistic model description, and linguistic control rules into a single adaptive fuzzy controller (the combined indirect/direct adaptive fuzzy controller).

- Convergence analysis of the adaptive fuzzy control systems.

The materials in this chapter are taken from Wang [1994a], where more simulation results can be found. Other adaptive fuzzy control approaches can be found in, for example, Spooner and Passino [1995], Johansen [1994], and Vandegrift, Lewis, Jagannathan, and Liu [1995].

24.4   Exercises

Exercise 24.1. Consider the indirect adaptive fuzzy control system in Fig. 23.2 and show that the tracking error of the closed-loop system satisfies the bound (24.41).

Exercise 24.2. Show that the tracking error of the direct adaptive fuzzy control system in Fig. 24.1 satisfies the bound (24.41).

Exercise 24.3. Suppose Informations 1 and 2 in Section 24.2 are replaced by a set of fuzzy IF-THEN rules describing the unknown function f. Combine these rules with Information 3 into an adaptive fuzzy controller.

Exercise 24.4. Suppose that the parameters in the membership functions $\mu_{A_i^{l_i}}$ in the fuzzy controller u_D of (24.6) are also adjustable parameters (so that the u_D

cannot be linear in the parameters θ as in (24.7)). Design a direct adaptive fuzzy controller in this case.

Exercise 24.5. Repeat the simulation in Fig. 24.3 with different initial conditions.

Exercise 24.6. Simulate the combined indirect/direct adaptive fuzzy controller for the systems in Examples 24.1 and 24.2.

Chapter 25

Advanced Adaptive Fuzzy Controllers I

In Chapters 23 and 24, we developed the basics of indirect, direct, and combined indirect/direct adaptive fuzzy controllers. The idea was to use the Lyapunov synthesis approach and the certainty equivalent principle. Although we did not provide much theoretical analysis, we proved in Theorem 24.1 that if all the variables involved are bounded and the minimum approximation error is squared integrable, the tracking error will converge to zero. Therefore, the key problems are: (i) how to guarantee the boundedness of all the variables, and (ii) how to make the minimum approximation error squared integrable. Since the fuzzy systems are universal approximators, we can make the minimum approximation error arbitrarily small by using more rules to construct the fuzzy systems. The main objective of this chapter is to solve the first problem. In Sections 25.1 and 25.2, we will use supervisory control and parameter projection to guarantee the boundedness of the states and the parameters, respectively.

25.1 State Boundedness By Supervisory Control

25.1.1 For Indirect Adaptive Fuzzy Control System

Consider the basic indirect adaptive fuzzy control system in Fig. 23.2; that is, the plant is (23.1)-(23.2), the controller is (23.7) with the fuzzy systems $\hat{f}$ and $\hat{g}$ given by (23.12) and (23.13), and the adaptation law is (23.27)-(23.28). Our task now is to make the state $\mathbf{x}$ bounded.

We use the supervisory control idea in Chapter 20 for this task. Specifically, we append a *supervisory controller* u_s on top of the basic controller u_I of (23.7), so that the controller now becomes

$$u = u_I + u_s \qquad (25.1)$$

The u_s should be designed to: (i) guarantee the boundedness of $\mathbf{x}$, and (ii) operate in a supervisory fashion; that is, u_s should equal zero when the state $\mathbf{x}$ is well inside the stable range and is nonzero only when $\mathbf{x}$ tends to leave the stable range. We now design this u_s.

Replacing $u = u_I$ by $u = u_I + u_s$, the error equation (23.18) becomes

$$\dot{\mathbf{e}} = \Lambda \mathbf{e} + \mathbf{b}[(\hat{f}(\mathbf{x}|\theta_f) - f(\mathbf{x})) + (\hat{g}(\mathbf{x}|\theta_g) - g(\mathbf{x}))u_I - g(\mathbf{x})u_s] \quad (25.2)$$

Define

$$V_e = \frac{1}{2}\mathbf{e}^T P \mathbf{e} \quad (25.3)$$

where $P > 0$ satisfies the Lyapunov equation (23.25). Since $\mathbf{e} = (y_m - x_1, \dot{y}_m - x_2, ..., y_m^{(n-1)} - x_n)^T$ and $y_m, \dot{y}_m, ..., y_m^{(n-1)}$ are assumed to be bounded, the boundedness of V_e implies the boundedness of $\mathbf{x} = (x_1, x_2, ..., x_n)^T$. Thus, our task becomes to design the u_s such that $V_e \leq \bar{V}$ is guaranteed, where $\bar{V}$ is a given constant determined from the bound for $\mathbf{x}$. From (25.3) we have $(V_e)^{1/2} \geq (\frac{\lambda_{Pmin}}{2})^{1/2}|\mathbf{e}| \geq (\frac{\lambda_{Pmin}}{2})^{1/2}(|\mathbf{x}| - |\mathbf{y_m}|)$, where λ_{Pmin} is the minimum eigenvalue of P and $\mathbf{y_m} = (y_m, \dot{y}_m, ..., y_m^{(n-1)})^T$. Hence, $V_e \leq \bar{V}$ is equivalent to $|\mathbf{x}| \leq |\mathbf{y_m}| + (\frac{2\bar{V}}{\lambda_{Pmin}})^{1/2}$. So if we want $|\mathbf{x}| \leq M_x$, where M_x is a constant, then we can choose

$$\bar{V} = \frac{\lambda_{Pmin}}{2}(M_x - \sup_{t \geq 0}|\mathbf{y_m}|)^2 \quad (25.4)$$

Since $V_e \geq 0$, a way to guarantee $V_e \leq \bar{V}$ is to design the u_s such that $\dot{V}_e < 0$ when $V_e \geq \bar{V}$. From (25.2) and (23.25) we have

$$\dot{V}_e = -\frac{1}{2}\mathbf{e}^T Q \mathbf{e} + \mathbf{e}^T P \mathbf{b}[(\hat{f} - f) + (\hat{g} - g)u_I - gu_s]$$

$$\leq -\frac{1}{2}\mathbf{e}^T Q \mathbf{e} + |\mathbf{e}^T P \mathbf{b}|[|\hat{f}| + |f| + |\hat{g}u_I| + |gu_I|] - \mathbf{e}^T P \mathbf{b} gu_s \quad (25.5)$$

In order to design the u_s such that the right-hand side of (25.5) is negative, we need to know the bounds of f and g; that is, we have to make the following assumption.

Assumption 25.1. We can determine functions $f^U(\mathbf{x}), g^U(\mathbf{x})$ and $g_L(\mathbf{x})$ such that $|f(\mathbf{x})| \leq f^U(\mathbf{x})$ and $0 < g_L(\mathbf{x}) \leq g(\mathbf{x}) \leq g^U(\mathbf{x})$ for all $\mathbf{x} \in R^n$.

Based on f^U, g^U and g_L and by observing (25.5), we choose the supervisory controller u_s as follows:

$$u_s = I^* sgn(\mathbf{e}^T P \mathbf{b})\frac{1}{g_L(\mathbf{x})}\left[|\hat{f}(\mathbf{x}|\theta_f)| + f^U(\mathbf{x}) + |\hat{g}(\mathbf{x}|\theta_g)u_I| + |g^U(\mathbf{x})u_I|\right] \quad (25.6)$$

where $I^* = 1$ if $V_e \geq \bar{V}$, $I^* = 0$ if $V_e < \bar{V}$, and $sgn(y) = 1(-1)$ if $y \geq 0 \ (< 0)$. Substituting (25.6) into (25.5) we have that if $V_e \geq \bar{V}$, then

$$\dot{V}_e \leq -\frac{1}{2}\mathbf{e}^T Q \mathbf{e} + |\mathbf{e}^T P \mathbf{b}|\left[|\hat{f}| + |f| + |\hat{g}u_I| + |gu_I| - \frac{g}{g_L}(|\hat{f}| + |f^U| + |\hat{g}u_I| + |g^U u_I|)\right]$$

$$\leq -\frac{1}{2}\mathbf{e}^T Q \mathbf{e} < 0 \quad (25.7)$$

where we assume $\mathbf{e} \neq 0$ since this is natural under $V_e \geq \bar{V}$. Consequently, using the control (25.1) with u_s given by (25.6), we can guarantee that $V_e \leq \bar{V}$, which, when $\bar{V}$ is chosen according to (25.4), guarantees $|\mathbf{x}| \leq M_x$ for any given constant M_x.

Because of the indicator function I^*, u_s is nonzero only when $V_e \geq \bar{V}$, therefore u_s is a supervisory controller. That is, if the closed-loop system with the fuzzy controller u_I of (23.7) is well-behaved in the sense that the error is within the constraint set (that is, $V_e \leq \bar{V}$ or equivalently $|\mathbf{x}| \leq M_x$), then the supervisory controller u_s is idle. On the other hand, if the system tends to be unstable (that is, $V_e \geq \bar{V}$), then the supervisory controller u_s takes action to force $V_e \leq \bar{V}$.

25.1.2 For Direct Adaptive Fuzzy Control System

Consider the basic direct adaptive fuzzy control system in Fig.24.1, where the process is modeled by (24.1)-(24.2), the controller is the fuzzy system (24.7), and the adaptation law is given by (24.16). Our task now is to design a *supervisory controller* u_s on top of the fuzzy controller u_D, such that the state $\mathbf{x}$ is guaranteed to be bounded.

Let the control be

$$u = u_D(\mathbf{x}|\theta) + u_s(\mathbf{x}) \tag{25.8}$$

and define the ideal control u^* as in (23.5) (with $g(\mathbf{x}) = b$). Substituting (25.8) into (24.1) and after rearrangement, we obtain the closed-loop error equation

$$\dot{\mathbf{e}} = \Lambda \mathbf{e} + \mathbf{b}[u^* - u_D(\mathbf{x}|\theta) - u_s(\mathbf{x})] \tag{25.9}$$

where Λ and $\mathbf{b}$ are the same as in (24.9). Defining the Lyapunov function candidate V_e as in (25.3) and using (25.9) and (23.25), we have

$$\dot{V}_e = -\frac{1}{2}\mathbf{e}^T Q\mathbf{e} + \mathbf{e}^T P\mathbf{b}[u^* - u_D(\mathbf{x}|\theta) - u_s(\mathbf{x})]$$

$$\leq -\frac{1}{2}\mathbf{e}^T Q\mathbf{e} + |\mathbf{e}^T P\mathbf{b}|(|u^*| + |u_D|) - \mathbf{e}^T P\mathbf{b}u_s(\mathbf{x}) \tag{25.10}$$

In order to design u_s such that $\dot{V}_e < 0$, we have to make the following assumption, which is essentially the same as Assumption 25.1.

Assumption 25.2. We can determine a function $f^U(\mathbf{x})$ and a constant b_L such that $|f(\mathbf{x})| \leq f^U(\mathbf{x})$ and $0 < b_L \leq b$.

By observing (25.10) and (23.5), we design the supervisory controller u_s as follows:

$$u_s(\mathbf{x}) = I^* sgn(\mathbf{e}^T P\mathbf{b}) \left[|u_D| + \frac{1}{b_L}(f^U + |y_m^{(n)}| + |\mathbf{k}^T \mathbf{e}|) \right] \tag{25.11}$$

where the indictor function I^* is the same as that in (25.6); that is, $I^* = 1$ if $V_e \geq \bar{V}$, $I^* = 0$ if $V_e < \bar{V}$, where $\bar{V}$ is determined according to the bound of $\mathbf{x}$ as in (25.4). Substituting (25.11) and (23.5) into (25.10) and considering the $I^* = 1$ case, we have

$$\dot{V}_e \le -\frac{1}{2}\mathbf{e}^T Q \mathbf{e} + |\mathbf{e}^T P \mathbf{b}| \left[\frac{1}{b}(|f| + |y_m^{(n)}| + |\mathbf{k}^T \mathbf{e}|) - \frac{1}{b_L}(f^U + |y_m^{(n)}| + |\mathbf{k}^T \mathbf{e}|) \right]$$

$$\le -\frac{1}{2}\mathbf{e}^T Q \mathbf{e} < 0 \tag{25.12}$$

Therefore, using the supervisory controller u_s, we can guarantee that $V_e \le \bar{V}$, from which we have $|\mathbf{x}| \le M_x$ if we choose $\bar{V}$ according to (25.4).

The design of supervisory controller for the combined indirect/direct adaptive fuzzy control system is left as an exercise.

25.2   Parameter Boundedness By Projection

25.2.1   For Indirect Adaptive Fuzzy Control System

Using the basic adaptation law (23.27)-(23.28), we cannot guarantee that the parameters θ_f and θ_g are bounded. If θ_f diverges to infinity, then the fuzzy system $\hat{f}(\mathbf{x}|\theta_f)$ will steadily increase and result in an unbounded control u_I; this is clearly unacceptable. Therefore, to develop a stable system (in the sense that all variables are bounded), we must modify the adaptation law such that the parameters are bounded.

Let the constrain sets Ω_f and Ω_g for θ_f and θ_g be defined as

$$\Omega_f = \{\theta_f \in R^{\Pi_{i=1}^n p_i} \mid |\theta_f| \le M_f\} \tag{25.13}$$

$$\Omega_g = \{\theta_g \in R^{\Pi_{i=1}^n q_i} \mid 0 < \epsilon \le |\theta_g| \le M_g\} \tag{25.14}$$

where p_i and q_i are as defined in Subsection 23.2.2, and M_f, ϵ and M_g are constants. We require $|\theta_g|$ to be bounded from below by $\epsilon > 0$ because from (23.7) we see that $\hat{g} = \theta_g^T \eta$ must be nonzero. We now modify the basic adaptation law (23.27)-(23.28) to guarantee $\theta_f \in \Omega_f$ and $\theta_g \in \Omega_g$.

The basic idea is the following: if the parameter vector is in the interior of the constraint set or on the boundary but moving toward the inside of the constraint set, then use the basic adaptation law (23.27)-(23.28); if the parameter vector is on the boundary of the constraint set but moving toward the outside of the constraint set, then project the gradient of the parameter vector onto the supporting hyperplane; see Fig.25.1. Specifically, the modified adaptation law with projection for the indirect adaptive fuzzy system is given as follows:

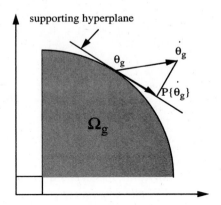

Figure 25.1. Illustration of the projection algorithm.

Adaptation Law with Projection:

- For θ_f, use

$$
\dot{\theta}_f = \begin{cases} -\gamma_1 \mathbf{e}^T P \mathbf{b} \xi(\mathbf{x}) & if \quad (|\theta_f| < M_f) \; or \; (|\theta_f| = M_f \; and \\ & \quad \mathbf{e}^T P \mathbf{b} \theta_f^T \xi(\mathbf{x}) \geq 0) \\ P\{-\gamma_1 \mathbf{e}^T P \mathbf{b} \xi(\mathbf{x})\} & if \quad (|\theta_f| = M_f \; and \; \mathbf{e}^T P \mathbf{b} \theta_f^T \xi(\mathbf{x}) < 0) \end{cases}
$$

$$(25.15)$$

where the projection operator $P\{*\}$ is defined as

$$
P\{-\gamma_1 \mathbf{e}^T P \mathbf{b} \xi(\mathbf{x})\} = -\gamma_1 \mathbf{e}^T P \mathbf{b} \xi(\mathbf{x}) + \gamma_1 \mathbf{e}^T P \mathbf{b} \frac{\theta_f \theta_f^T \xi(\mathbf{x})}{|\theta_f|^2}
$$

$$(25.16)$$

- For θ_g, use

  · Whenever an element θ_{gi} of θ_g equals ϵ, use

$$
\dot{\theta}_{gi} = \begin{cases} -\gamma_2 \mathbf{e}^T P \mathbf{b} \eta_i(\mathbf{x}) u_I & if \quad \mathbf{e}^T P \mathbf{b} \eta_i(\mathbf{x}) u_I < 0 \\ 0 & if \quad \mathbf{e}^T P \mathbf{b} \eta_i(\mathbf{x}) u_I \geq 0 \end{cases}
$$

$$(25.17)$$

where $\eta_i(\mathbf{x})$ is the *ith* component of $\eta(\mathbf{x})$.

· Otherwise, use

$$
\dot{\theta}_g = \begin{cases} -\gamma_2 \mathbf{e}^T P \mathbf{b} \eta(\mathbf{x}) u_I & if \quad (|\theta_g| < M_g) \; or \; (|\theta_g| = M_g \; and \\ & \qquad \mathbf{e}^T P \mathbf{b} \theta_g^T \eta(\mathbf{x}) u_I \geq 0) \\ P\{-\gamma_2 \mathbf{e}^T P \mathbf{b} \eta(\mathbf{x}) u_I\} & if \quad (|\theta_g| = M_g \; and \; \mathbf{e}^T P \mathbf{b} \theta_g^T \eta(\mathbf{x}) u_I < 0) \end{cases}
$$

(25.18)

where the projection operator $P\{*\}$ is defined as

$$
P\{-\gamma_2 \mathbf{e}^T P \mathbf{b} \eta(\mathbf{x}) u_I\} = -\gamma_2 \mathbf{e}^T P \mathbf{b} \eta(\mathbf{x}) u_I + \gamma_2 \mathbf{e}^T P \mathbf{b} \frac{\theta_g \theta_g^T \eta(\mathbf{x}) u_I}{|\theta_g|^2}
$$

(25.19)

The following theorem shows that the modified adaptation law (25.15)-(25.19) guarantees that $\theta_f \in \Omega_f$ and $\theta_g \in \Omega_g$.

Theorem 25.1. Let the constraint sets Ω_f and Ω_g be defined in (25.13) and (25.14). If the initial values of the parameters satisfy $\theta_f(0) \in \Omega_f$ and $\theta_g(0) \in \Omega_g$, then the adaptation law (25.15)-(25.19) guarantees that $\theta_f(t) \in \Omega_f$ and $\theta_g(t) \in \Omega_g$ for all $t \geq 0$.

Proof: To prove $|\theta_f| \leq M_f$, let $V_f = \frac{1}{2} \theta_f^T \theta_f$. If the condition in the first line of (25.15) is true, we have either $|\theta_f| < M_f$ or $\dot{V}_f = -\gamma_1 \mathbf{e}^T P \mathbf{b} \theta_f^T \xi(\mathbf{x}) \leq 0$ when $|\theta_f| = M_f$; hence, we have $|\theta_f| \leq M_f$ in this case. If the condition in the second line of (25.15) is true, we have $|\theta_f| = M_f$ and $\dot{V}_f = -\gamma_1 \mathbf{e}^T P \mathbf{b} \theta_f^T \xi(\mathbf{x}) + \gamma_1 \mathbf{e}^T P \mathbf{b} \frac{\theta_f^T \theta_f}{|\theta_f|^2} \theta_f^T \xi(\mathbf{x}) = 0$; hence, $|\theta_f| \leq M_f$ in this case. Since the initial $|\theta_f(0)| \leq M_f$, we have $|\theta_f(t)| \leq M_f$ for all $t \geq 0$.

Using the same method, we can prove that $|\theta_g(t)| \leq M_g$ for all $t \geq 0$. To show $|\theta_g| \geq \epsilon$, we see from (25.17) that if $\theta_{gi} = \epsilon$, then $\dot{\theta}_{gi} \geq 0$; hence, we always have $\theta_{gi} \geq \epsilon$ which guarantees $|\theta_g| \geq \epsilon$. $\square$

25.2.2   For Direct Adaptive Fuzzy Control System

The idea is the same. Let the constraint set for the θ in the direct adaptive fuzzy controller (24.7) be

$$
\Omega_D = \{\theta \in R^{\Pi_{i=1}^n m_i} \mid |\theta| \leq M_D\}
$$

(25.20)

The modified adaptation law with projection is

$$
\dot{\theta} = \begin{cases} \gamma \mathbf{e}^T p_n \xi(\mathbf{x}) & if \quad (|\theta| < M_D) \; or \; (|\theta| = M_D \; and \; \mathbf{e}^T p_n \theta^T \xi(\mathbf{x}) \geq 0) \\ P\{\gamma \mathbf{e}^T p_n \xi(\mathbf{x})\} & if \quad (|\theta| = M_D \; and \; \mathbf{e}^T p_n \theta^T \xi(\mathbf{x}) < 0) \end{cases}
$$

(25.21)

where the projection operator is defined as

$$
P\{\gamma \mathbf{e}^T p_n \xi(\mathbf{x})\} = \gamma \mathbf{e}^T p_n \xi(\mathbf{x}) - \gamma \mathbf{e}^T p_n \frac{\theta \theta^T \xi(\mathbf{x})}{|\theta|^2}
$$

(25.22)

The following theorem shows the boundedness of the parameters.

Theorem 25.2. Let the constraint set Ω_D be defined in (25.20). If the initial parameter $\theta(0) \in \Omega_D$, then the adaptation law (25.21) guarantees that $\theta(t) \in \Omega_D$ for all $t \geq 0$.

The proof of this theorem is the same as the proof of Theorem 25.1 and is left as an exercise.

25.3   Stable Direct Adaptive Fuzzy Control System

25.3.1   Stability and Convergence Analysis

Adding the supervisory controller (25.11) to the basic adaptive fuzzy control system in Fig. 24.1 and using the modified adaptation law (25.21), we obtain the advanced direct adaptive fuzzy control system that is shown in Fig.25.2. The following theorem shows the properties of this adaptive fuzzy control system.

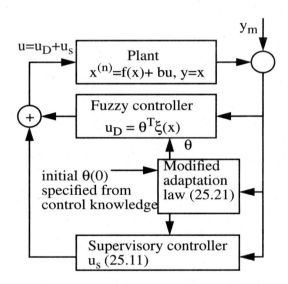

Figure 25.2. Advanced direct adaptive fuzzy control system with supervisory controller and modified adaptation law.

Theorem 25.3. Consider the adaptive fuzzy control system in Fig.25.2; that is, the plant under control is in the form of (24.1)-(24.2), the controller $u = u_D(\mathbf{x}|\theta) + u_s(\mathbf{x})$ with the fuzzy system $u_D(\mathbf{x}|\theta)$ designed as in (24.6) or (24.7) and the supervisory controller $u_s(\mathbf{x})$ designed as in (25.11), and the adaptation law is given by (25.21)-(25.22). This adaptive fuzzy control system is guaranteed to have the following properties:

(a) All the parameters and states are bounded, that is,

$$|\theta(t)| \le M_D \tag{25.23}$$
$$|\mathbf{x}(t)| \le M_x \tag{25.24}$$

where M_D and M_x are constants, and the $\bar{V}$ that determined the I^* in u_s is chosen according to (25.4).

(b) The tracking error $\mathbf{e}$ is bounded by the minimum approximation error w (defined in (24.11)) according to

$$\int_0^t |\mathbf{e}(\tau)|^2 d\tau \le a + b \int_0^t |w(\tau)|^2 d\tau, \ \forall t \ge 0 \tag{25.25}$$

where a and b are constants.

(c) If w is squared integrable, that is, if $\int_0^\infty |w(t)|^2 dt < \infty$, then $lim_{t\to\infty}|\mathbf{e}(t)| = 0$.

Proof: (a) (25.23) follows from Theorem 25.2, and (25.24) is obtained from (25.12) (which shows $V_e \le \bar{V}$) and (25.4) (which shows that $V_e \le \bar{V}$ is equivalent to $|\mathbf{x}(t)| \le M_x$).

(b) With the supervisory controller $u_s(\mathbf{x}|\theta)$ added, the error equation (24.12) becomes

$$\dot{\mathbf{e}} = \Lambda \mathbf{e} + \mathbf{b}(\theta^* - \theta)^T \xi(\mathbf{x}) - \mathbf{b}w - \mathbf{b}u_s(\mathbf{x}) \tag{25.26}$$

Considering the Lyapunov function candidate V of (24.13) and using (25.26), we have

$$\dot{V} = -\frac{1}{2}\mathbf{e}^T Q\mathbf{e} + \frac{b}{\gamma}(\theta^* - \theta)^T[\gamma \mathbf{e}^T p_n \xi(\mathbf{x}) - \dot{\theta}] - \mathbf{e}^T p_n bw - \mathbf{e}^T P\mathbf{b}u_s(\mathbf{x}) \tag{25.27}$$

Substituting the adaptation law (25.21) into (25.27) and letting $I_\theta^* = 1$ indicate that the condition in the second line of (25.21) is true and $I_\theta^* = 0$ indicate the other cases, we have

$$\dot{V} = -\frac{1}{2}\mathbf{e}^T Q\mathbf{e} + I_\theta^* \mathbf{e}^T p_n b \frac{(\theta^* - \theta)^T \theta\theta^T \xi(\mathbf{x})}{|\theta|^2} - \mathbf{e}^T p_n bw - \mathbf{e}^T P\mathbf{b}u_s(\mathbf{x}) \tag{25.28}$$

We now show that the second term in the right-hand side of (25.28) is nonpositive. If $I_\theta^* = 0$, the conclusion is trivial. Let $I_\theta^* = 1$, which means that $|\theta| = M_D$ and $\mathbf{e}^T p_n \theta^T \xi(\mathbf{x}) < 0$, we have $(\theta^* - \theta)^T\theta = \frac{1}{2}(|\theta^*|^2 - |\theta|^2 - |\theta - \theta^*|^2) < 0$ since $|\theta| = M_D \ge |\theta^*|$. Therefore, (25.28) is simplified to

$$\dot{V} \le -\frac{1}{2}\mathbf{e}^T Q\mathbf{e} - \mathbf{e}^T p_n bw - \mathbf{e}^T P\mathbf{b}u_s(\mathbf{x}) \tag{25.29}$$

Since the supervisory controller $u_s(\mathbf{x})$ of (25.11) has the same sign as $\mathbf{e}^T P \mathbf{b}$, we obtain from (25.29) that

$$\dot{V} \leq -\frac{1}{2}\mathbf{e}^T Q \mathbf{e} - \mathbf{e}^T p_n b w$$

$$\leq -\frac{\lambda_{Qmin} - 1}{2}|\mathbf{e}|^2 - \frac{1}{2}(|\mathbf{e}|^2 - 2\mathbf{e}^T p_n b w + |p_n b w|^2) + \frac{1}{2}|p_n b w|^2$$

$$\leq -\frac{\lambda_{Qmin} - 1}{2}|\mathbf{e}|^2 + \frac{1}{2}|p_n b w|^2 \qquad\qquad (25.30)$$

where λ_{Qmin} is the minimum eigenvalue of Q and we assume $\lambda_{Qmin} > 1$. Integrating both sides of (25.30), we obtain

$$\int_0^t |\mathbf{e}(\tau)|^2 d\tau \leq \frac{2}{\lambda_{Qmin} - 1}(|V(0)| + |V(t)|) + \frac{1}{\lambda_{Qmin} - 1}|p_n b|^2 \int_0^t |w(\tau)|^2 d\tau$$

$$(25.31)$$

Define $a = \frac{2}{\lambda_{Qmin}-1}(|V(0)| + sup_{t\geq 0}|V(t)|)$ and $b = \frac{1}{\lambda_{Qmin}-1}|p_n b|^2$, (25.31) becomes (25.25).

(c) If $w \in L_2$, then from (25.25) we have $\mathbf{e} \in L_2$. Because all the variables in the right-hand side of (25.26) are bounded, we have $\dot{\mathbf{e}} \in L_\infty$. Using Barbalat's Lemma (Sastry and Bodson [1989]: if $\mathbf{e} \in L_2 \cap L_\infty$ and $\dot{\mathbf{e}} \in L_\infty$, then $lim_{t\to\infty}|\mathbf{e}(t)| = 0$), we have $lim_{t\to\infty}|\mathbf{e}(t)| = 0$. $\square$

25.3.2   Simulations

Example 25.1.   Consider the same situation as in Example 24.1, except that the supervisory controller u_s of (25.11) and the modified adaptation law (25.21) is used. We still use the six fuzzy sets as shown in Fig. 24.2 and choose the initial $\theta_i(0)'s$ randomly in $[-2, 2]$. The difference is that we require $|x| \leq M_x = 1.5$ and $|\theta| \leq M_D = 10$. With initial $x(0) = 1$, the simulation result is shown in Fig. 25.3. From Fig. 25.3 we see that the supervisory controller indeed forced the state to be within the constraint set. $\square$

Example 25.2.   The same as Example 24.2 except that the supervisory controller is added and the adaptation law is modified. Consider the same condition as in Fig. 24.6, except that we require $|\mathbf{x}| \leq M_x = 3$. Fig. 25.4 shows the simulation result for this case. We see again that the supervisory controller did its job. $\square$

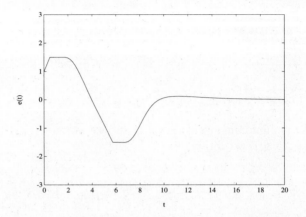

Figure 25.3. Closed-loop system state $x(t) = e(t)$ for Example 25.1.

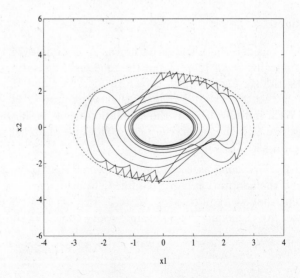

Figure 25.4. Closed-loop system trajectory in the $x_1 - x_2$ phase plane for Example 25.2.

25.4   Summary and Further Readings

In this chapter we have demonstrated the following:

- How to design the supervisory controller to guarantee the boundedness of the states.

- How to modifiy the basic adaptation laws through projection such that the parameters are guaranteed to be bounded.

- The stable direct adaptive fuzzy control scheme and its properties.

Projection algorithms were studied in Luenberger [1969] and Luenberger [1984]. There are other methods to guarantee the boundedness of the parameters, such as dead-zone (Narendra and Annaswamy [1989] and Sastry and Bodson [1989]), δ-modification (Ioannou and Kokotovic [1983]), and ϵ-modification (Narendra and Annaswamy [1989]). The stable direct adaptive fuzzy controller in this chapter was taken from Wang [1993].

25.5   Exercises

Exercise 25.1. Design a supervisory controller for the combined indirect/direct adaptive fuzzy control system developed in Section 24.2, such that the states of the closed-loop system are bounded. Prove that your supervisory controller achieves the objective.

Exercise 25.2. Modify the adaptation law by introducing projection for the combined indirect/direct adaptive fuzzy control system in Section 24.2, such that the parameters are bounded. Prove that your adaptation law achieves the objective.

Exercise 25.3. Show that (25.16) is indeed a projection operator; that is, prove that:

(a) $P\{-\gamma_1 \mathbf{e}^T P \mathbf{b} \xi(\mathbf{x})\}$ is orthogonal to $P\{-\gamma_1 \mathbf{e}^T P \mathbf{b} \xi(\mathbf{x})\} - (-\gamma_1 \mathbf{e}^T P \mathbf{b} \xi(\mathbf{x}))$, and

(b) $P\{-\gamma_1 \mathbf{e}^T P \mathbf{b} \xi(\mathbf{x})\}$ lies on the supporting hyperplane of Ω_f at the point θ_f.

Exercise 25.4. Prove Theorem 25.2.

Exercise 25.5. Repeat the simulation in Example 25.1 with $x(0) = 1.4$.

Exercise 25.6. Repeat the simulation in Example 25.2 with $M_x = 4$.

Chapter 26

Advanced Adaptive Fuzzy
Controllers II

26.1 Stable Indirect Adaptive Fuzzy Control System

26.1.1 Stability and Convergence Analysis

Continuing with Chapter 25, we add the supervisory controller $u_s(\mathbf{x})$ of (25.6) to the basic indirect adaptive fuzzy controller u_I of (23.7) and change the adaptation law to (25.15)-(25.19). This advanced indirect adaptive fuzzy control system is shown in Fig. 26.1. The following Theorem shows the stability and convergence properties of this system.

Theorem 26.1. Consider the indirect adaptive fuzzy control system in Fig. 26.1; that is, the plant is (23.1)-(23.2), the control $u = u_I + u_s$ where u_I is given by (23.7) and u_s is given by (25.6), and the adaptation law is (25.15)-(25.19). This adaptive fuzzy control system is guaranteed to have the following properties:

   (a) All the parameters and states are within the constraint sets, that is,

$$\theta_f \in \Omega_f, \ \theta_g \in \Omega_g \tag{26.1}$$

$$|\mathbf{x}| \le M_x \tag{26.2}$$

where Ω_f and Ω_g are defined in (25.13) and (25.14), respectively, and the $\bar{V}$ in the I^* of u_s is chosen according to (25.4).

   (b) The tracking error $\mathbf{e}$ satisfies (25.25), where w is defined in (23.21) and a, b are constants.

   (c) If w is squared integrable, then $lim_{t \to \infty}|\mathbf{e}(t)| = 0$.

   The proof of this Theorem follows the same arguments as in the proof of Theorem 25.3, and is left as an exercise.

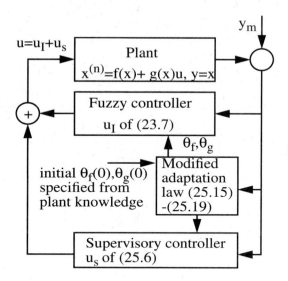

Figure 26.1. Advanced indirect adaptive fuzzy control system with supervisory controller and modified adaptation law.

26.1.2   Nonlinear Parameterization

In the fuzzy controller of the indirect adaptive fuzzy control systems in Figs. 23.2 and 26.1, the fuzzy systems $\hat{f}(\mathbf{x}|\theta_f)$ and $\hat{g}(\mathbf{x}|\theta_g)$ are linearly parameterized, as shown in (23.12) and (23.13). To achieve this linear parameterization, the IF-part membership functions must be fixed and the adaptation laws only adjust the centers θ_f and θ_g of the THEN-part fuzzy sets. In order for the fuzzy systems $\hat{f}(\mathbf{x}|\theta_f)$ and $\hat{g}(\mathbf{x}|\theta_g)$ to be well-behaved at any $\mathbf{x}$ in the domain of interest, the IF-part membership functions have to cover the whole domain. Therefore, the numbers of rules in the fuzzy systems $\hat{f}(\mathbf{x}|\theta_f)$ and $\hat{g}(\mathbf{x}|\theta_g)$ increase exponentially with the dimension of $\mathbf{x}$. One way to reduce the number of rules is to allow the IF-part membership functions also to change during the adaptation procedure, so that the same rule can be used to cover different regions of the state space at different times.

Specifically, we choose $\hat{f}(\mathbf{x}|\theta_f)$ and $\hat{g}(\mathbf{x}|\theta_g)$ to be the fuzzy systems in the form of (9.6) (fuzzy systems with product inference engine, singleton fuzzifier, center-average defuzzifier, and Gaussian membership functions), that is,

$$\hat{f}(\mathbf{x}|\theta_f) = \frac{\sum_{l=1}^{M} \bar{y}_f^l [\prod_{i=1}^{n} exp(-(\frac{x_i - \bar{x}_{fi}^l}{\sigma_{fi}})^2)]}{\sum_{l=1}^{M} [\prod_{i=1}^{n} exp(-(\frac{x_i - \bar{x}_{fi}^l}{\sigma_{fi}})^2)]} \qquad (26.3)$$

$$\hat{g}(\mathbf{x}|\theta_g) = \frac{\sum_{l=1}^{M} \bar{y}_g^l [\prod_{i=1}^{n} exp(-(\frac{x_i - \bar{x}_{gi}^l}{\sigma_{gi}})^2)]}{\sum_{l=1}^{M} [\prod_{i=1}^{n} exp(-(\frac{x_i - \bar{x}_{gi}^l}{\sigma_{gi}})^2)]} \qquad (26.4)$$

where θ_f and θ_g are the collections of the parameters $(\bar{y}_f^l, \bar{x}_{fi}^l, \sigma_{fi}^l)$ and $(\bar{y}_g^l, \bar{x}_{gi}^l, \sigma_{gi}^l)$, respectively. Let the optimal parameters θ_f^* and θ_g^* be defined as in (23.19) and (23.20). In order to use the same approach as in Subsection 23.2.3 to design the adaptation law, we take the Taylor series expansions of $\hat{f}(\mathbf{x}|\theta_f^*)$ and $\hat{g}(\mathbf{x}|\theta_g^*)$ around θ_f and θ_g and obtain

$$\hat{f}(\mathbf{x}|\theta_f) - \hat{f}(\mathbf{x}|\theta_f^*) = (\theta_f - \theta_f^*)^T \left[\frac{\partial \hat{f}(\mathbf{x}|\theta_f)}{\partial \theta_f} \right] + O(|\theta_f - \theta_f^*|^2) \qquad (26.5)$$

$$\hat{g}(\mathbf{x}|\theta_g) - \hat{g}(\mathbf{x}|\theta_g^*) = (\theta_g - \theta_g^*)^T \left[\frac{\partial \hat{g}(\mathbf{x}|\theta_g)}{\partial \theta_g} \right] + O(|\theta_g - \theta_g^*|^2) \qquad (26.6)$$

where $O(|\theta_f - \theta_f^*|^2)$ and $O(|\theta_g - \theta_g^*|^2)$ are higher-order terms.

Substituting (26.5)-(26.6) into (23.22), we obtain the error equation

$$\dot{\mathbf{e}} = \Lambda \mathbf{e} + \mathbf{b} \left\{ (\theta_f - \theta_f^*)^T \frac{\partial \hat{f}(\mathbf{x}|\theta_f)}{\partial \theta_f} + (\theta_g - \theta_g^*)^T \frac{\partial \hat{g}(\mathbf{x}|\theta_g)}{\partial \theta_g} u_I + v \right\} \qquad (26.7)$$

where

$$v = w + O(|\theta_f - \theta_f^*|^2) + O(|\theta_g - \theta_g^*|^2) \qquad (26.8)$$

and w is defined in (23.21). Comparing (26.7) with (23.23) we see that we can use the same approach in Subsection 23.2.3 to design the adaptation law—just replacing $\xi(\mathbf{x})$ by $\frac{\partial \hat{f}(\mathbf{x}|\theta_f)}{\partial \theta_f}$, $\eta(\mathbf{x})$ by $\frac{\partial \hat{g}(\mathbf{x}|\theta_g)}{\partial \theta_g}$, and w by v; that is, the adaptaion law in this case is

$$\dot{\theta}_f = -\gamma_1 \mathbf{e}^T P \mathbf{b} \left[\frac{\partial \hat{f}(\mathbf{x}|\theta_f)}{\partial \theta_f} \right] \qquad (26.9)$$

$$\dot{\theta}_g = -\gamma_2 \mathbf{e}^T P \mathbf{b} \left[\frac{\partial \hat{g}(\mathbf{x}|\theta_g)}{\partial \theta_g} \right] u_I \qquad (26.10)$$

The (advanced) adaptation law with projection can be developed in the same way as in the linear parameterization cases.

To implement the adaptation law (26.9)-(26.10), we must know how to compute $\frac{\partial \hat{f}(\mathbf{x}|\theta_f)}{\partial \theta_f}$ and $\frac{\partial \hat{g}(\mathbf{x}|\theta_g)}{\partial \theta_g}$. From Chapter 13 we know that these two derivatives can be computed using the back-propagation algorithm. Specifically, $\frac{\partial \hat{f}(\mathbf{x}|\theta_f)}{\partial \theta_f}$ is computed from

$$\frac{\partial \hat{f}}{\partial \bar{y}_f^l} = \frac{z_f^l}{b_f} \qquad (26.11)$$

$$\frac{\partial \hat{f}}{\partial \bar{x}^l_{fi}} = \frac{\bar{y}^l_f - \hat{f}}{b_f} z^l_f \frac{2(x_i - \bar{x}^l_{fi})}{(\sigma^l_{fi})^2} \tag{26.12}$$

$$\frac{\partial \hat{f}}{\partial \sigma^l_{fi}} = \frac{\bar{y}^l_f - \hat{f}}{b_f} z^l_f \frac{2(x_i - \bar{x}^l_{fi})^3}{(\sigma^l_{fi})^3} \tag{26.13}$$

where

$$z^l_f = \prod_{i=1}^{n} exp(-(\frac{x_i - \bar{x}^l_{fi}}{\sigma^l_{fi}})^2) \tag{26.14}$$

$$b_f = \sum_{l=1}^{M} z^l_f \tag{26.15}$$

$\frac{\partial \hat{g}(\mathbf{X}|\theta_g)}{\partial \theta_g}$ can be computed using the same algorithm (26.11)-(26.15) with f replaced by g.

26.2 Adaptive Fuzzy Control of General Nonlinear Systems

The adaptive fuzzy controllers developed up to this point were designed for nonlinear systems in the canonical form (23.1)-(23.2). In practice, however, many nonlinear systems may not be represented in the canonical form. In general, a single-input-single-output continuous-time nonlinear system is described by

$$\dot{\mathbf{x}} = F(\mathbf{x}, u), \quad y = h(\mathbf{x}) \tag{26.16}$$

where $\mathbf{x} \in R^n$ is the state vector, $u \in R$ and $y \in R$ are the input and output of the system, respectively, and F and h are nonlinear functions. In this section, we consider the general nonlinear system (26.16); our objective is to make the output $y(t)$ track a desired trajectory $y_m(t)$.

Comparing (26.16) with the canonical form (23.1)-(23.2), we see that a difficulty with this general model is that the output y is only indirectly related to the input u, through the state variable $\mathbf{x}$ and the nonlinear state equation; on the other hand, with the canonical form the output (which equals the first state variable) is directly related to the input. Therefore, inspired by the results of Chapters 23-25, we might guess that the difficulty of the tracking control for the general nonlinear system (26.16) can be reduced if we can find *a direct relation between the system output y and the control input u*. Indeed, this idea constitutes the intuitive basis for the so-called *input-output linearization* approach to nonlinear controller design (Isidori [1989]).

Input-output linearization is an approach to nonlinear controller design that has attracted a great deal of interest in the nonlinear control community in recent years. This approach differs entirely from conventional linearization (for example, Jacobian linearization) in that linearization is achieved by exact state transformations and feedback rather than by linear approximations of the nonlinear dynamics. We now briefly describe the basic concepts of linearization (Subsection 26.2.1), show how to design adaptive fuzzy controllers for the general nonlinear system (26.16) based on the input-output linearization concept (Subsection 26.2.2), and apply these adaptive fuzzy controllers to the ball-and-beam system (Subsection 26.2.3).

26.2.1   Intuitive Concepts of Input-Output Linearization

The basic idea of input-output linearization can be summarized as follows: *differentiate the output y repeatly until the input u appears, then specify u in such a way that the nonlinearity is canceled, and finally design a controller based on linear control.* We now illustrate this basic idea through an example.

Consider the third-order system

$$\dot{x}_1 = sin(x_2) + (x_2 + 1)x_3 \tag{26.17}$$
$$\dot{x}_2 = x_1^5 + x_3 \tag{26.18}$$
$$\dot{x}_3 = x_1^2 + u \tag{26.19}$$
$$y = x_1 \tag{26.20}$$

To generate a direct relation between y and u, we differentiate y

$$\dot{y} = \dot{x}_1 = sin(x_2) + (x_2 + 1)x_3 \tag{26.21}$$

Since $\dot{y}$ is still not directly related to u, we differentiate it again

$$\ddot{y} = (cosx_2 + x_3)(x_1^5 + x_3) + (x_2 + 1)x_1^2 + (x_2 + 1)u \tag{26.22}$$

Clearly, (26.22) represents a direct relationship between y and u. If we choose

$$u = \frac{1}{x_2 + 1}(v - f_1) \tag{26.23}$$

where $f_1 \equiv (cosx_2 + x_3)(x_1^5 + x_3) + (x_2 + 1)x_1^2$, then we have

$$\ddot{y} = v \tag{26.24}$$

If we view v as a new input, then the orignal nonlinear system (26.17)-(26.20) is *linearized* to the linear system (26.24). This linearization procedure is shown in Fig. 26.2. Now if we choose the new control input

$$v = \ddot{y}_m + k_1 \dot{e} + k_2 e \tag{26.25}$$

where $e \equiv y_m - y$, then the closed-loop system is characterized by

$$\ddot{e} + k_1 \dot{e} + k_2 e = 0 \tag{26.26}$$

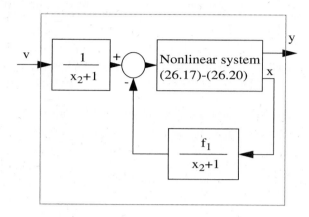

Linearized system

Figure 26.2. Diagram of the linearized system for the
nonlinear system (26.17)-(26.20).

If we choose k_1 and k_2 such that all roots of $s^2 + k_1 s + k_2 = 0$ are in the open left-half complex plane, then we have $lim_{t \to \infty} e(t) = 0$, which is our control objective.

If we need to differentiate the output of a system r times to generate a direct relationship between the output y and the input u, this system is said to have *relative degree r*. Thus, the system (26.17)-(26.20) has relative degree 2. It can be shown formally that for any controllable system of order n, it will take at most n differentiations of any output for the control input to appear; that is, the relative degree of any $n'th$-order controllable system is less than or equal to n.

At this point one might feel that the tracking control problem for the nonlinear system (26.17)-(26.20) has been solved with control law (26.23) and (26.25). However, one must realize that (26.26) only accounts for part of the whole system because it has only order 2, while the whole system has order 3. Therefore, a part of the system dynamics has been rendered "unobservable" in the input-output linearization. This part of the system is called the *internal dynamics*, because it cannot be seen from the external input-output relationship (26.24). For the preceding example, the internal dynamics are represented by

$$\dot{x}_3 = x_1^2 + \frac{1}{x_2 + 1}(\ddot{y}_m + k_1 \dot{e} + k_2 e - f_1) \qquad (26.27)$$

which is obtained by substituting (26.23) and (26.25) into (26.19). If these internal dynamics are stable in the sense that the state x_3 is bounded, our tracking control problem has indeed been solved. Otherwise, we have to redesign the control law.

26.2.2   Design of Adaptive Fuzzy Controllers Based on Input-Output Linearization

From the last subsection we see that the input-output linearization approach requires that the mathematical model of the system is known, because otherwise the differentiation procedure cannot be performed. In our problem, however, the nonlinear functions F and h in (26.16) are assumed to be unknown. How can we generalize the approach of input-output linearization to our problem?

First, inspired by the certainty equivalent controller in Chapter 23, one may think to replace the F and h by fuzzy systems and then develop an adaptation law to adjust the parameters of the fuzzy systems to make them approximate F and h. Indeed, this is a valid idea. However, this approach will result in a very complicated adaptive control system. The reason is that, although the original fuzzy systems for approximating F and h are linear in their parameters, the differentiations cause these parameters to appear in a nonlinear fashion in later stages of the differentiation procedure. Therefore, we do not take this approach.

Our approach is based on the following consideration: from the last subsection we see that the control design (26.23) is based only on the final system (26.22) in the differentiation procedure, the intermediate system (26.21) is not directly used; therefore, instead of approximating the F and h by fuzzy systems, we approximate the nonlinear functions in the final equation of the differentiation procedure by fuzzy systems. We then can develop an adaptation law to adjust the parameters of the fuzzy systems to make y track y_m. To give the details in a formal way, we need the following assumption.

Assumption 26.1. We assume that: (i) the nonlinear system (26.16) has relative degree r, (ii) the control u appears linearly with respect to $y^{(r)}$, that is,

$$y^{(r)} = f(\mathbf{x}) + g(\mathbf{x})u \qquad (26.28)$$

where f and g are unknown functions and $g(\mathbf{x}) \neq 0$, and (iii) the internal dynamics of the system with the following adaptive fuzzy controller are stable.

Design of Adaptive Fuzzy Controller Based on Input-Output Linearization:

- **Step 1.** Determine the relative degree r of the nonlinear system (26.16) based on physical intuitions. Specifically, we analyze the physical meanings of $y, \dot{y}, \ddot{y}$, ..., and determine the $y^{(r)}$ that is directly related to u. We will show how to do this for the ball-and-beam system in the next subsection.

- **Step 2.** Choose the fuzzy systems $\hat{f}(\mathbf{x}|\theta_f)$ and $\hat{g}(\mathbf{x}|\theta_g)$ in the form of (9.6), that is,

$$\hat{f}(\mathbf{x}|\theta_f) = \theta_f^T \xi(\mathbf{x}) \qquad (26.29)$$

$$\hat{g}(\mathbf{x}|\theta_g) = \theta_g^T \xi(\mathbf{x}) \qquad (26.30)$$

- **Step 3**. Design the controller as

$$u = \frac{1}{\hat{g}(\mathbf{x}|\theta_g)}[-\hat{f}(\mathbf{x}|\theta_f) + y_m^{(r)} + \mathbf{k}^T \mathbf{e}] \tag{26.31}$$

where $\mathbf{k} = (k_r, ..., k_1)^T$ is such that all roots of $s^r + k_1 s^{r-1} + \cdots + k_r = 0$ are in the open left-half complex plane, and $\mathbf{e} = (e, \dot{e}, ..., e^{(r-1)})^T$ with $e = y_m - y$.

- **Step 4**. Use the adaptation law

$$\dot{\theta}_f = -\gamma_1 \mathbf{e}^T P \mathbf{b} \xi(\mathbf{x}) \tag{26.32}$$
$$\dot{\theta}_g = -\gamma_2 \mathbf{e}^T P \mathbf{b} \eta(\mathbf{x}) u \tag{26.33}$$

to adjust the parameters, where P and $\mathbf{b}$ are defined as in Chapter 23. The overall control system is the same as in Fig.23.2 except that the plant is changed to (26.16) and n is changed to r.

The controller (26.31) and adaptation law (26.32)-(26.33) are obtained by using the same Lyapunov synthesis approach as in Chapter 23. This can be done because Assumption 26.1 ensures that controlling (26.16) is equivalent to controlling (26.28).

Using the same ideas as in Chapter 25, we can add a supervisory controller to the fuzzy controller (26.31) to guarantee the boundedness of $\mathbf{y} = (y, \dot{y}, ..., y^{(r-1)})^T$. We also can use the projection algorithm to modify the adaptation law (26.32)-(26.33) to guarantee the boundedness of θ_f and θ_g. If linguistic information for f and g is available, we can incorporate it into the initial $\theta_f(0)$ and $\theta_g(0)$ in the same way as in Chapter 23.

The preceding adaptive fuzzy controller is an indirect adaptive fuzzy controller. We also can develop a direct adaptive fuzzy controller using the same idea as in Chapter 24. The resulting controller is

$$u = \theta^T \xi(\mathbf{x}) \tag{26.34}$$

with adaptation law

$$\dot{\theta} = \gamma \mathbf{e}^T \mathbf{p}_r \xi(\mathbf{x}) \tag{26.35}$$

where $\mathbf{p}_r$ is defined as in Chapter 24. Similarly, we can also develop a combined indirect/direct adaptive fuzzy controller for the general nonlinear system.

26.2.3 Application to the Ball-and-Beam System

The ball-and-beam system is illustrated in Fig. 16.7 and is characterized by (16.4) and (16.5). In this subsection, we use the adaptive fuzzy controllers developed in the last subsection to control the ball position $y = x_1$ to track the trajectory $y_m(t) = sin(t)$.

We begin with Step 1, that is, determining the relative degree r of the ball-and-beam system based on physical intuitions. First, we realize that the control u equals the acceleration of the beam angle θ; thus, our goal is to determine which derivative of y is directly related to the acceleration of θ. Clearly, the ball position $y = x_1 = r$ and the ball speed along the beam $\dot{y}$ are not directly related to u. Based on Newton's Law, the acceleration of the ball position, $\ddot{y}$, is propositional to $sin(\theta)$, which is not directly related to $u = \ddot{\theta}$. Therefore, $\ddot{y}$ is not directly related to u. Because $\ddot{y}$ is propositional to $sin(\theta)$, $y^{(3)}$ is directly related to $\dot{\theta}$, but not directly related to $u = \ddot{\theta}$. Finally, we see that $y^{(4)}$ is directly related to u. Therefore, the relative degree of the ball-and-beam system equals 4. [1]

In Step 2, we define three fuzzy sets $\mu_N(x_i) = 1/(1 + exp(5(x_i + 1)))$, $\mu_Z(x_i) = exp(-x_i^2)$ and $\mu_P(x_i) = 1/(1 + exp(-5(x_i - 1)))$ for all x_1 to x_4. Therefore, the dimension of θ_f, θ_g and $\xi(\mathbf{x})$ equals $3^4 = 81$. Because there is no linguistic information about f and g, the initial $\theta_f(0)$, and $\theta_g(0)$ were chosen randomly in the interval $[-2, 2]$.

In Step 3, we choose $\mathbf{k} = (1, 4, 6, 4)^T$, and in Step 4 we choose $\gamma_1 = 2$ and $\gamma_2 = 0.2$. Figs. 26.3 and 26.4 show the $y(t)$ (solid line) using this adaptive fuzzy controller along with the desired trajectory $y_m(t)$ (dashed line) for initial conditions $\mathbf{x}(0) = (1, 0, 0, 0)^T$ and $\mathbf{x}(0) = (-1, 0, 0, 0)^T$, respectively.

We also simulated the direct adaptive fuzzy controller (26.34) and (26.35). We chose the $\xi(\mathbf{x})$ to be the same as in the preceding indirect adaptive fuzzy controller. We chose $\gamma = 10$. Figs. 26.5 and 26.6 show the $y(t)$ (solid line) using this direct adaptive fuzzy controller along with the desired trajectory $y_m(t)$ for initial conditions $\mathbf{x}(0) = (1, 0, 0, 0)^T$ and $\mathbf{x}(0) = (-0.4, 0, 0, 0)^T$, respectively. Comparing Figs. 26.3-26.4 with Figs. 26.5-26.6 we see that the direct adaptive fuzzy controller gave better performance than the indirect adaptive fuzzy controller for this example.

---

[1] We did not give a mathematically rigorous definition of relative degree in Subsection 26.2.1; therefore, the relative degree determined in this fashion can only be viewed as an intuitive relative degree. For the rigorous definition of relative degree, see Isidori [1989] and Slotine and Li [1991].

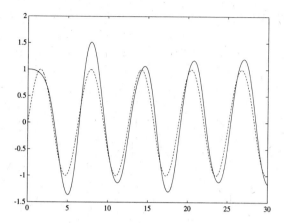

Figure 26.3. The output $y(t)$ (solid line) using the indirect adaptive fuzzy controller (26.31)-(26.33) and the desired trajectory $y_m(t)$ (dashed line) for the ball-and-beam system with the initial condition $\mathbf{x}(0) = (1, 0, 0, 0)^T$.

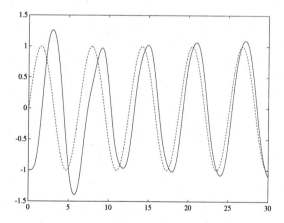

Figure 26.4. The output $y(t)$ (solid line) using the indirect adaptive fuzzy controller (26.31)-(26.33) and the desired trajectory $y_m(t)$ (dashed line) for the ball-and-beam system with the initial condition $\mathbf{x}(0) = (-1, 0, 0, 0)^T$.

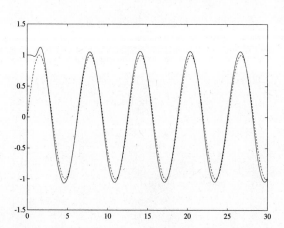

Figure 26.5. The output $y(t)$ (solid line) using the direct adaptive fuzzy controller (26.34)-(26.35) and the desired trajectory $y_m(t)$ (dashed line) for the ball-and-beam system with the initial condition $\mathbf{x}(0) = (1,0,0,0)^T$.

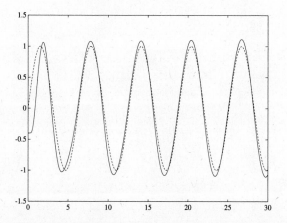

Figure 26.6. The output $y(t)$ (solid line) using the direct adaptive fuzzy controller (26.34)-(26.35) and the desired trajectory $y_m(t)$ (dashed line) for the ball-and-beam system with the initial condition $\mathbf{x}(0) = (-0.4,0,0,0)^T$.

26.3 Summary and Further Readings

In this chapter we have demonstrated the following:

- How to design adaptation laws if the IF-part membership functions of the fuzzy systems also are allowed to change during the adaptation procedure.

- The basic idea of input-output linearization.

- How to design adaptive fuzzy controllers for the general nonlinear systems, based on the input-output linearization concept.

Nonlinear control theory should be the principal tool in the further development of fuzzy control. There are many good books on nonlinear control, for example, Isidori [1989] provided a rigorous treatment for nonlinear control and Slotine and Li [1991] is more readable. The materials in this chapter are taken from Wang [1994a].

26.4 Exercises

Exercise 26.1. Prove Theorem 26.1.

Exercise 26.2. Suppose that the direct adaptive fuzzy controller $u_D(\mathbf{x}|\theta)$ in (24.9) is a fuzzy system in the form of (9.6) with $a_i^l = 1$. Design an adaptation law for the parameters θ using the Lyapunov synthesis approach.

Exercise 26.3. Consider the nonlinear system

$$\dot{x}_1 = x_1^2 x_2 \qquad (26.36)$$
$$\dot{x}_2 = 3x_2 + u \qquad (26.37)$$
$$y = -2x_1 - x_2 \qquad (26.38)$$

Design a state feedback controller such that the equilibrium $\mathbf{x} = 0$ of the closed-loop system is locally asympototically stable.

Exercise 26.4. For the system

$$\dot{x}_1 = sinx_2 + \sqrt{t+1}x_2 \qquad (26.39)$$
$$\dot{x}_2 = \alpha_1 x_1^4 cosx_2 + \alpha_2 x_1 x_2 sinx_2 + u \qquad (26.40)$$

design an adaptive controller to track an arbitrary desired trajectory $x_{d1}(t)$. Assume that the state $(x_1, x_2)^T$ is measured, that $x_{d1}(t), \dot{x}_{d1}(t), \ddot{x}_{d1}(t)$ are all known, and that α_1, α_2 are unknown constants.

Exercise 26.5. Design a direct adaptive fuzzy controller for the general nonlinear system (26.16) and simulate it for the ball-and-beam system.

Part VI

Miscellaneous Topics

We pointed out in the Preface that in writing this book we first established the structure that a reasonable theory of fuzzy systems and fuzzy control should follow, and then filled in the details. This structure is constituted by Chapters 1-26. Because fuzzy systems and fuzzy control is a large and diversified field, some important topics were unavoidably not included in this structure. In this final part of the book (Chapters 27-31), we will study a number of important topics in fuzzy systems and fuzzy control that were not covered in Chapters 1-26. These topics are equally important as those in Chapters 1-26.

In Chapter 27, we will study perhaps the most important method in fuzzy approaches to pattern recognition—the fuzzy c-means algorithm. We will show the details of the algorithm and study its convergence properties. In Chapter 28, we will study fuzzy relation equations that are potentially very useful in fuzzy systems and fuzzy control. We will show how to obtain exact and approximate solutions to different fuzzy relation equations. In Chapter 29, we will introduce the basic arithmetic operations for fuzzy numbers, including fuzzy addition, fuzzy subtraction, fuzzy multiplication and fuzzy division. These fuzzy operations are very useful in fuzzy decision making. In Chapter 30, we will study the most important topic in fuzzy decision making—fuzzy linear programming. We will justify why fuzzy linear programming is needed and show how to solve a number of fuzzy linear programming problems. Finally, in Chapter 31, we will briefly review the basics in possibility theory and conclude the book with a discussion on fuzziness versus probability.

Chapter 27

The Fuzzy C-Means Algorithm

27.1 Why Fuzzy Models for Pattern Recognition?

Pattern recognition is a field concerned with machine recognition of meaningful regularities in noisy or complex environments. In simpler words, pattern recognition is the search for structures in data. For example, Fig. 27.1 shows four cases of data structures in the plane. Observing Fig.27.1, we see that the data in each of the four cases should be classified into two groups, but the definition of a "group" is different. Specifically, the group in Fig. 27.1(a) should be classified according to the distance between data points (that is, data points with short distances among themselves should be grouped together), the group in Fig. 27.1(b) should be recognized according to the connectivity of the data points (that is, data points well connected together should be classified into the same group), and the groups in Fig. 27.1(c)-(d) should be defined from a mixture of distance-based and linkage-based criteria. In pattern recognition, a group of data is called a *cluster*.

In practice, the data are usually not well distributed, therefore the "regularities" or "structures" may not be precisely defined. That is, pattern recognition is, by its very nature, an inexact science. To deal with the ambiguity, it is helpful to introduce some "fuzziness" into the formulation of the problem. For example, the boundary between clusters could be fuzzy rather than crisp; that is, a data point could belong to two or more clusters with different degrees of membership. In this way, the formulation is closer to the real-world problem and therefore better performance may be expected. This is the first reason for using fuzzy models for pattern recognition: the problem by its very nature requires fuzzy modeling (in fact, fuzzy modeling means more flexible modeling—by extending the zero-one membership to the membership in the interval [0,1], more flexibility is introduced).

The second reason for using fuzzy models is that the formulated problem may be easier to solve computationally. This is due to the fact that a nonfuzzy model often results in an exhaustive search in a huge space (because some key variables can only take two values 0 and 1), whereas in a fuzzy model all the variables are continuous, so that derivatives can be computed to find the right direction for the

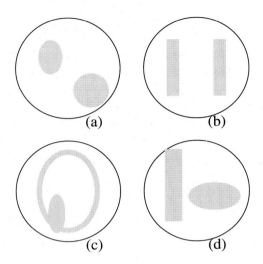

Figure 27.1. Possible data structures.

search. This will become clear in Section 27.3 when we introduce the fuzzy c-means algorithm.

A key problem in pattern recognition is to find clusters from a set of data points. In the literature, a number of fuzzy clustering algorithms were proposed. In this chapter, we study the most famous fuzzy clustering algorithm: the fuzzy c-means algorithm proposed by Bezdek [1981]. Interesting readers are referred to Bezdek and Pal [1992] for other methods. Pattern recognition using fuzzy models is a rich and currently very active research field; this chapter serves only as a short primer.

27.2   Hard and Fuzzy c-Partitions

Suppose that we are given a set of data $X = \{\mathbf{x}_1, \mathbf{x}_2, ..., \mathbf{x}_n\}$, where $\mathbf{x}_k$ can be any element, for example, $\mathbf{x}_i \in R^p$. Let $P(X)$ be the power set of X, that is, the set of all the subsets of X. A *hard c-partition of X* is the family $\{A_i \in P(X) : 1 \le i \le c\}$ such that $\bigcup_{i=1}^{c} A_i = X$ and $A_i \cap A_j = \emptyset$ for $1 \le i \ne j \le c$. Each A_i is viewed as a cluster, so $\{A_1, ..., A_c\}$ partitions X into c clusters.

The hard c-partition can be reformulated through the characteristic (membership) functions of the element $\mathbf{x}_k$ in A_i. Specifically, define

$$u_{ik} = \begin{cases} 1, & \mathbf{x}_k \in A_i \\ 0, & \mathbf{x}_k \notin A_i \end{cases} \qquad (27.1)$$

where $\mathbf{x}_k \in X, A_i \in P(X), i = 1, 2, ..., c$ and $k = 1, 2, ..., n$. Clearly, $u_{ik} = 1$ means that $\mathbf{x}_k$ belongs to cluster A_i. Given the value of u_{ik}, we can uniquely determine a hard c-partition of X, and vice versa. The u_{ik}'s should satisfy the following three conditions:

$$u_{ik} \in \{0, 1\}, \ 1 \le i \le c, \ 1 \le k \le n \qquad (27.2)$$

$$\sum_{i=1}^{c} u_{ik} = 1, \ \forall k \in \{1, 2, ..., n\} \qquad (27.3)$$

$$0 < \sum_{k=1}^{n} u_{ik} < n, \ \forall i \in \{1, 2, ..., c\} \qquad (27.4)$$

(27.2) and (27.3) together mean that each $\mathbf{x}_k \in X$ should belong to one and only one cluster. (27.4) requires that each cluster A_i must contain at least one and at most $n - 1$ data points. Collecting u_{ik} with $1 \le i \le c$ and $1 \le k \le n$ into a $c \times n$ matrix U, we obtain the matrix representation for hard c-partition, defined as follows.

Definition 27.1. *Hard c-Partition.* Let $X = \{\mathbf{x}_1, ..., \mathbf{x}_n\}$ be any set, V_{cn} be the set of real $c \times n$ matrices $U = [u_{ik}]$, and c be an integer with $2 \le c < n$. Then *hard c-partition space for X* is the set

$$M_c = \{U \in V_{cn} \mid (27.2) - (27.4) \ are \ true\} \qquad (27.5)$$

We now consider an example of hard c-partition.

Example 27.1. Let X be the set of three cars:

$$X = \{x_1 = Ford, x_2 = Toyota, x_3 = Chrysler\} \qquad (27.6)$$

If $c = 2$, then under the constraints (27.2)-(27.4), there are three hard c-partitions of X:

$$U_1 = \begin{bmatrix} 1 & 1 & 0 \\ 0 & 0 & 1 \end{bmatrix}, U_2 = \begin{bmatrix} 1 & 0 & 0 \\ 0 & 1 & 1 \end{bmatrix}, U_3 = \begin{bmatrix} 1 & 0 & 1 \\ 0 & 1 & 0 \end{bmatrix} \qquad (27.7)$$

Constraints (27.3) and (27.4) rule out matrices like

$$\begin{bmatrix} 1 & 1 & 0 \\ 1 & 0 & 1 \end{bmatrix}, \begin{bmatrix} 1 & 1 & 1 \\ 0 & 0 & 0 \end{bmatrix} \qquad (27.8)$$

respectively. If our objective is to partition X into US cars and Japanese cars, then U_3 is the most appropriate partition. However, as we discussed in Chapter 2, the distinction between US and Japanese cars is not crisp, because many parts of US cars are imported from Japan and some Japanese cars are manufactured in the US.

Clearly, a way to solve this problem is to allow the u_{ik}'s to take any value in the interval $[0, 1]$. $\square$

Another problem with hard c-partition is that the space M_c is too large. In fact,

$$|M_c| = \frac{1}{c!}\left[\sum_{j=1}^{c}\left(\begin{array}{c}c\\j\end{array}\right)(-1)^{c-j}j^n\right] \qquad (27.9)$$

is the number of distinct ways to partition X into c nonempty subsets. If, for example, $c = 10$ and $n = 25$, there are roughly 10^{18} distinct 10-partitions of the 25 points. This problem is due to the discrete nature of the characteristic function u_{ik}. Although discrete u_{ik}'s result in a finite space M_c, the number of elements in M_c is so large that the search for the "optimal" partition becomes a forbidding task. If we change u_{ik} to a continuous variable that can take any value in the interval $[0, 1]$, then we can compute the derivatives of some objective function with respect to u_{ik}. Using these derivatives, we could find the best search direction so that the search for optimal partition would be greatly simplified.

Due to the above two reasons (conceptual appropriateness and computational simplicity), we introduce the concept of fuzzy c-partition, as follows.

Definition 27.2. *Fuzzy c-Partition.* Let X, V_{cn} and c be as in Definition 27.1. Then *fuzzy c-partition space for X* is the set

$$M_{fc} = \{U \in V_{cn} \mid u_{ik} \in [0, 1], 1 \le i \le c, 1 \le k \le n; (27.3) \text{ is true}\} \qquad (27.10)$$

u_{ik} is the membership value of $\mathbf{x}_k$ belonging to cluster A_i. Note that condition (27.4) is not included in defining M_{fc}; this is called the *degenerate fuzzy c-partition* in Bezdek [1981].

Consider Example 27.1 again. Using fuzzy c-partition, a more reasonable partition of X might be

$$U = \left[\begin{array}{ccc} 0.9 & 0.2 & 0.9 \\ 0.1 & 0.8 & 0.1 \end{array}\right] \qquad (27.11)$$

27.3   Hard and Fuzzy c-Means Algorithms

27.3.1   Objective Function Clustering and Hard c-Means Algorithm

How to choose the "optimal" partition from the space M_c or M_{fc}? There are three types of methods: hierarchical methods, graph-theoretic methods, and objective function methods. In the hierarchical methods, merging and splitting techniques are used to construct new clusters based on some measure of similarity; the result is a hierarchy of nested clusters. In the graph-theoretic methods, X is regarded as a node set which are connected by edges according to a measure similarity; the criterion for clustering is typically some measure of connectivity between groups of nodes. In the objective function methods, an objective function measuring the "desirability" of clustering candidates is established for each c, and local minima of the objective function are defined as optimal clusters. The objective function methods allow the most precise formulation of the clustering criterion; we will adapt this approach in this chapter.

The most extensively studied objective function is the *overall within-group sum of squared errors*, defined as

$$J_W(U,V) = \sum_{k=1}^{n} \sum_{i=1}^{c} u_{ik} ||\mathbf{x}_k - \mathbf{v}_i||^2 \qquad (27.12)$$

where $U = [u_{ik}] \in M_c$ or M_{fc}, $V = (\mathbf{v}_1, ..., \mathbf{v}_c)$ with $\mathbf{v}_i$ being the center of cluster A_i defined by

$$\mathbf{v}_i = \frac{\sum_{k=1}^{n} u_{ik} \mathbf{x}_k}{\sum_{k=1}^{n} u_{ik}} \qquad (27.13)$$

Clearly, $\mathbf{v}_i$ is the average (for hard c-partition) or weighted average (for fuzzy c-partition) of all the points in cluster A_i. From now on, we assume that $\mathbf{x}_k, \mathbf{v}_i \in R^p$. If U is a hard c-partition, then $J_W(U,V)$ of (27.12) can be rewritten as

$$J_W(U,V) = \sum_{i=1}^{c} (\sum_{\mathbf{x}_k \in A_i} ||\mathbf{x}_k - \mathbf{v}_i||^2) \qquad (27.14)$$

which explains why $J_W(U,V)$ is called the overall within-group sum of squared errors. Since $u_{ik} ||\mathbf{x}_k - \mathbf{v}_i||^2$ is the squared error incurred by representing $\mathbf{x}_k$ with $\mathbf{v}_i$, it is also a measure of local density. $J_W(U,V)$ would be small if the points in each hard cluster A_i adhere tightly to their cluster center $\mathbf{v}_i$.

Finding the optimal pair (U,V) for J_W is not an easy task. The difficulty stems from the size of M_c, which is finite but huge (see (27.9)). One of the most popular algorithms for finding the approximate minima of J_W is the following hard c-means algorithm (also called ISODATA algorithm).

Hard c-Means Algorithm:

- **Step 1.** Suppose that we are given n data points $X = \{\mathbf{x}_1, ..., \mathbf{x}_n\}$ with $\mathbf{x}_i \in R^p$. Fix c, $2 \leq c < n$, and initialize $U^{(0)} \in M_c$.

- **Step 2.** At iteration l, $l = 0, 1, 2, ...$, compute the c mean vectors

$$\mathbf{v}_i^{(l)} = \frac{\sum_{k=1}^{n} \mathbf{x}_k u_{ik}^{(l)}}{\sum_{k=1}^{n} u_{ik}^{(l)}} \qquad (27.15)$$

where $[u_{ik}^{(l)}] = U^{(l)}$, and $i = 1, 2, ..., c$.

- **Step 3.** Update $U^{(l)}$ to $U^{(l+1)} = [u_{ik}^{(l+1)}]$ using

$$u_{ik}^{(l+1)} = \begin{cases} 1 & ||\mathbf{x}_k - \mathbf{v}_i^{(l)}|| = \min_{1 \le j \le c}(||\mathbf{x}_k - \mathbf{v}_j^{(l)}||) \\ 0 & otherwise \end{cases} \qquad (27.16)$$

- **Step 4.** Compare $U^{(l)}$ with $U^{(l+1)}$: if $||U^{(l+1)} - U^{(l)}|| < \epsilon$ for a small constant ϵ, stop; otherwise, set $l = l + 1$ and go to Step 2.

The hard c-means algorithm is quite reasonable from an intuitive point of view: guess c hard clusters (Step 1), find their centers (Step 2), reallocate cluster memberships to minimize squared errors between the data and the current centers (Step 3), and stop when looping ceases to lower J_W significantly (Step 4). Since the hard space M_c is discrete, the notion of local minima is not defined for J_W. The necessity of computing $\{\mathbf{v}_i^{(l)}\}$ with (27.15) can be established by setting the gradients of J_W with respect to each $\mathbf{v}_i$ equal to zero.

Example 27.2. Suppose that X consists of the 15 points in R^2 shown in Fig. 27.2. These data points look like a butterfly, where $\mathbf{x}_1$ to $\mathbf{x}_7$ form the left wing, $\mathbf{x}_9$ to $\mathbf{x}_{15}$ form the right wing, and $\mathbf{x}_8$ is a bridge between the two wings. With $c = 2$ and

$$U^{(0)} = \begin{bmatrix} 1 & 1 & 1 & 1 & 1 & 0 & 0 & 0 & 0 & 0 & 0 & 0 & 0 & 0 & 0 \\ 0 & 0 & 0 & 0 & 0 & 1 & 1 & 1 & 1 & 1 & 1 & 1 & 1 & 1 & 1 \end{bmatrix} \qquad (27.17)$$

the hard c-means algorithm stops at $l = 3$ with

$$U^{(3)} = U^{(4)} = \begin{bmatrix} 1 & 1 & 1 & 1 & 1 & 1 & 1 & 0 & 0 & 0 & 0 & 0 & 0 & 0 & 0 \\ 0 & 0 & 0 & 0 & 0 & 0 & 0 & 1 & 1 & 1 & 1 & 1 & 1 & 1 & 1 \end{bmatrix} \qquad (27.18)$$

$U^{(3)}$ shows that $\mathbf{x}_1$ to $\mathbf{x}_7$ are grouped into one cluster A_1 and $\mathbf{x}_8$ to $\mathbf{x}_{15}$ are grouped into the other cluster A_2. Note that A_1 and A_2 cannot be symmetric with respect to $\mathbf{x}_8$ because $\mathbf{x}_8$ must belong entirely to either A_1 or A_2. Since the data in X are themselves perfectly symmetrical, the unsymmetry of the clusters is intuitively unappealing. A way to solve this problem is to use the fuzzy c-means algorithm, which we will introduce next. □

27.3.2 The Fuzzy c-Means Algorithm

For the fuzzy c-means algorithm, the objective is to find $U = [u_{ik}] \in M_{fc}$ and $V = (\mathbf{v}_1, ..., \mathbf{v}_c)$ with $\mathbf{v}_i \in R^p$ such that

$$J_m(U, V) = \sum_{k=1}^{n} \sum_{i=1}^{c} (u_{ik})^m ||\mathbf{x}_k - \mathbf{v}_i||^2 \qquad (27.19)$$

is minimized, where $m \in (1, \infty)$ is a weighting constant. We first establish a necessary condition for this minimization problem, and then propose the fuzzy c-means algorithm based on the condition.

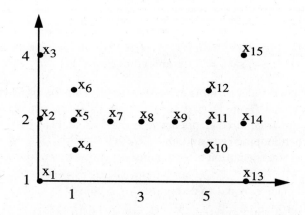

Figure 27.2. The butterfly.

Theorem 27.1. Let $X = \{\mathbf{x}_1, ..., \mathbf{x}_n\}$, $\mathbf{x}_i \in R^p$, be a given set of data. Fix $c \in \{2, 3, ..., n-1\}$ and $m \in (1, \infty)$, and assume that $\|\mathbf{x}_k - \mathbf{v}_i\| \neq 0$ for all $1 \leq k \leq n$ and $1 \leq i \leq c$. Then $U = [u_{ik}]$ and $V = (\mathbf{v}_1, ..., \mathbf{v}_c)$ is a local minimum for $J_m(U, V)$ only if

$$u_{ik} = \frac{1}{\sum_{j=1}^c \left(\frac{\|\mathbf{x}_k - \mathbf{v}_i\|}{\|\mathbf{x}_k - \mathbf{v}_j\|}\right)^{\frac{2}{m-1}}}, \quad 1 \leq i \leq c, \ 1 \leq k \leq n \tag{27.20}$$

and

$$\mathbf{v}_i = \frac{\sum_{k=1}^n (u_{ik})^m \mathbf{x}_k}{\sum_{k=1}^n (u_{ik})^m}, \quad 1 \leq i \leq c \tag{27.21}$$

Proof: To show (27.20), assume that the $\mathbf{v}_i$'s are fixed. Then the problem becomes minimizing J_m with respect to u_{ik} under the constraint (27.3). Using the Lagrange multiplier method, we obtain that the problem is equivalent to minimizing

$$L(U, \lambda) = \sum_{k=1}^n \sum_{i=1}^c (u_{ik})^m \|\mathbf{x}_k - \mathbf{v}_i\|^2 - \sum_{k=1}^n \lambda_k \left(\sum_{i=1}^c u_{ik} - 1\right) \tag{27.22}$$

without constraints. The necessary condition for this problem is

$$\frac{\partial L(U, \lambda)}{\partial u_{ik}} = [m(u_{ik})^{m-1} \|\mathbf{x}_k - \mathbf{v}_i\|^2 - \lambda_k] = 0 \tag{27.23}$$

$$\frac{\partial L(U, \lambda)}{\partial \lambda_k} = \sum_{i=1}^c u_{ik} - 1 = 0 \tag{27.24}$$

From (27.23), we have

$$u_{ik} = \left(\frac{\lambda_k}{m \|\mathbf{x}_k - \mathbf{v}_i\|^2} \right)^{\frac{1}{m-1}} \qquad (27.25)$$

Substituting (27.25) into (27.24), we have

$$\left(\frac{\lambda_k}{m} \right)^{\frac{1}{m-1}} = \frac{1}{\sum_{i=1}^{c} \left(\frac{1}{\|\mathbf{x}_k - \mathbf{v}_i\|^2} \right)^{\frac{1}{m-1}}} \qquad (27.26)$$

Substituting (27.26) into (27.25), we obtain (27.20).

To show (27.21), assume that u_{ik}'s are fixed. Then this is an unconstraint minimization problem and the necessary condition is

$$\frac{\partial J_m(U, V)}{\partial \mathbf{v}_i} = -\sum_{k=1}^{n} 2(u_{ik})^m (\mathbf{x}_k - \mathbf{v}_i) = 0 \qquad (27.27)$$

from which we get (27.21). $\square$

The fuzzy c-means algorithm is based on the necessary condition (27.20) and (27.21).

Fuzzy c-Means Algorithm:

- **Step 1**. For given data set $X = \{\mathbf{x}_1, ..., \mathbf{x}_n\}$, $\mathbf{x}_i \in R^p$, fix $c \in \{2, 3, ..., n-1\}$, $m \in (1, \infty)$, and initialize $U^{(0)} \in M_{fc}$.

- **Step 2**. At iteration l, $l = 0, 1, 2, ...$, compute the c mean vectors

$$\mathbf{v}_i^{(l)} = \frac{\sum_{k=1}^{n} (u_{ik}^{(l)})^m \mathbf{x}_k}{\sum_{k=1}^{n} (u_{ik}^{(l)})^m}, \quad 1 \le i \le c \qquad (27.28)$$

- **Step 3**. Update $U^{(l)} = [u_{ik}^{(l)}]$ to $U^{(l+1)} = [u_{ik}^{(l+1)}]$ using

$$u_{ik}^{(l+1)} = \frac{1}{\sum_{j=1}^{c} \left(\frac{\|\mathbf{x}_k - \mathbf{v}_i^{(l)}\|}{\|\mathbf{x}_k - \mathbf{v}_j^{(l)}\|} \right)^{\frac{2}{m-1}}}, \quad 1 \le i \le c, \ 1 \le k \le n \qquad (27.29)$$

- **Step 4**. If $\|U^{(l+1)} - U^{(l)}\| < \epsilon$, stop; otherwise, set $l = l + 1$ and go to Step 2.

We now apply the fuzzy c-means algorithm to the butterfly example.

Example 27.3. Let $X = \{\mathbf{x}_1, ..., \mathbf{x}_{15}\}$ be the butterfly data in Fig. 27.2. With $c = 2, m = 1.25, \epsilon = 0.01$, and

$$U^{(0)} = \begin{bmatrix} 0.854 & 0.146 & 0.854 & 0.854 & \cdots & 0.854 \\ 0.146 & 0.854 & 0.146 & 0.146 & \cdots & 0.146 \end{bmatrix}_{2 \times 15} \qquad (27.30)$$

the fuzzy c-means algorithm terminated at $l = 5$ with the membership values $u_{1k}^{(5)}$ shown in Fig.27.3. We see that the data in the right and left wings are well classified, while the bridge $\mathbf{x}_8$ belongs to both clusters to almost the same degree; this is intuitively appealing. □

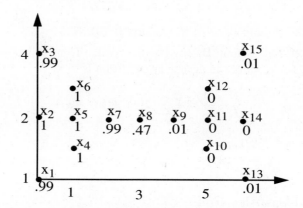

Figure 27.3. Membership values of the butterfly data points using the fuzzy c-means algorithm.

Next, we analyze the convergence properties of the fuzzy c-means algorithm.

27.4 Convergence of the Fuzzy c-Means Algorithm

Theorem 27.1 shows that (27.20) and (27.21) establish the necessary condition for (U, V) to be a local minimum of J_m. In the fuzzy c-means algorithm, the $\mathbf{v}_i$ and u_{ik} are iteratively computed according to this necessary condition. Therefore, it is not clear whether the algorithm will converge to a local minimum of J_m. In fact, this is a rather complicated issue. In this section, we prove only a fundamental property of the fuzzy c-means algorithm: the objective function $J_m(U, V)$ will decrease or keep the same through the iterations; that is, $J_m(U^{(l+1)}, V^{(l+1)}) \leq J_m(U^{(l)}, V^{(l)})$.

To prove this property, we first show that if V in $J_m(U, V)$ is fixed, then (27.20) is also a sufficient condition to compute the local minimum of J_m (Lemma 27.1). Similarly, if U in $J_m(U, V)$ is fixed, then (27.21) is also a sufficient condition for minimizing J_m (Lemma 27.2).

Lemma 27.1. Let $\phi(U) = J_m(U, V)$, where $V \in R^{p \times c}$ is fixed, and $||\mathbf{x}_k - \mathbf{v}_i|| \neq$

0 for all $1 \leq k \leq n$ and $1 \leq i \leq c$. Then $U = [u_{ik}]$ is a local minimum of $\phi(U)$ if and only if U is computed via (27.20).

Proof: The only if part was proven in Theorem 27.1. To show the sufficiency, we examine $H(U)$ — the $cn \times cn$ Hessian of the Lagrangian of $\phi(U)$ evaluated at the U given by (27.20). From (27.23), we have

$$h_{st,ik}(U) = \frac{\partial}{\partial u_{st}} \left[\frac{\partial \phi(U)}{\partial u_{ik}} \right] = \begin{cases} m(m-1)(u_{st})^{m-2} \|\mathbf{x}_t - \mathbf{v}_s\|^2 & \text{if } s = i, t = k \\ 0 & \text{otherwise} \end{cases}$$

$$(27.31)$$

where u_{st} is computed from (27.20). Thus, $H(U) = [h_{st,ik}(U)]$ is a diagonal matrix. Since $m > 1$ and $\|\mathbf{x}_t - \mathbf{v}_s\| \neq 0$ for all $1 \leq t \leq n$ and $1 \leq s \leq c$, we have $m(m-1)(u_{st})^{m-2} \|\mathbf{x}_t - \mathbf{v}_s\|^2 > 0$. Therefore, the Hessian $H(U)$ is positive definite and consequently (27.20) is also a sufficient condition for minimizing $\phi(U)$. □

Lemma 27.2. Let $\psi(V) = J_m(U, V)$, where $U \in M_{fc}$ is fixed, $\|\mathbf{x}_k - \mathbf{v}_i\| \neq 0$ for $1 \leq k \leq n$ and $1 \leq i \leq c$, and $m > 1$. Then $V = [\mathbf{v}_i]$ is a local minimum of $\psi(V)$ if and only if V is computed via (27.21).

Proof: The necessity was proven in Theorem 27.1. To show the sufficiency, we have from (27.27) that

$$\frac{\partial}{\partial \mathbf{v}_j} \left[\frac{\partial \psi(V)}{\partial \mathbf{v}_i} \right] = \begin{cases} \sum_{k=1}^{n} 2(u_{jk})^m & \text{if } j = i \\ 0 & \text{otherwise} \end{cases}$$

$$(27.32)$$

Therefore, the Hessian is positive definite and consequently (27.21) is a sufficient condition for minimizing $\psi(V)$. □

From Lemmas 27.1 and 27.2, we can prove our main result: $J_m(U^{(l+1)}, V^{(l+1)}) \leq J_m(U^{(l)}, V^{(l)})$.

Theorem 27.2. Let $\mathbf{v}_i^{(l)}, u_{ik}^{(l)}, l = 0, 1, 2, ...,$ be the sequence generated from the fuzzy c-means algorithm (27.28) and (27.29). If $m > 1$ and $\|\mathbf{x}_k^{(l)} - \mathbf{v}_i^{(l)}\| \neq 0$ for all $k = 1, 2, ..., n, i = 1, 2, ..., c$, and $l = 0, 1, 2, ...,$ then we have

$$J_m(U^{(l+1)}, V^{(l+1)}) \leq J_m(U^{(l)}, V^{(l)})$$

$$(27.33)$$

for $l = 0, 1, 2, ...$

Proof: Since $V^{(l)}$ is computed from (27.28) for fixed U, we have from Lemma 27.2 that

$$J_m(U^{(l+1)}, V^{(l+1)}) \leq J_m(U^{(l+1)}, V^{(l)})$$

$$(27.34)$$

Since $U^{(l+1)}$ is computed from (27.29) for fixed V, we have from Lemma 27.1 that

$$J_m(U^{(l+1)}, V^{(l)}) \leq J_m(U^{(l)}, V^{(l)})$$

$$(27.35)$$

Combining (27.34) and (27.35), we obtain (27.33). □

Let Ω be the set

$$\Omega = \{(U^*, V^*) \in M_{fc} \times R^{p \times c} \mid J_m(U^*, V^*) \leq J_m(U, V^*), \forall U \in M_{fc};$$
$$J_m(U^*, V^*) < J_m(U^*, V), V \neq V^*\} \quad (27.36)$$

Then it was proven in Bezdek, Hathaway, Sabin and Tucker [1987] that the fuzzy c-means algorithm either terminates at a point in Ω, or a subsequence exists that converges to a point in Ω. For details, the readers are referred to Bezdek, Hathaway, Sabin, and Tucker [1987].

27.5 Summary and Further Readings

In this chapter we have demonstrated the following:

- The motivations of using fuzzy models for pattern recognition.

- The definitions of hard and fuzzy c-partitions and the steps of the hard c-means algorithm.

- The detailed steps of the fuzzy c-means algorithm and its theoretical justification.

- The convergence properties of the fuzzy c-means algorithm.

The original book Bezdek [1981] is still the best source to learn the fuzzy c-means algorithm. A number of important papers on the fuzzy c-means algorithm and other algorithms in fuzzy approaches to pattern recognition are collected in the edited book Bezdek and Pal [1992]. For classical methods for pattern recognition, see Duda and Hart [1973].

27.6 Exercises

Exercise 27.1. Show that the number of distinct ways to partition n elements into c nonempty subsets is given by (27.9).

Exercise 27.2. Let $U \in M_{fc}$ and

$$F(U) = \sum_{k=1}^{n} \sum_{i=1}^{c} \frac{(u_{ik})^2}{n} \quad (27.37)$$

Show that:

(a) $(1/c) \leq F(U) \leq 1$ for any $U \in M_{fc}$.

(b) $F(U) = 1$ if and only if U is hard.

(c) $F(U) = (1/c)$ if and only if $u_{ik} = (1/c)\forall i, k$.

Exercise 27.3. Let $\mathbf{x}_1, ..., \mathbf{x}_n$ be n d-dimensional samples and Σ be any non-singular d-by-d matrix. Show that the vector $\mathbf{x}$ that minimizes

$$\sum_{k=1}^{n}(\mathbf{x}_k - \mathbf{x})^T \Sigma^{-1}(\mathbf{x}_k - \mathbf{x}) \tag{27.38}$$

is the sample mean $\frac{1}{n}\sum_{k=1}^{n}\mathbf{x}_k$.

Exercise 27.4. Verify that (27.16) is necessary to minimize $J_W(U, V)$ of (27.12) for fixed V.

Exercise 27.5. Prove that $J_W(U, V)$ of (27.12) decreases monotonically as c increases.

Exercise 27.6. Repeat Example 27.3 with the initial partition

$$U^{(0)} = \begin{bmatrix} (0.8)_{1\times 8} & (0.2)_{1\times 7} \\ (0.2)_{1\times 8} & (0.8)_{1\times 7} \end{bmatrix} \tag{27.39}$$

where $(x)_{1\times r}$ is a $1 \times r$ vector with all its elements equal to x.

Exercise 27.7. Apply the fuzzy c-means algorithm to the butterfly data with $c = 3, m = 1.25, \epsilon = 0.01$, and

$$U^{(0)} = \begin{bmatrix} (0.4)_{1\times 8} & (0.6)_{1\times 7} \\ (0.3)_{1\times 8} & (0.7)_{1\times 7} \\ (0.3)_{1\times 8} & (0.7)_{1\times 7} \end{bmatrix} \tag{27.40}$$

Exercise 27.8. Develop a design method for fuzzy systems from input-output pairs based on the fuzzy c-means algorithm.

Chapter 28

Fuzzy Relation Equations

28.1 Introduction

Given fuzzy set A in the input space U and fuzzy relation Q in the input-output product space $U \times V$, the compositional rule of inference (Chapter 6) gives a fuzzy set B in V as

$$\mu_B(y) = \sup_{x \in U}(\mu_A(x) \star \mu_Q(x,y)) \qquad (28.1)$$

Let $\circ$ denote the sup-star composition, then (28.1) can be rewritten as

$$B = A \circ Q \qquad (28.2)$$

If we view the fuzzy relation Q as a pure fuzzy system (see Chapter 1), then (28.1) or (28.2) tells us how to compute the system's output B given its input A and the system itself Q. Sometimes, it is of interest to consider the following two problems:

- Given the system Q and its output B, determine the corresponding input A. This is similar to the deconvolution or equalization problems in signal processing and communications, so we call this problem the *"fuzzy deconvolution problem."*

- Given the system's input A and output B, determine the system Q. This is similar to the system identification problem in control, so we call this problem the *"fuzzy identification problem."*

Therefore, solving the fuzzy relation equation (28.2) means solving the above two problems. The objective of this chapter is to study how to obtain the solutions to the two problems.

28.2 Solving the Fuzzy Relation Equations

In this section, we first introduce a useful operator—the φ-operator, and study its properties. Then, we prove a few lemmas from which we obtain solutions to the

354

fuzzy identification and fuzzy deconvolution problems.

Definition 28.1. The φ-operator is a two-place operator $\varphi : [0,1] \times [0,1] \rightarrow [0,1]$ defined by

$$a\varphi b = \sup[c \in [0,1]|a \star c \leq b] \qquad (28.3)$$

where $\star$ denotes t-norm operator.

Clearly, for different t-norms $\star$ we have different φ-operators. For the $\star$ specified as minimum, the φ-operator becomes the so-called α-operator:

$$a\alpha b = \sup[c \in [0,1]|min(a,c) \leq b] = \begin{cases} 1 & if \ a \leq b \\ b & if \ a > b \end{cases} \qquad (28.4)$$

We now show some useful properties of the φ-operator.

Lemma 28.1. Let the φ-operator be defined by (28.3). Then the following inequalities are true:

$$a\varphi max(b,c) \geq max(a\varphi b, a\varphi c) \qquad (28.5)$$
$$a \star (a\varphi b) \leq b \qquad (28.6)$$
$$a\varphi(a \star b) \geq b \qquad (28.7)$$

where $a,b,c \in [0,1]$ (usually, a,b,c are membership functions).

Proof: From (28.3) it is clear that $a\varphi b$ is a nondecreasing function of the second argument b, that is, $a\varphi b_1 \geq a\varphi b_2$ if $b_1 \geq b_2$. Hence, $a\varphi max(b,c) \geq a\varphi b$ and $a\varphi max(b,c) \geq a\varphi c$; this gives (28.5). (28.6) is a direct conclusion of the definition (28.3). Finally, from $a\varphi(a \star b) = \sup[c \in [0,1]|a \star c \leq a \star b]$ and the nondecreasing property of t-norm, we obtain (28.7). $\square$

Now consider the "fuzzy identification problem," that is, solving the fuzzy relation equation (28.2) for Q given A and B. We will show that a particular solution is $\hat{Q} = A\varphi B$. To prove this result, we need the following two lemmas.

Lemma 28.2. Let A be a fuzzy set in U and Q be a fuzzy relation in $U \times V$. Then,

$$Q \subseteq A\varphi(A \circ Q) \qquad (28.8)$$

where $A\varphi(A \circ Q)$ is a fuzzy relation in $U \times V$ with membership function $\mu_A(x)\varphi\mu_{A \circ Q}(y)$.

Proof: Since $a\varphi b$ is a nondecreasing function of b, we have

$$\mu_{A\varphi(A \circ Q)}(x,y) = \mu_A(x)\varphi[\sup_{x \in U}(\mu_A(x) \star \mu_Q(x,y))]$$
$$\geq \mu_A(x)\varphi[\mu_A(x) \star \mu_Q(x,y)] \qquad (28.9)$$

Using (28.7) we get

$$\mu_{A\varphi(A \circ Q)}(x,y) \geq \mu_Q(x,y) \qquad (28.10)$$

which gives (28.8). $\square$

Lemma 28.3. Let A and B be fuzzy sets in U and V, respectively. Then,

$$A \circ (A\varphi B) \subseteq B \qquad (28.11)$$

where $A\varphi B$ is a fuzzy relation in $U \times V$ with membership function $\mu_A(x)\varphi\mu_B(y)$.

Proof: From (28.6) we have

$$\mu_{A\circ(A\varphi B)}(y) = \sup_{x \in U}[\mu_A(x) \star (\mu_A(x)\varphi\mu_B(y))]$$
$$\leq \sup_{x \in U} \mu_B(y)$$
$$= \mu_B(y) \qquad (28.12)$$

which is equivalent to (28.11). $\square$

We are now ready to determine a solution to the "fuzzy identification problem."

Theorem 28.1. Let $\mathcal{Q}$ be the set of all solutions Q of the fuzzy relation equation (28.2) given A and B. If $\mathcal{Q}$ is non-empty, then the largest element of $\mathcal{Q}$ (in the sense of set-theoretic inclusion) is given by

$$\hat{Q} = A\varphi B \qquad (28.13)$$

Proof: Let Q be an arbitrary element of $\mathcal{Q}$, then $B = A \circ Q$ and from Lemma 28.2 we have $Q \subseteq A\varphi B = \hat{Q}$. Since t-norms are nondecreasing functions, we have $B = A \circ Q \subseteq A \circ \hat{Q}$. But from Lemma 28.3 we have $A \circ \hat{Q} = A \circ (A\varphi B) \subseteq B$, hence $B = A \circ \hat{Q}$, which means $\hat{Q}$ is an element of $\mathcal{Q}$. Since $Q \subseteq \hat{Q}$ and Q is an arbitrary element of $\mathcal{Q}$, we conclude that $\hat{Q}$ is the largest element of $\mathcal{Q}$. $\square$

Next, we consider the "fuzzy deconvolution problem," that is, solving the fuzzy relation equation (28.2) for A given B and Q. Again, we first prove two lemmas from which we determine a possible solution.

Lemma 28.4. Let B be a fuzzy set in V and Q be a fuzzy relation in $U \times V$. Then,

$$(Q\varphi B) \circ Q \subseteq B \qquad (28.14)$$

where the composition $(Q\varphi B) \circ Q$ is a fuzzy set in V with membership function $\sup_{x \in U}[\mu_{Q\varphi B}(x,y) \star \mu_Q(x,y)]$.

Proof: From (28.6) we get

$$\mu_{(Q\varphi B)\circ Q}(y) = \sup_{x \in U}[(\mu_Q(x,y)\varphi\mu_B(y)) \star \mu_Q(x,y)]$$
$$\leq \sup_{x \in U}[\mu_B(y)]$$
$$= \mu_B(y) \qquad (28.15)$$

which gives (28.14). $\square$

Lemma 28.5. For fuzzy set A in U and fuzzy relation Q in $U \times V$, the following is true:

$$A \subseteq Q\varphi(A \circ Q) \tag{28.16}$$

where $Q\varphi(A \circ Q)$ is a fuzzy set in U with membership function $\sup_{y \in V}[\mu_Q(x,y)\varphi\mu_{A \circ Q}(y)]$.

Proof: Using (28.7) and the fact that $a\varphi b$ is a nondecreasing function of b, we have

$$\begin{aligned}
\mu_{Q\varphi(A \circ Q)}(x,y) &= \sup_{y \in V}\{\mu_Q(x,y)\varphi[\sup_{x \in U}(\mu_A(x) \star \mu_Q(x,y))]\} \\
&\geq \sup_{y \in V}[\mu_Q(x,y)\varphi(\mu_A(x) \star \mu_Q(x,y))] \\
&\geq \sup_{y \in V}[\mu_A(x)] = \mu_A(x) \tag{28.17}
\end{aligned}$$

which gives (28.16). $\square$

We are now ready to give a compact formula of the solution to the "fuzzy deconvolution problem."

Theorem 28.2. Let $\mathcal{A}$ be the set of all solutions A of the fuzzy relation equation (28.2) given B and Q. If $\mathcal{A}$ is non-empty, then the greatest element of $\mathcal{A}$ is given by

$$\hat{A} = Q\varphi B \tag{28.18}$$

where $Q\varphi B$ is a fuzzy set in U with membership function $\sup_{y \in V}[\mu_Q(x,y)\varphi\mu_B(y)]$.

Proof: Let A be an arbitrary element of $\mathcal{A}$, so $B = A \circ Q$ and from Lemma 28.5 we have $A \subseteq Q\varphi(A \circ Q) = Q\varphi B = \hat{A}$, which implies $B = A \circ Q \subseteq \hat{A} \circ Q$. From Lemma 28.4 we have $\hat{A} \circ Q = (Q\varphi B) \circ Q \subseteq B$, hence $\hat{A} \circ Q = B$, which means that $\hat{A}$ is an element of $\mathcal{A}$. Since A is an arbitrary element in $\mathcal{A}$ and we have shown $A \subseteq \hat{A}$, $\hat{A}$ is the largest element of $\mathcal{A}$. $\square$

A fundamental assumption in Theorems 28.1 and 28.2 is that the solutions to the problems exist. In many cases, however, an exact solution may not exist. Therefore, $\hat{Q} = A\varphi B$ and $\hat{A} = Q\varphi B$ may not be the solutions; see the following example.

Example 28.1. Let $U = \{x_1, x_2\}$, $V = \{y_1, y_2\}$, and

$$A = 0.4/x_1 + 0.3/x_2 \tag{28.19}$$
$$B = 0.5/y_1 + 0.4/y_2 \tag{28.20}$$
$$Q = 0.3/(x_1,y_1) + 0.2/(x_1,y_2) + 0.1/(x_2,y_1) + 0/(x_2,y_2) \tag{28.21}$$

Let the φ-operator be the α-operator (28.4), then

$$\hat{Q} = A\varphi B = 1/(x_1,y_1) + 1/(x_1,y_2) + 1/(x_2,y_1) + 1/(x_2,y_2) \tag{28.22}$$

Using *min* for the t-norm in the sup-star composition, we have

$$A \circ \hat{Q} = 0.4/y_1 + 0.4/y_2 \neq B \tag{28.23}$$

Hence, $\hat{Q} = A\varphi B$ is not a solution to the problem. Similarly, using α-operator for φ and min for the t-norm, we have

$$\hat{A} = Q\varphi B = 1/x_1 + 1/x_2 \qquad (28.24)$$

and

$$\hat{A} \circ Q = 0.3/y_1 + 0.2/y_2 \neq B \qquad (28.25)$$

□

Under what conditions do the solutions to the "fuzzy identification and deconvolution problems" exist? We answer this question in the next section.

28.3   Solvability Indices of the Fuzzy Relation Equations

Since B and $A \circ Q$ are fuzzy sets in V, whether the fuzzy relation equation (28.2) is solvable is equivalent to whether the two fuzzy sets B and $A \circ Q$ are equal. Therefore, we must first study how to measure the equality of two fuzzy sets.

28.3.1   Equality Indices of Two Fuzzy Sets

Let C and D be two fuzzy sets in U. Then, a natural measurement of the difference between C and D is

$$d_p(C, D) = (\int_U |C(x) - D(x)|^p dx)^{1/p}, \ p \geq 1 \qquad (28.26)$$

For $p = 1$ one has the *Hamming distance*, and $p = 2$ yields the *Euclidean distance*. The *equality index* of C and D is defined as

$$eq_p(C, D) = 1 - d_p(C, D) \qquad (28.27)$$

Clearly, $eq_t(C, D) = 1$ if and only if $C = D$, therefore $eq_p(C, D)$ is qualified as an equality index. Since the computation of $d_p(C, D)$ involves the complicated integration, it is helpful to explore other measures of equality of two fuzzy sets.

Another approach to the comparison of fuzzy sets is to use more set-theoretically oriented tools. First, from the definition of the φ-operator (28.3) and the boundary condition of t-norm (see Chapter 3) we have that $C \subseteq D$ if and only if $\inf_{x \in U}[\mu_C(x)\varphi\mu_D(x)] = 1$. Since $C = D$ is equivalent to $C \subseteq D$ *and* $D \subseteq C$, we can define the *equality index* of C and D as

$$eq_t(C, D) = \inf_{x \in U}[\mu_C(x)\varphi\mu_D(x)] \star \inf_{x \in U}[\mu_D(x)\varphi\mu_C(x)] \qquad (28.28)$$

Clearly, $eq_t(C, D) = 1$ if and only if $C = D$, therefore $eq_t(C, D)$ is qualified as an equality index. Using the facts that $a\varphi b = 1$ if and only if $a \leq b$ and that the second

$\inf_{x \in U}$ in (28.28) can be moved in the front (due to the continuity of t-norm), we can rewrite the equality index as

$$eq_t(C, D) = \inf_{x \in U} [(\mu_C(x) \vee \mu_D(x)) \varphi (\mu_C(x) \wedge \mu_D(x))] \qquad (28.29)$$

where $\vee$ denotes maximum, and $\wedge$ denotes minimum.

We are now ready to define the solvability indices for the fuzzy relation equation.

28.3.2   The Solvability Indices

Using the equality index of two fuzzy sets, it is natural to define the solvability index of the fuzzy relation equation (28.2) as

$$\xi = eq_t(A \circ Q, B) \qquad (28.30)$$

However, since either A or Q is unknown, ξ cannot be computed. From Theorems 28.1 and 28.2 we know that a possible solution to the "fuzzy identification (deconvolution) problem" is $\hat{Q} = A \varphi B$ ($\hat{A} = Q \varphi B$), a way to define the solvability indices is to replace the Q in (28.30) by $\hat{Q}$ and A by $\hat{A}$. Specifically, we have the following definition.

Definition 28.2. The *solvability index of the "fuzzy identification problem"* is defined as

$$\xi_I = eq_t[A \circ (A \varphi B), B] \qquad (28.31)$$

and the *solvability index of the "fuzzy deconvolution problem"* is defined as

$$\xi_D = eq_t[(Q \varphi B) \circ Q, B] \qquad (28.32)$$

Clearly, $\xi_I = 1$ ($\xi_D = 1$) not only means that the "fuzzy identification (deconvolution) problem" is solvable but also implies that $\hat{Q} = A \varphi B$ ($\hat{A} = Q \varphi B$) is a solution. Since we use these special solutions in defining ξ_I and ξ_D, one may wonder whether $\xi_I \neq 1$ ($\xi_D \neq 1$) implies that the "fuzzy identification (deconvolution) problem" is not solvable because we may have $\hat{Q}'$ ($\hat{A}'$) other than $A \varphi B$ ($Q \varphi B$) such that $eq_t(A \circ \hat{Q}', B) = 1$ ($eq_t(\hat{A}' \circ Q, B) = 1$). Fortunately, this $\hat{Q}'$ ($\hat{A}'$) does not exist, because from Theorems 28.1 and 28.2 we know that if $\hat{Q} = A \varphi B$ ($\hat{A} = Q \varphi B$) is not a solution to the "fuzzy identification (deconvolution) problem," then no other solution exists. Therefore, $\xi_I = 1$ ($\xi_D = 1$) if and only if the "fuzzy identification (deconvolution) problem" is solvable.

Our next task is to simplify ξ_I and ξ_D to make them easy to compute. First, we need the following lemma.

Lemma 28.6. For $a, b \in [0, 1]$, it holds that

$$a \star (a \varphi b) = min(a, b) \qquad (28.33)$$
$$b \varphi (a \star (a \varphi b)) = b \varphi a \qquad (28.34)$$

Proof: If $a \leq b$, then $a\varphi b = 1$ and $a \star (a\varphi b) = a = min(a,b)$. If $a > b$, then from the definition of φ-operator we have $a \star (a\varphi b) = a \star sup[c \in [0,1]|a \star c \leq b] = sup[a \star c|a \star c \leq b] = b = min(a,b)$; this proves (28.33). For (28.34), we have from (28.33) that $b\varphi(a \star (a\varphi b)) = b\varphi[min(a,b)] = b\varphi a$, which is obvious for $a \leq b$, but in the $b < a$ case we have $b\varphi b = 1 = b\varphi a$. $\square$

The following theorem shows that ξ_I and ξ_D are simple functions of the heights of A, B and Q.

Theorem 28.3. Let $hgh(A), hgh(B)$ and $hgh(Q)$ be the heights of A, B and Q, respectively (recall that $hgh(A) = sup_{x \in U} \mu_A(x)$). Then,

$$\xi_I = hgh(B)\varphi hgh(A) \tag{28.35}$$

$$\xi_D = hgh(B)\varphi hgh(Q) \tag{28.36}$$

Proof: Since $A \circ (A\varphi B) \subseteq B$ (Lemma 28.3), we have from (28.31) and (28.29) that

$$\xi_I = \inf_{y \in V} \{\mu_B(y)\varphi[\mu_{A \circ (A\varphi B)}(y)]\}$$

$$= \inf_{y \in V} \{\mu_B(y)\varphi[\sup_{x \in U}(\mu_A(x) \star (\mu_A(x)\varphi\mu_B(y)))]\} \tag{28.37}$$

From the monotonicity of φ in the second argument, we get

$$\xi_I = \inf_{y \in V} \sup_{x \in U} \{\mu_B(y)\varphi[\mu_A(x) \star (\mu_A(x)\varphi\mu_B(y))]\} \tag{28.38}$$

Using (28.34) we have

$$\xi_I = \inf_{y \in V} \sup_{x \in U} [\mu_B(y)\varphi\mu_A(x)]$$

$$= \inf_{y \in V} [\mu_B(y)\varphi \sup_{x \in U} \mu_A(x)] \tag{28.39}$$

Since φ is a monotonic nonincreasing function of the first argument, the $\inf_{y \in V}$ in (28.29) is achieved when $\mu_B(y)$ equals its maximum value. Hence,

$$\xi_I = \sup_{y \in V} [\mu_B(y)]\varphi \sup_{x \in U} [\mu_A(x)] \tag{28.40}$$

The proof of (28.36) is left as an exercise. $\square$

We can now go back to Example 28.1 and see why an exact solution does not exist. For the A, B and Q in (28.19)-(28.21), we have $hgh(A) = 0.4, hgh(B) = 0.5$ and $hgh(Q) = 0.3$, therefore $\xi_I = 0.5\varphi 0.4 \neq 1$ and $\xi_D = 0.5\varphi 0.3 \neq 1$. If an exact solution does not exist, what we can do is to determine approximate solutions; this is the topic of the next section.

28.4 Approximate Solution—A Neural Network Approach

From Theorem 28.3 and the definition of the φ-operator we see that if the height of B is larger than the height of A or Q, exact solutions do not exist. In these cases, a natural way is to determine A or Q such that

$$d_2(B, A \circ Q) = \int_V \left\{ \mu_B(y) - \sup_{x \in U}[\mu_A(x) \star \mu_Q(x, y)] \right\}^2 dy \qquad (28.41)$$

is minimized; this gives the least-squares solution. Here we consider only the "fuzzy deconvolution problem;" the "fuzzy identification problem" can be solved in a similar fashion. To make the minimization of (28.41) trackable, we do the following two things: (i) parameterize $\mu_A(x)$ as the Gaussian function (we also can use triangular or trapezoidal function)

$$\mu_A(x) = exp\left(-\frac{1}{2}\frac{||x - \bar{x}||^2}{\sigma^2} \right) \qquad (28.42)$$

where $\bar{x}$ and σ are free parameters, and (ii) make samples x_i and y_j ($i = 1, 2, ..., N, j = 1, 2, ..., M$) over the domains U and V, respectively, and approximate (28.41) by

$$\sum_{j=1}^{M} \left\{ \mu_B(y_j) - \max_{1 \leq i \leq N}[\mu_A(x_i) \star \mu_Q(x_i, y_j)] \right\}^2 \qquad (28.43)$$

Therefore, the problem becomes determining the parameters $\bar{x}$ and σ, such that (28.43) is minimized.

We now develop a neural network approach to this problem. First, we represent $\mu_A(x)\mu_Q(x, y)$ as the network shown in Fig. 28.1, where $a = exp[-\frac{1}{2}(\frac{||x-\bar{x}||}{\sigma})^2]$, $b = \mu_Q(x, y)$, and $z = ab$. For fixed $y = y_j$, we compute $z_1, z_2, ..., z_N$ for $x = x_1, x_2, ..., x_N$, and determine $z_{i*} = max[z_1, ..., z_N]$. Clearly, z_{i*} equals the $\max_{1 \leq i \leq N}[\mu_A(x_i)\mu_Q(x_i, y_j)]$ in (28.43). We then adjust the parameters $\bar{x}$ and σ to minimize the squared error

$$e_j^2 = \left\{ \mu_Y(y_j) - \max_{1 \leq i \leq N}[\mu_X(x_i)\mu_Q(x_i, y_j)] \right\}^2 \qquad (28.44)$$

using a gradient descent algorithm. The details of this method is given as follows.

A Neural Network Approach to the "Fuzzy Deconvolution Problem"

- **Step 1**: Make samples x_i ($i = 1, 2, ..., N$) over the domain U and samples y_j ($j = 1, 2, ..., M$) over the domain V. If U and V are bounded, we usually make the samples uniformly distributed over U and V. Determine initial guesses $\bar{x}(0)$ of $\bar{x}$ and $\sigma(0)$ of σ. Sometimes, human experts can provide these initial guesses. If not, choose $\bar{x}(0)$ randomly in U and choose σ to be a reasonable positive number, for example, $\sigma = 1$. Initialize $j = 1$.

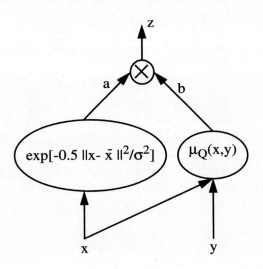

Figure 28.1. Network representation of $\mu_A(x)\mu_Q(x,y)$.

- **Step 2**: Fix $y = y_j$ in the network of Fig. 28.1, and for $x = x_1, x_2, ..., x_N$ compute forward along the network to determine the corresponding $z_1, z_2, ..., z_N$. Find i^* such that $z_{i^*} = max[z_1, z_2, ..., z_N]$.

- **Step 3**: Update the parameters $\bar{x}$ and σ using the gradient descent algorithm:

$$\bar{x}(k+1) = \bar{x}(k) - \frac{\alpha}{2}\frac{\partial e_j^2}{\partial \bar{x}}$$

$$= \bar{x}(k) + \alpha e_j b^* a^* \frac{x_{i^*} - \bar{x}(k)}{(\sigma(k))^2}, \qquad (28.45)$$

$$\sigma(k+1) = \sigma(k) - \frac{\alpha}{2}\frac{\partial e_j^2}{\partial \sigma}$$

$$= \sigma(k) + \alpha e_j b^* a^* \frac{||x_{i^*} - \bar{x}(k)||^2}{(\sigma(k))^3}, \qquad (28.46)$$

where

$$a^* = exp[-\frac{1}{2}(\frac{||x_{i^*} - \bar{x}(k)||}{\sigma(k)})^2], \qquad (28.47)$$

$$b^* = \mu_Q(x_{i^*}, y_j), \qquad (28.48)$$

e_j^2 is given by (28.44), and $k = 0, 1, 2, ...$

- **Step 4**: Go to Step 2 with $y = y_{j+1}$, and repeat for $y = y_1, y_2, ..., y_M, y_1, ...,$ until the differences $|\bar{x}(k+1) - \bar{x}(k)|$ and $|\sigma(k+1) - \sigma(k)|$ are smaller than a small threshold for a number of $k's$ greater than M.

The training algorithm (28.45) and (28.46) are obtained by using the chain rule: $\frac{\partial e_j^2}{\partial \bar{x}} = 2e_j \frac{\partial e_j}{\partial \bar{x}} = -2e_j \frac{\partial z_{j^*}}{\partial \bar{x}} = -2e_j b^* \frac{\partial a^*}{\partial \bar{x}} = -2e_j b^* a^* \frac{x_{j^*} - \bar{x}(k)}{(\sigma(k))^2}$; similar for $\frac{\partial e_j^2}{\partial \sigma}$. It can be proven that the above training algorithm guarantees that the total matching error $\sum_{j=1}^{M} e_j^2$ will decrease after every M steps of training. Because it is required that $\sigma > 0$, in the implementation we need to choose a small constant $\epsilon > 0$ such that if $\sigma(k+1) > \epsilon$ then keep $\sigma(k+1)$ as computed by (28.46), and if $\sigma(k+1) \leq \epsilon$ then set $\sigma(k+1) = \epsilon$. In this way we can guarantee $\sigma(k) \geq \epsilon$ for all k.

We now simulate the algorithm for an example.

Example 28.2. Consider a fuzzy relation that is characterized by the two fuzzy IF-THEN rules:

$$Ru^{(1)}: \quad IF \ x \ is \ small, \ THEN \ y \ is \ near \ 1, \quad (28.49)$$
$$Ru^{(2)}: \quad IF \ x \ is \ large, \ THEN \ y \ is \ near \ 0, \quad (28.50)$$

where "small," "large," "near 1," and "near 0" are fuzzy sets with membership functions: $\mu_{small}(x) = \frac{1}{1+e^{10(x-1)}}, \mu_{large}(x) = \frac{1}{1+e^{10(-x+3)}}, \mu_{near1}(y) = e^{-10(y-1)^2}$ and $\mu_{near0}(y) = e^{-10y^2}$, respectively. Using Mamdani's product implication, the $Ru^{(i)}$ (i=1,2) are fuzzy relations in $U \times V$ with membership functions

$$\mu_{Ru^{(1)}}(x, y) = \mu_{small}(x)\mu_{near1}(y) \quad (28.51)$$
$$\mu_{Ru^{(2)}}(x, y) = \mu_{large}(x)\mu_{near0}(y) \quad (28.52)$$

The final fuzzy relation Q is

$$\mu_Q(x, y) = \max[\mu_{Ru^{(1)}}, \mu_{Ru^{(2)}}]$$
$$= \max \left[\frac{e^{-10(y-1)^2}}{1 + e^{10(x-1)}}, \frac{e^{-10y^2}}{1 + e^{10(-x+3)}} \right] \quad (28.53)$$

Now, for simplicity, we assume that $\sigma \equiv 1$ in (28.42), and our task is to determine $\bar{x}$ for different $\mu_B(y)'s$. We consider three cases of $\mu_B(y)$: (i) $\mu_B(y) = e^{-10(y-0.5)^2}$ (that is, y is near 0.5), (ii) $\mu_B(y) = e^{-10(y-1)^2}$ (that is, y is near 1), and (iii) $\mu_B(y) = e^{-10y^2}$ (that is, y is near 0). We choose the domains $U = [0,4]$ and $V = [0,1]$. We uniformly sample U and V with $N = 17$ and $M = 9$ points, respectively, that is, $(x_1, x_2, x_3, ..., x_{17}) = (0, 0.25, 0.5, ..., 4)$, and $(y_1, y_2, y_3, ..., y_9) = (0, 0.125, 0.25, ..., 1)$. Figs.28.2-28.4 show the convergence procedures of $\bar{x}(k)$ for different initial conditions for the three $\mu_B(y)$ cases (i)-(iii), respectively, where the horizontal axis represents the k in the training algorithm, and the vertical axis represents the $\bar{x}(k)$. From Figs. 28.2-28.4 we see that the training procedure converged in about 10 steps for all the three cases.

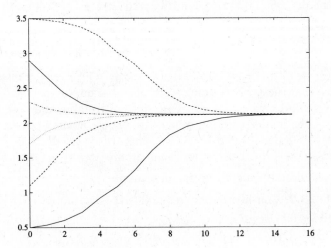

Figure 28.2. Convergence procedure of $\bar{x}(k)$ for $\mu_Y(y) = e^{-10(y-0.5)^2}$.

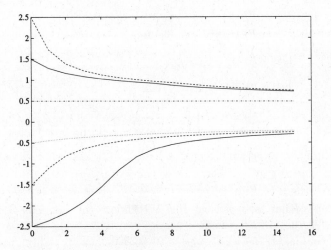

Figure 28.3. Convergence procedure of $\bar{x}(k)$ for $\mu_Y(y) = e^{-10(y-1)^2}$.

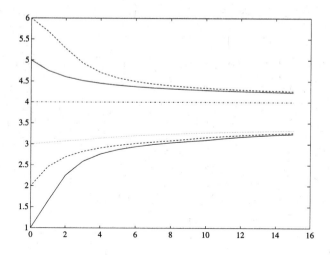

Figure 28.4. Convergence procedure of $\bar{x}(k)$ for $\mu_Y(y) = e^{-10y^2}$.

From Fig. 28.2 we see that in this case all the $\bar{x}(k)'s$ for different initial $\bar{x}(0)'s$ converged to the same value near 2. This is consistent with our intuition because from (28.49) and (28.50) we expect that if y is near 0.5, that is, in the middle of 0 and 1, then x should be in the middle of "small" and "large" which, by observing the membership functions μ_{small} and μ_{large}, is somewhere around 2. Figures 28.3 and 28.4 show that for "y is near 1" and "y is near 0", the $\bar{x}(k)$ converged to a region of values for different initial conditions, not to a single point. This also is intuitively appealing because from (28.49) we see that if y is near 1, then x should be small, which in Fig. 28.3 means somewhere between -0.3 and 0.8. Similarly, from (28.50) we see that if y is near 0, then x should be large, which in Fig. 28.4 means somewhere between 3.2 and 4.3. In summary, all the solutions in the three cases are consistent with our intuition. $\square$

28.5 Summary and Further Readings

In this chapter we have demonstrated the following:

- The definition of "fuzzy deconvolution and fuzzy identification problems."

- How to obtain the particular solutions to the problems.

- The definition and computation of the solvability indices.

- How to transform the problems of solving the fuzzy relation equations into a neural network framework and how to develop a training algorithm for computing the approximate solutions.

A special book on fuzzy relation equations is Nola, Sessa, Pedrycz, and Sanchez [1989]. The book Pedrycz [1993] also covered the fuzzy relation equations extensively. A good tutorial paper for fuzzy relation equations is Pedrycz [1991]. The neural network approach in this chapter was taken from Wang [1993b].

28.6 Exercises

Exercise 28.1. Solve the following "fuzzy deconvolution problems" for the max-min composition.

$$(a) \ A \circ \begin{bmatrix} .9 & .8 & .7 \\ .8 & .9 & .8 \\ .7 & .8 & .9 \end{bmatrix} = [.6 \ .5 \ .4] \tag{28.54}$$

$$(b) \ A \circ \begin{bmatrix} .5 & .7 & 0 & .2 \\ .4 & .6 & 1 & 0 \\ .2 & .4 & .5 & .6 \\ 0 & .2 & 0 & .8 \end{bmatrix} = [.5 \ .5 \ .4 \ .2] \tag{28.55}$$

$$(c) \ A \circ \begin{bmatrix} .5 & .4 & .6 & .7 \\ .2 & 0 & .6 & .8 \\ .2 & .4 & .6 & .7 \\ 0 & .3 & .3 & 1 \end{bmatrix} = [.5 \ .5 \ .4 \ .2] \tag{28.56}$$

$$(d) \ A \circ \begin{bmatrix} .5 & 0 & .3 & 0 \\ .4 & 1 & .3 & 0 \\ 0 & .1 & 1 & .1 \\ .4 & .3 & .3 & .5 \end{bmatrix} = \begin{bmatrix} .5 & .3 & .3 & .1 \\ .5 & .4 & .4 & .2 \end{bmatrix} \tag{28.57}$$

Exercise 28.2. Solve the following "fuzzy identification problems" for the max-min composition.

$$(a) \ [1 \ .3 \ .5] \circ Q = [.6 \ .5 \ .4] \tag{28.58}$$

$$(b) \ [.1 \ .3 \ .5 \ .7] \circ Q = [.6 \ .5 \ .4 \ .3] \tag{28.59}$$

$$(c) \ [.6 \ .4 \ .6 \ .8] \circ Q = [.1 \ .5 \ .4 \ .2] \tag{28.60}$$

$$(d) \ \begin{bmatrix} .6 & .2 & .3 & .9 \\ .1 & .3 & .5 & .7 \end{bmatrix} \circ Q = \begin{bmatrix} .1 & .5 & .4 & .2 \\ .2 & .2 & .4 & .4 \end{bmatrix} \tag{28.61}$$

Exercise 28.3. Prove (28.36) in Theorem 28.3.

Exercise 28.4. Let P, Q and R be fuzzy relations in $U \times V, V \times W$ and $U \times W$, respectively, and $\mathcal{F}(U \times V)$ be the set of all fuzzy relations in $U \times V$. Define the solvability index of the fuzzy relation equation $P \circ Q = R$ as

$$\delta = \sup_{P \in \mathcal{F}(U \times V)} eq_t(P \circ Q, R) \tag{28.62}$$

Show that

$$\delta \leq \inf_{z \in W} \left[\sup_{x \in U} \mu_R(x, z) \varphi \sup_{y \in V} \mu_Q(y, z) \right] \tag{28.63}$$

Exercise 28.5. Develop a neural network approach for solving the "fuzzy identification problem."

Chapter 29

Fuzzy Arithmetic

29.1 Fuzzy Numbers and the Decomposition Theorem

In the fuzzy IF-THEN rules, the atomic fuzzy proposition, x_i *is* A_i^l, is often an imprecise characterization of numbers in the real line R, for example, "close to 5," "approximately 7," etc. Since fuzzy sets defined in R appear quite often in many applications, they deserve an in-depth study. Roughly speaking, a fuzzy number is a fuzzy set in R; but in order to perform meaningful arithmetic operations, we add some constraints. Specifically, we have the following definition.

Definition 29.1. Let A be a fuzzy set in R. A is called a *fuzzy number* if: (i) A is normal, (ii) A is convex, (iii) A has a bounded support, and (iv) all α-cuts of A are closed intervals of R.

We require a fuzzy number to be normal because our conception of real numbers close to r is fully satisfied by r itself, hence we should have $\mu_A(r) = 1$. The convexity and boundedness conditions allow us to define meaningful arithmetic operations for fuzzy numbers.

Two special classes of fuzzy numbers often are used in practice; they are triangular and trapezoidal fuzzy numbers. A *triangular fuzzy number* A is a fuzzy set in R with the triangular membership function:

$$\mu_A(x) = \mu_A(x; a, b, c) = \begin{cases} (x-a)/(b-a) & if \quad a \le x < b \\ (c-x)/(c-b) & if \quad b \le x \le c \\ 0 & if \quad x > c \ or \ x < a \end{cases} \tag{29.1}$$

Similarly, if a fuzzy set A in R has the trapezoidal membership function:

$$\mu_A(x) = \mu_A(x; a, b, c, d) = \begin{cases} (x-a)/(b-a) & if \quad a \le x < b \\ 1 & if \quad b \le x \le c \\ (d-x)/(d-c) & if \quad c < x \le d \\ 0 & if \quad x > d \ or \ x < a \end{cases} \tag{29.2}$$

it is called *trapezoidal fuzzy number*. It is easy to check that the triangular and trapezoidal fuzzy numbers satisfy the four conditions in Definition 29.1.

368

A useful approach to study the arithmetic operations of fuzzy numbers is to use the α-cuts of fuzzy numbers. Recall that the α-cut of a fuzzy set A in R is defined as

$$A_\alpha = \{x \in R | \mu_A(x) \geq \alpha\} \tag{29.3}$$

Now we show that a fuzzy set can be uniquely determined by its α-cuts. To make precise representation, we define fuzzy set $\tilde{A}_\alpha$ in R with the membership function

$$\mu_{\tilde{A}_\alpha}(x) = \alpha I_{A_\alpha}(x) \tag{29.4}$$

where $I_{A_\alpha}(x)$ is the indicator function of the (crisp) set A_α, that is, $I_{A_\alpha}(x) = 1$ if $x \in A_\alpha$ and $I_{A_\alpha}(x) = 0$ if $x \in R - A_\alpha$. The following theorem, known as the Decomposition Theorem, states that the fuzzy set A equals the union of the fuzzy sets $\tilde{A}_\alpha$ for all $\alpha \in [0, 1]$.

Theorem 29.1. (Decomposition Theorem) Let A and $\tilde{A}_\alpha$ be fuzzy sets in R with $\tilde{A}_\alpha$ defined by (29.4). Then,

$$A = \bigcup_{\alpha \in [0,1]} \tilde{A}_\alpha \tag{29.5}$$

where $\bigcup$ denotes the standard fuzzy union (that is, sup over $\alpha \in [0, 1]$).

Proof: Let x be an arbitrary point in R and $a = \mu_A(x)$. Then,

$$\mu_{\bigcup_{\alpha \in [0,1]} \tilde{A}_\alpha}(x) = \sup_{\alpha \in [0,1]} \mu_{\tilde{A}_\alpha}(x)$$

$$= \max\left[\sup_{\alpha \in [0,a]} \mu_{\tilde{A}_\alpha}(x), \sup_{\alpha \in (a,1]} \mu_{\tilde{A}_\alpha}(x) \right] \tag{29.6}$$

For each $\alpha \in (a, 1]$, we have $\mu_A(x) = a < \alpha$ and thus $x \notin A_\alpha$ which implies $\mu_{\tilde{A}_\alpha}(x) = 0$. If $\alpha \in [0, a]$, then $\mu_A(x) = a \geq \alpha$ so that $x \in A_\alpha$ and from (29.4) we have $\mu_{\tilde{A}_\alpha}(x) = \alpha$. Hence,

$$\mu_{\bigcup_{\alpha \in [0,1]} \tilde{A}_\alpha}(x) = \sup_{\alpha \in [0,a]} \alpha = a = \mu_A(x) \tag{29.7}$$

Since x is arbitrary, (29.7) implies (29.5). $\square$

From Theorem 29.1 we see that if we can determine the α-cuts of a fuzzy set for all $\alpha \in [0, 1]$, we can specify the fuzzy set itself. Therefore, determining a fuzzy set is equivalent to determining its α-cuts for all $\alpha \in [0, 1]$. In the next two sections, we will use the α-cuts to define the arithmetic operations of fuzzy numbers.

29.2 Addition and Subtraction of Fuzzy Numbers

29.2.1 The α-Cut Method

Let A and B be two fuzzy numbers and $A_\alpha = [a_\alpha^-, a_\alpha^+], B_\alpha = [b_\alpha^-, b_\alpha^+]$ be their α-cuts. Then, the *addition of A and B*, $A + B$, is a fuzzy number with its α-cuts defined by

$$(A + B)_\alpha = [a_\alpha^- + b_\alpha^-, a_\alpha^+ + b_\alpha^+] \qquad (29.8)$$

for every $\alpha \in [0,1]$. Since a fuzzy set is completely determined by its α-cuts (Theorem 29.1), the fuzzy number $A + B$ is well-defined by its α-cuts (29.8).

Similarly, the *subtraction of A and B, A − B*, is a fuzzy number with its α-cuts defined by

$$(A - B)_\alpha = [\min(a_\alpha^- - b_\alpha^-, a_\alpha^+ - b_\alpha^+), \max(a_\alpha^- - b_\alpha^-, a_\alpha^+ - b_\alpha^+)] \qquad (29.9)$$

for every $\alpha \in [0,1]$.

Example 29.1. Compute the addition and subtraction of the triangular fuzzy numbers A and B whose membership functions are

$$\mu_A(x) = \mu_A(x; -1, 0, 1) \qquad (29.10)$$
$$\mu_B(x) = \mu_B(x; -1, 1, 3) \qquad (29.11)$$

For a given $\alpha \in [0,1]$, the α-cuts of A and B are

$$A_\alpha = [\alpha - 1, 1 - \alpha], \; B_\alpha = [2\alpha - 1, 3 - 2\alpha] \qquad (29.12)$$

Hence,

$$(A + B)_\alpha = [3\alpha - 2, 4 - 3\alpha] \qquad (29.13)$$

Since $3\alpha - 2$ and $4 - 3\alpha$ are linear functions of α, $A + B$ is also a triangular fuzzy number. By setting $\alpha = 0$ and $\alpha = 1$ in (29.13), we obtain

$$\mu_{A+B}(x) = \mu_{A+B}(x; -2, 1, 4) \qquad (29.14)$$

which is plotted in Fig. 29.1. Similarly, we have

$$(A - B)_\alpha = [\alpha - 2, -\alpha] \qquad (29.15)$$

and

$$\mu_{A-B}(x) = \mu_{A-B}(x; -2, -1, 0) \qquad (29.16)$$

which is shown in Fig. 29.1. □

Another way to compute the addition and subtraction of fuzzy numbers is to use the extension principle.

29.2.2 The Extension Principle Method

Using the extension principle (Chapter 4), we can define the addition of fuzzy numbers A and B as

$$\mu_{A+B}(z) = \sup_{x+y=z} \min[\mu_A(x), \mu_B(y)] \qquad (29.17)$$

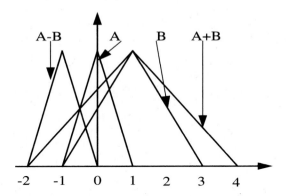

Figure 29.1. Addition and subtraction of the two fuzzy numbers in Example 29.1.

where $z \in R$ and the $\sup_{x+y=z}$ is taken over all $x, y \in R$ with $x + y = z$. Similarly, we define

$$\mu_{A-B}(z) = \sup_{x-y=z} \min[\mu_A(x), \mu_B(y)] \qquad (29.18)$$

We now prove that this definition is equivalent to the α-cut method.

Theorem 29.2. The extension principle method (29.17) and (29.18) is equivalent to the α-cut method (29.8) and (29.9); that is, they give the same fuzzy numbers $A + B$ and $A - B$.

Proof: First, note that the fuzzy sets $A + B$ and $A - B$ defined by (29.17) and (29.18) are fuzzy numbers (the proof is left as an exercise). Hence, according to Theorem 29.1 we only need to show that the α-cuts of $A + B$ and $A - B$ defined by (29.17) and (29.18) are given by (29.8) and (29.9), respectively. From (29.17) we see that for any $x \in A_\alpha = [a_\alpha^-, a_\alpha^+]$ and $y \in B_\alpha = [b_\alpha^-, b_\alpha^+]$, we have $\mu_{A+B}(z) \geq \alpha$ if $z = x+y$; hence, $[a_\alpha^- + b_\alpha^-, a_\alpha^+ + b_\alpha^+] \subseteq (A+B)_\alpha$ because for any $z \in [a_\alpha^- + b_\alpha^-, a_\alpha^+ + b_\alpha^+]$ we can find $x \in [a_\alpha^-, a_\alpha^+]$ and $y \in [b_\alpha^-, b_\alpha^+]$ such that $z = x + y$. Conversely, if $z \in R - [a_\alpha^- + b_\alpha^-, a_\alpha^+ + b_\alpha^+]$ and for any decomposition $z = x + y$, we have either $x \notin [a_\alpha^-, a_\alpha^+]$ or $y \notin [b_\alpha^-, b_\alpha^+]$ so that $\mu_{A+B}(z) = 0 < \alpha$ according to (29.17). This means that for any z with $\mu_{A+B}(z) \geq \alpha$, it must be true that $z \in [a_\alpha^- + b_\alpha^-, a_\alpha^+ + b_\alpha^+]$; that is, $(A+B)_\alpha \subseteq [a_\alpha^- + b_\alpha^-, a_\alpha^+ + b_\alpha^+]$. Hence, we have (29.8). (29.9) can be proven in the same manner. $\square$

Since a real number a is a special fuzzy number (with its membership function $\mu_a(x) = 1$ if $x = a$ and $\mu_a(x) = 0$ if $x \neq a$), we have from (29.17) that $A + a$ is a fuzzy number with membership function

$$\mu_{A+a}(z) = \mu_A(z - a) \tag{29.19}$$

That is, $\mu_{A+a}(z)$ is obtained by shifting $\mu_A(z)$ a distance of a. Similarly, $A - a$ is a fuzzy number with membership function

$$\mu_{A-a}(z) = \mu_A(z + a) \tag{29.20}$$

Viewing 0 as a special fuzzy number, we have from (29.18) that

$$\mu_{-B}(z) = \mu_{0-B}(z) = \mu_B(-z) \tag{29.21}$$

Finally, we consider an example of computing the addition and subtraction of two fuzzy sets defined over a finite number of points.

Example 29.2. Compute the addition and subtraction of the fuzzy sets:

$$A = 0.1/1 + 0.5/2 + 1/3 + 0.7/4 + 0.3/5 \tag{29.22}$$
$$B = 0.2/1 + 1/2 + 0.5/3 \tag{29.23}$$

Using (29.17) and (29.18), we have

$$A + B = 0.1/2 + 0.2/3 + 0.5/4 + 1/5 + 0.7/6 + 0.5/7 + 0.3/8 \tag{29.24}$$
$$A - B = 0.1/-2 + 0.5/-1 + 0.5/0 + 1/1 + 0.7/6 + 0.3/3 + 0.2/4 \tag{29.25}$$

$\square$

29.3 Multiplication and Division of Fuzzy Numbers

29.3.1 The α-Cut Method

In classical interval analysis, the multiplication of intervals $[a, b]$ and $[c, d]$ is defined as

$$[a, b] \cdot [c, d] = [\min(ac, ad, bc, bd), \max(ac, ad, bc, bd)] \tag{29.26}$$

If $0 \notin [c, d]$, then the division of $[a, b]$ and $[c, d]$ is defined as

$$[a, b]/[c, d] = [a, b] \cdot [\frac{1}{c}, \frac{1}{d}]$$
$$= [\min(a/c, a/d, b/c, b/d), \max(a/c, a/d, b/c, b/d)] \tag{29.27}$$

We now define the multiplication and division of fuzzy numbers using their α-cuts and (29.26)-(29.27).

Let A and B be fuzzy numbers and $A_\alpha = [a_\alpha^-, a_\alpha^+]$ and $B_\alpha = [b_\alpha^-, b_\alpha^+]$ be their α-cuts. Then, the *multiplication of A and B*, $A \cdot B$, is a fuzzy number with its α-cuts defined by

$$(A \cdot B)_\alpha = [\min(a_\alpha^- b_\alpha^-, a_\alpha^- b_\alpha^+, a_\alpha^+ b_\alpha^-, a_\alpha^+ b_\alpha^+), \max(a_\alpha^- b_\alpha^-, a_\alpha^- b_\alpha^+, a_\alpha^+ b_\alpha^-, a_\alpha^+ b_\alpha^+)] \tag{29.28}$$

for every $\alpha \in [0,1]$. If $0 \notin [b_\alpha^-, b_\alpha^+]$ for all $\alpha \in [0,1]$, then the *division of A and B*, A/B, is a fuzzy number with its α-cuts defined by

$$(A/B)_\alpha = [\min(a_\alpha^-/b_\alpha^-, a_\alpha^-/b_\alpha^+, a_\alpha^+/b_\alpha^-, a_\alpha^+/b_\alpha^+), \max(a_\alpha^-/b_\alpha^-, a_\alpha^-/b_\alpha^+, a_\alpha^+/b_\alpha^-, a_\alpha^+/b_\alpha^+)]$$
$$(29.29)$$

for every $\alpha \in [0,1]$.

Example 29.3. Consider the fuzzy number A in Example 29.1 and B defined by

$$\mu_B(x) = \mu_B(x; 0.5, 1, 3) \tag{29.30}$$

We now compute $A \cdot B$ and A/B. First, note that the α-cuts $A_\alpha = [a_\alpha^-, a_\alpha^+] = [\alpha - 1, 1 - \alpha]$ and $B_\alpha = [b_\alpha^-, b_\alpha^+] = [(\alpha+1)/2, 3 - 2\alpha]$. Hence,

$$\min(a_\alpha^- b_\alpha^-, a_\alpha^- b_\alpha^+, a_\alpha^+ b_\alpha^-, a_\alpha^+ b_\alpha^+) = a_\alpha^- b_\alpha^+ = -2\alpha^2 + 5\alpha - 3 \tag{29.31}$$

$$\max(a_\alpha^- b_\alpha^-, a_\alpha^- b_\alpha^+, a_\alpha^+ b_\alpha^-, a_\alpha^+ b_\alpha^+) = a_\alpha^+ b_\alpha^+ = 2\alpha^2 - 5\alpha + 3 \tag{29.32}$$

and

$$(A \cdot B)_\alpha = [-2\alpha^2 + 5\alpha - 3, 2\alpha^2 - 5\alpha + 3] \tag{29.33}$$

From (28.33) we have

$$\mu_{A \cdot B}(x) = \frac{5 - (1 + 8|x|)^{1/2}}{4} \tag{29.34}$$

which is illustrated in Fig. 29.2. Similarly, we have

$$\min(a_\alpha^-/b_\alpha^-, a_\alpha^-/b_\alpha^+, a_\alpha^+/b_\alpha^-, a_\alpha^+/b_\alpha^+) = a_\alpha^-/b_\alpha^- = (\alpha - 1)/(3 - 2\alpha) \tag{29.35}$$

$$\max(a_\alpha^-/b_\alpha^-, a_\alpha^-/b_\alpha^+, a_\alpha^+/b_\alpha^-, a_\alpha^+/b_\alpha^+) = a_\alpha^+/b_\alpha^- = 2(1 - \alpha)/(1 + \alpha) \tag{29.36}$$

and

$$(A/B)_\alpha = [(\alpha - 1)/(3 - 2\alpha), 2(1 - \alpha)/(1 + \alpha)] \tag{29.37}$$

From (29.37) we have

$$\mu_{A/B}(x) = \begin{cases} (3x + 1)/(2x + 1) & for \quad x < 0 \\ (2 - x)/(2 + x) & for \quad x \geq 0 \end{cases} \tag{29.38}$$

which is plotted in Fig. 29.2. $\square$

29.3.2   The Extension Principle Method

Similar to addition and subtraction of fuzzy numbers, we can use the extension principle to define $A \cdot B$ and A/B. Specifically, the fuzzy numbers $A \cdot B$ and A/B are defined by

$$\mu_{A \cdot B}(z) = \sup_{xy=z} \min[\mu_A(x), \mu_B(y)] \tag{29.39}$$

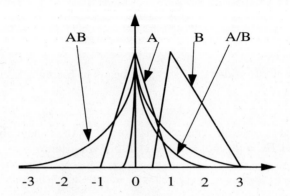

Figure 29.2. Multiplication and division of the two fuzzy numbers in Example 29.3.

and

$$\mu_{A/B}(z) = \sup_{x/y=z} \min[\mu_A(x), \mu_B(y)] \qquad (29.40)$$

respectively. It can be shown that (29.39) and (29.40) give the same fuzzy numbers as defined by (29.28) and (29.29), respectively.

If $B = b$ is a crisp number, then from (29.39) we have

$$\mu_{bA}(z) = \mu_A(z/b) \qquad (29.41)$$

This gives the multiplication of a real number and a fuzzy number.

29.4   Fuzzy Equations

Let A, B and X be fuzzy numbers. It is interesting to solve the *fuzzy equations*:

$$A + X = B \qquad (29.42)$$

and

$$A \cdot X = B \qquad (29.43)$$

where A, B are known, and X is unknown. That is, the problem is to determine fuzzy number X such that (29.42) or (29.43) is true for given fuzzy numbers A and B.

We first study how to solve (29.42). Let $A_\alpha = [a_\alpha^-, a_\alpha^+], B_\alpha = [b_\alpha^-, b_\alpha^+]$ and $X_\alpha = [x_\alpha^-, x_\alpha^+]$. According to the Decomposition Theorem, (29.42) is true if and only if

$$[a_\alpha^-, a_\alpha^+] + [x_\alpha^-, x_\alpha^+] = [b_\alpha^-, b_\alpha^+] \tag{29.44}$$

for all $\alpha \in (0, 1]$. Hence, a potential solution is given by

$$X_\alpha = [x_\alpha^-, x_\alpha^+] = [b_\alpha^- - a_\alpha^-, b_\alpha^+ - a_\alpha^+] \tag{29.45}$$

In order for the X_α in (29.45) to be qualified as α-cuts, the following two conditions must be satisfied:

- (i) $b_\alpha^- - a_\alpha^- \le b_\alpha^+ - a_\alpha^+$ for every $\alpha \in (0, 1]$
- (ii) If $\alpha \le \beta$, then $b_\alpha^- - a_\alpha^- \le b_\beta^- - a_\beta^- \le b_\beta^+ - a_\beta^+ \le b_\alpha^+ - a_\alpha^+$

Condition (i) guarantees that $[b_\alpha^- - a_\alpha^-, b_\alpha^+ - a_\alpha^+]$ is an interval, and condition (ii) ensures that $[b_\alpha^- - a_\alpha^-, b_\alpha^+ - a_\alpha^+]$ are nested intervals so that they are qualified as α-cuts of a fuzzy number. Therefore, if conditons (i) and (ii) are satisfied, then using the Decomposition Theorem we obtain the solution as

$$X = \bigcup_{\alpha \in (0,1]} \tilde{X}_\alpha \tag{29.46}$$

where $\tilde{X}_\alpha$ are fuzzy sets defined by

$$\mu_{\tilde{X}_\alpha}(x) = \alpha I_{[b_\alpha^- - a_\alpha^-, b_\alpha^+ - a_\alpha^+]}(x) \tag{29.47}$$

Example 29.4. Let A and B be the triangular fuzzy numbers in Example 29.1. Then, from (29.12) we have $a_\alpha^- = \alpha - 1, a_\alpha^+ = 1 - \alpha, b_\alpha^- = 2\alpha - 1$ and $b_\alpha^+ = 3 - 2\alpha$. Hence, $b_\alpha^- - a_\alpha^- = \alpha$ and $b_\alpha^+ - a_\alpha^+ = 2 - \alpha$. Since $\alpha \le 2 - \alpha$ for $\alpha \in (0, 1]$ and $\alpha \le \beta \le 2 - \beta \le 2 - \alpha$ if $\alpha \le \beta \le 1$, conditions (i) and (ii) are satisfied. Therefore,

$$X_\alpha = [\alpha, 2 - \alpha] \tag{29.48}$$

from which we obtain the solution X as the triangular fuzzy number

$$\mu_X(x) = \mu_X(x; 0, 1, 2) \tag{29.49}$$

$\square$

Next, we study the fuzzy equation (29.43). For simplicity, assume that A and B are fuzzy numbers in R^+. Then (29.43) is equivalent to

$$[a_\alpha^-, a_\alpha^+] \cdot [x_\alpha^-, x_\alpha^+] = [b_\alpha^-, b_\alpha^+] \tag{29.50}$$

for all $\alpha \in (0, 1]$. If the following two conditions are satisfied:

- (i) $b_\alpha^-/a_\alpha^- \leq b_\alpha^+/a_\alpha^+$ for every $\alpha \in (0,1]$

- (ii) $\alpha \leq \beta$ implies $b_\alpha^-/a_\alpha^- \leq b_\beta^-/a_\beta^- \leq b_\beta^+/a_\beta^+ \leq b_\alpha^+/a_\alpha^+$

then the solution X of (29.43) is given through its α-cuts

$$X_\alpha = [x_\alpha^-, x_\alpha^+] = [b_\alpha^-/a_\alpha^-, b_\alpha^+/a_\alpha^+] \tag{29.51}$$

Example 29.5. Consider the fuzzy equation (29.43) with $\mu_A(x) = \mu_A(x; 3, 4, 5)$ and $\mu_B(x) = \mu_B(x; 12, 20, 32)$. Then $A_\alpha = [\alpha+3, 5-\alpha]$ and $B_\alpha = [8\alpha+12, 32-12\alpha]$. It is easy to verify that the two conditions above are satisfied, hence

$$X_\alpha = \left[\frac{8\alpha + 12}{\alpha + 3}, \frac{32 - 12\alpha}{5 - \alpha} \right] \tag{29.52}$$

from which we obtain the solution as

$$\mu_X(x) = \begin{cases} 0 & \text{if} \quad x \leq 4 \text{ and } x \geq 32/5 \\ \frac{12-3x}{x-8} & \text{if} \quad 4 < x \leq 5 \\ \frac{32-5x}{12-x} & \text{if} \quad 5 \leq x \leq 32/5 \end{cases} \tag{29.53}$$

□

29.5   Fuzzy Ranking

In fuzzy multiple attribute decision making (Chen and Hwang [1991]), the final scores of alternatives are represented by fuzzy numbers. In order to make a crisp choice among the alternatives, we need a method for comparing fuzzy numbers. However, unlike the real numbers in R that can be linearly ordered by $\leq$, fuzzy numbers cannot be linearly ordered. Intuitively, this is easy to understand. Consider the fuzzy numbers A and B in Fig. 29.3. Some people feel that B is larger than A because the center of B is larger than the center of A, while other people do not agree because B spreads widely over the small numbers while A is concentrated on large numbers. Clearly, the problem itself is ill-posed (that is, no crisp answer), and no fuzzy ranking method is perfect. In the literature, a large number of methods were proposed for ranking fuzzy numbers, with each method good for certain situations and bad for other situations. In Chen and Hwang [1992], some 20 fuzzy ranking methods were reviewed. In this section, we consider only two types of methods: the α-cut-based methods, and the Hamming distance-based methods.

First, let us examize under what conditions can we confidently rank fuzzy numbers. If the boundaries of the α-cuts $A_\alpha = [a_\alpha^-, a_\alpha^+]$ and $B_\alpha = [b_\alpha^-, b_\alpha^+]$ satisfy $a_\alpha^- \leq b_\alpha^-$ and $a_\alpha^+ \leq b_\alpha^+$ for all $\alpha \in (0,1]$, then we can reasonably say that A is smaller than B. This gives us the first α-cuts-based method.

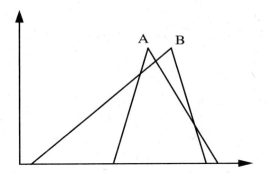

Figure 29.3. Ranking fuzzy numbers is an ill-posed problem.

- **α-cuts-based method 1**: Let A and B be fuzzy numbers with α-cuts $A_\alpha = [a_\alpha^-, a_\alpha^+]$ and $B_\alpha = [b_\alpha^-, b_\alpha^+]$. We say A is smaller than B, denoted by $A \leq B$, if $a_\alpha^- \leq b_\alpha^-$ and $a_\alpha^+ \leq b_\alpha^+$ for all $\alpha \in (0, 1]$.

The advantage of this method is that the conclusion is less controversial. Its disadvantage is that it applies only to some obvious situations. For example, the A and B in Fig. 29.3 cannot be compared according this method, because $b_\alpha^- \leq a_\alpha^-$ for small α but $b_\alpha^- \geq a_\alpha^-$ for large α. Hence, a weaker ranking method was proposed, as follows:

- **α-cuts-based method 2**: We say $A \leq B$ if $a_\alpha^+ \leq b_\alpha^+$ for all $\alpha \in (c, 1]$, where c is a constant that is usually larger than 0.5.

Clearly, this method emphasizes the numbers with large membership values. According to this method, we have $A \leq B$ for the A and B in Fig. 29.3.

Next, we consider the Hamming distance-based methods. The basic idea is to find a fuzzy number C such that $A \leq C$ and $B \leq C$ according to the (quite reasonable but restrictive) α-cuts-based method 1. Then, if the Hamming distance between C and A is larger than the Hamming distance between C and B, we conclude that $A \leq B$. Clearly, the choice of C is not unique. Here we choose the C to be the fuzzy number $MAX(A, B)$ which is defined by

$$\mu_{MAX(A,B)}(z) = \sup_{\max(x,y)=z} \min[\mu_A(x), \mu_B(y)] \qquad (29.54)$$

Note that $MAX(A, B)$ is different from the fuzzy set with membership function $\max(\mu_A, \mu_B)$. It is easy to show that $A \leq MAX(A, B)$ and $B \leq MAX(A, B)$ according to the α-cuts-based method 1. Let the Hamming distance $d_1(C, D)$ be defined as in (28.26) with $p = 1$, then we have the following fuzzy ranking method:

- **Hamming distance-based method**: We say $A \leq B$ if $d_1(MAX(A, B), A) \geq d_1(MAX(A, B), B)$.

Example 29.6. Consider the fuzzy numbers

$$\mu_A(x) = \mu_A(x; 1, 2, 5) \tag{29.55}$$
$$\mu_B(x) = \mu_A(x; 2, 3, 4) \tag{29.56}$$

From (29.54) we have

$$\mu_{MAX(A,B)}(x) = \begin{cases} x - 2 & if \quad 2 \leq x < 3 \\ 4 - x & if \quad 3 \leq x < 3.5 \\ \frac{5-x}{3} & if \quad 3.5 \leq x \leq 5 \\ 0 & if \quad x < 1 \; or \; x > 5 \end{cases} \tag{29.57}$$

which is plotted in heavy lines in Fig. 29.4. We can easily find that

$$d_1(MAX(A, B), A) = \int_R |\mu_{MAX(A,B)}(x) - \mu_A(x)| dx = 1 \tag{29.58}$$

and

$$d_1(MAX(A, B), B) = \int_R |\mu_{MAX(A,B)}(x) - \mu_B(x)| dx = 0.25 \tag{29.59}$$

Hence, the Hamming distance-based method gives $A \leq B$. If we use the α-cuts-based method 2 with $c = 0.5$, we have $A \leq B$. Clearly, the α-cuts-based method 1 does not apply in this case. $\square$

29.6   Summary and Further Readings

In this chapter we have demonstrated the following:

- Fuzzy numbers and their addition, subtraction, multiplication, and division.

- Equivalence of the α-cut and extension principle methods for fuzzy number operations.

- Representation of a fuzzy set by its α-cuts (the Decomposition Theorem).

- How to solve the fuzzy equations.

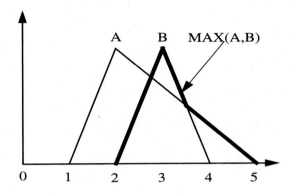

Figure 29.4. Fuzzy numbers for Example 29.6.

- The α-cuts-based and Hamming distance-based methods for ranking fuzzy numbers.

A special book on fuzzy arithmetic is Kaufmann and Gupta [1985]. A variety of fuzzy ranking methods were reviewed in Chen and Hwang [1992].

29.7 Exercises

Exercise 29.1. Let A be a fuzzy set in R. Show that A is a fuzzy number if and only if there exists a closed interval $[a, b] \neq \phi$ such that

$$\mu_A(x) = \begin{cases} 1 & for \ x \in [a, b] \\ l(x) & for \ x \in (-\infty, a) \\ r(x) & for \ x \in (b, \infty) \end{cases} \tag{29.60}$$

where $l(x)$ is a function from $(-\infty, a)$ to $[0, 1]$ that is monotonic increasing, continuous from the right, and such that $l(x) = 0$ for $x \in (-\infty, w_1)$; $r(x)$ is a function from (b, ∞) to $[0, 1]$ that is monotonic decreasing, continuous from the left, and such that $r(x) = 0$ for $x \in (w_2, \infty)$.

Exercise 29.2. For any fuzzy set A, show that

$$A_\alpha = \bigcap_{\beta < \alpha} A_\beta \tag{29.61}$$

Exercise 29.3. Prove that the fuzzy sets $A + B$ and $A - B$ defined by (29.17) and (29.18) are fuzzy numbers.

Exercise 29.4. Compute $A + B, A - B, A \cdot B, A/B$ and $MAX(A, B)$ for the fuzzy numbers A and B given as follows:

(a) $\mu_A(x) = \mu_A(x; -1, 1, 3), \mu_B(x) = \mu_B(x; 1, 3, 5)$

(b) $\mu_A(x) = \mu_A(x; -2, 0, 2), \mu_B(x) = \mu_B(x; 2, 4, 6)$

Exercise 29.5. Show that the α-cut method and the extension principle method for multiplication and division of fuzzy numbers are equivalent.

Exercise 29.6. Let A and B be the fuzzy numbers in Exercise 29.4 ((a) or (b)) and C be a triangular fuzzy number with $\mu_C(x) = \mu_C(x; 6, 8, 10)$. Solve the following equations for X:

(a) $A + X = B$

(b) $B \cdot X = C$

Exercise 29.7. Rank the three fuzzy numbers $\mu_A(x) = \mu_A(x; 0, 1, 3, 4), \mu_B(x) = \mu_B(x; 3, 4, 5)$, and $\mu_C(x) = \mu_C(x; 4, 5, 6)$ using the methods in Section 29.5.

Exercise 29.8. Show that $A \le MAX(A, B)$ and $B \le MAX(A, B)$ according to the α-cuts-based method 1, where $MAX(A, B)$ is defined by (29.54).

Chapter 30

Fuzzy Linear Programming

30.1 Classification of Fuzzy Linear Programming Problems

Linear programming is the most natural mechanism for formulating a vast array of problems with modest effort. The popularity of linear programming is mainly due to two reasons: (i) many practical problems can be formulated as linear programming problems, and (ii) there are efficient methods (for example, the Simplex method, see Luenberger [1984]) for solving the linear programming problems. The classical linear programming problem is to find the values of unknown variables such that a linear objective function is maximized under the constraints represented by linear inequalities or equations. Specifically, the standard linear programming problem is

$$
\begin{aligned}
maximize \quad & \mathbf{cx} \\
subject\ to \quad & A\mathbf{x} \le \mathbf{b} \\
& \mathbf{x} \ge 0
\end{aligned}
\tag{30.1}
$$

where $\mathbf{x} = (x_1, ..., x_n)^T \in R^n$ are the *decision variables* to be determined, $\mathbf{c} = (c_1, ..., c_n)$ are called *objective coefficients*, and $A = [a_{ij}] \in R^{m \times n}$ is called *constraint matrix* with its elements a_{ij} called *constraint coefficients*, and $\mathbf{b} = (b_1, ..., b_m)^T$ are called *resources*.

In many practical situations, it is quite restrictive to require the objective function and the constraints to be specified in precise, crisp terms. In order to understand where and how fuzziness is applied in the linear programming problem, let us consider a simple example.

Example 30.1 (Lai and Hwang [1992]). A toy company makes two kinds of toy dolls. Doll A is a high quality toy with a $0.4 per unit profit and doll B is of lower quality with a $0.3 per unit profit. Suppose that x_1 doll A and x_2 doll B are produced each day, so the profit is $0.4x_1 + 0.3x_2$. Although doll A can make a higher profit per unit, it requires twice as many labor hours as doll B. If the total available labor hours are 500 hours per day, then we have the constraint $2x_1 + x_2 \le 500$. Additionally, the supply of material is sufficient for only 400 dolls per day (both A and B combined), hence we have one more constraint $x_1 + x_2 \leq 400$. In summary, the company manager formulates the production scheduling problem as follows:

$$\begin{aligned} maximize \quad & 0.4x_1 + 0.3x_2 \quad (profit) \\ subject\ to \quad & 2x_1 + x_2 \leq 500 \quad (labor\ hours) \\ & x_1 + x_2 \leq 400 \quad (materials) \\ & x_1, x_2 \geq 0 \end{aligned} \qquad (30.2)$$

However, the total available labor hours and materials may not be that precise, because the company manager can ask workers to work overtime and require additional materials from suppliers. Therefore, some tolerance would be put on the constraints in (30.2). For example, when the actual labor hours $2x_1 + x_2$ is less than 500, we say the constraint $2x_1 + x_2 \leq 500$ is absolutely satisfied; when $2x_1 + x_2$ is larger than 600, we say the constraint $2x_1 + x_2 \leq 500$ is completely violated; and, when $2x_1 + x_2$ lies between 500 and 600, we use a linear monotonic decreasing function to represent the degree of satisfaction. Similarly, we can define a membership function to characterize the degree of satisfaction for the constraint $x_1 + x_2 \leq 400$. Fig.30.1 shows these membership functions. In conclusion, the first kind of fuzziness appears in the specification of resources; we call this problem *linear programming with fuzzy resources*.

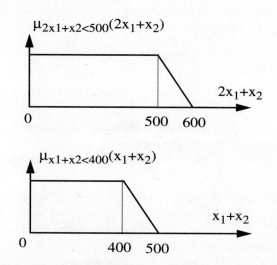

Figure 30.1. The membership functions for labor hours and materials in Example 30.1.

The second kind of fuzziness appears in the specification of the objective coefficients 0.4 and 0.3. Since the market changes constantly, it may not be certain that the per unit profit of doll A and doll B are $0.4 and $0.3, respectively. The numbers could only be viewed as the most possible values. Therefore, it is reasonable to represent the objective coefficients as fuzzy numbers; this gives us the second type of fuzzy linear programming problem—*linear programming with fuzzy objective coefficients*.

The third kind of fuzziness goes to the constraint coefficients. Because of the inconsistence of human workers, the ratio of labor hours in making doll A and doll B could only be viewed as around 2. Similarly, doll A and doll B may require slightly different amount of materials. Therefore, it is reasonable to represent the coefficients by fuzzy numbers; this gives the third fuzzy linear programming problem—*linear programming with fuzzy constraint coefficients*. □

In summary of the analysis in Example 30.1, we have the following three types of fuzzy linear programming problems:

- **Linear programming with fuzzy resources:**

$$
\begin{aligned}
maximize \quad & \mathbf{cx} \\
subject\ to \quad & A\mathbf{x} \tilde{\leq} \mathbf{b} \\
& \mathbf{x} \geq 0
\end{aligned}
\tag{30.3}
$$

where the fuzzy inequality $\tilde{\leq}$ is characterized by the membership functions like those in Fig. 30.1 (it will be precisely defined in Section 30.2).

- **Linear programming with fuzzy objective coefficients:**

$$
\begin{aligned}
maximize \quad & \tilde{\mathbf{c}}\mathbf{x} \\
subject\ to \quad & A\mathbf{x} \leq \mathbf{b} \\
& \mathbf{x} \geq 0
\end{aligned}
\tag{30.4}
$$

where $\tilde{\mathbf{c}} = (\tilde{c}_1, ..., \tilde{c}_n)$ is a vector of fuzzy numbers.

- **Linear programming with fuzzy constraint coefficients:**

$$
\begin{aligned}
maximize \quad & \mathbf{cx} \\
subject\ to \quad & \tilde{A}\mathbf{x} \leq \mathbf{b} \\
& \mathbf{x} \geq 0
\end{aligned}
\tag{30.5}
$$

where $\tilde{A} = [\tilde{a}_{ij}]$ is a matrix consisting of fuzzy numbers.

Of course, the combinations of these three problems give more types of fuzzy linear programming problems. However, if we know how to solve these three problems, other problems can be solved in a similar manner. In the next three sections, we will develop some basic approaches to solving these three problems, respectively.

30.2   Linear Programming with Fuzzy Resources

Consider the linear programming problem with fuzzy resources given by (30.3). Let $t_i(>0)$ be the *tolerance* of the i'th resource b_i, then the fuzzy inequality $(A\mathbf{x})_i \tilde{\leq} b_i$ is specified as $(A\mathbf{x})_i \leq b_i + \theta t_i$, where $\theta \in [0,1]$. In other words, the fuzzy constraint $(A\mathbf{x})_i \tilde{\leq} b_i$ is defined as a fuzzy set i with membership function

$$\mu_i(\mathbf{x}) = \begin{cases} 1 & if \quad (A\mathbf{x})_i < b_i \\ 1 - [(A\mathbf{x})_i - b_i]/t_i & if \quad b_i \leq (A\mathbf{x})_i \leq b_i + t_i \\ 0 & if \quad (A\mathbf{x})_i > b_i + t_i \end{cases} \qquad (30.6)$$

Therefore, the problem becomes to find $\mathbf{x}$ such that $\mathbf{cx}$ and $\mu_i(\mathbf{x})$ for $i = 1, 2, ..., m$ are maximized. This is a multiple objective optimization problem.

Werners [1987] proposed the following method to solve this problem. First, solve the following two standard linear programming problems:

$$\begin{aligned} maximize \quad & \mathbf{cx} \\ subject\ to \quad & (A\mathbf{x})_i \leq b_i, \ i = 1, 2, ..., m \\ & \mathbf{x} \geq 0 \end{aligned} \qquad (30.7)$$

$$\begin{aligned} maximize \quad & \mathbf{cx} \\ subject\ to \quad & (A\mathbf{x})_i \leq b_i + t_i, \ i = 1, 2, ..., m \\ & \mathbf{x} \geq 0 \end{aligned} \qquad (30.8)$$

Let $\mathbf{x}^0$ and $\mathbf{x}^1$ be the solutions of (30.7) and (30.8), respectively, and define $z^0 = \mathbf{cx}^0$ and $z^1 = \mathbf{cx}^1$. Then, the following membership function is defined to characterize the degree of optimality:

$$\mu_0(\mathbf{x}) = \begin{cases} 1 & if \quad \mathbf{cx} > z^1 \\ 1 - \frac{z^1 - \mathbf{cx}}{z^1 - z^0} & if \quad z^0 \leq \mathbf{cx} \leq z^1 \\ 0 & if \quad \mathbf{cx} < z^0 \end{cases} \qquad (30.9)$$

Clearly, when $\mathbf{cx} \geq z^1$ we have $\mu_0(\mathbf{x}) = 1$, which gives us maximum degree of optimality, when $\mathbf{cx} \leq z^0$ we have $\mu_0(\mathbf{x}) = 0$, which gives minimum degree of optimality, and when $\mathbf{cx}$ lies between z^1 and z^0 the degree of optimality changes from 1 to 0.

Since the constraints and objective function are represented by the membership functions (30.6) and (30.9), respectively, we can use the max-min method to solve this multiple objective optimization problem. Specifically, the problem becomes:

$$\max_{\mathbf{x} \geq 0} \min[\mu_0(\mathbf{x}), \mu_1(\mathbf{x}), ..., \mu_m(\mathbf{x})] \qquad (30.10)$$

or equivalently

$$\begin{aligned}
maximize \quad & \alpha \\
subject\ to \quad & \mu_0(\mathbf{x}) \geq \alpha \\
& \mu_i(\mathbf{x}) \geq \alpha,\ i = 1, 2, ..., m \\
& \alpha \in [0, 1],\ \mathbf{x} \geq 0
\end{aligned} \tag{30.11}$$

Substituting (30.6) and (30.9) into (30.11), we conclude that the fuzzy resource linear programming problem (30.3) can be solved by solving the following standard linear programming problem:

$$\begin{aligned}
maximize \quad & \alpha \\
subject\ to \quad & \mathbf{cx} \geq z^1 - (1 - \alpha)(z^1 - z^0) \\
& (A\mathbf{x})_i \leq b_i + (1 - \alpha)t_i,\ i = 1, 2, ..., m \\
& \alpha \in [0, 1],\ \mathbf{x} \geq 0
\end{aligned} \tag{30.12}$$

Example 30.1 (Lai and Hwang [1992]). Consider the following product-mix selection problem:

$$\begin{aligned}
maximize \quad & 4x_1 + 5x_2 + 9x_3 + 11x_4 && (profit) \\
subject\ to \quad & g_1(\mathbf{x}) = x_1 + x_2 + x_3 + x_4 \tilde{\leq} 15 && (man - weeks) \\
& g_2(\mathbf{x}) = 7x_1 + 5x_2 + 3x_3 + 2x_4 \tilde{\leq} 80 && (material\ Y) \\
& g_3(\mathbf{x}) = 3x_1 + 5x_2 + 10x_3 + 15x_4 \tilde{\leq} 100 && (material\ Z) \\
& x_1, x_2, x_3, x_4 \geq 0
\end{aligned} \tag{30.13}$$

where the tolerances for man-weeks, materials Y and Z are $t_1 = 5, t_2 = 40$ and $t_3 = 30$, respectively. Solving (30.7) and (30.8), we obtain $z^0 = 99.29$ and $z^1 = 130$. From (30.12) we have that the problem is equivalent to

$$\begin{aligned}
minimize \quad & \theta \\
subject\ to \quad & z = 4x_1 + 5x_2 + 9x_3 + 11x_4 \geq 130 - 30.71\theta \\
& g_1(\mathbf{x}) = x_1 + x_2 + x_3 + x_4 \leq 15 + 5\theta \\
& g_2(\mathbf{x}) = 7x_1 + 5x_2 + 3x_3 + 2x_4 \leq 80 + 40\theta \\
& g_3(\mathbf{x}) = 3x_1 + 5x_2 + 10x_3 + 15x_4 \leq 100 + 30\theta \\
& x_1, x_2, x_3, x_4 \geq 0,\ \theta \in [0, 1]
\end{aligned} \tag{30.14}$$

where $\theta = 1 - \alpha$. The solution gives $z^* = 114.65$ at $\theta = 0.5$. $\square$

30.3 Linear Programming with Fuzzy Objective Coefficients

Consider the linear programming problem with fuzzy objective coefficients given by (30.4). For simplicity and without loss of much generality, we assume that the $\tilde{c}_i$'s are triangular fuzzy numbers with membership functions $\mu_{\tilde{c}_i}(x; c_i^-, c_i^0, c_i^+)$. Symbolically, let $\tilde{c}_i = (c_i^-, c_i^0, c_i^+)$. Then (30.4) becomes

$$
\begin{aligned}
maximize \quad & (\mathbf{c}^-\mathbf{x}, \mathbf{c}^0\mathbf{x}, \mathbf{c}^+\mathbf{x}) \\
subject\ to \quad & A\mathbf{x} \le \mathbf{b} \\
& \mathbf{x} \ge 0
\end{aligned}
\tag{30.15}
$$

where $\mathbf{c}^- = (c_1^-, ..., c_n^-)$, $\mathbf{c}^0 = (c_1^0, ..., c_n^0)$ and $\mathbf{c}^+ = (c_1^+, ..., c_n^+)$. This is a multiple objective linear programming problem. A number of approaches were proposed in the literature to solve this problem (see Lai and Hwang [1992]); we now consider two approaches.

The first approach is to simply combine the three objectives into a single objective function. For example, $\mathbf{c}^-\mathbf{x}, \mathbf{c}^0\mathbf{x}$ and $\mathbf{c}^+\mathbf{x}$ can be combined into the so-called most-likely criterion $\frac{(4\mathbf{c}^0+\mathbf{c}^-+\mathbf{c}^+)\mathbf{x}}{6}$ (Lai and Hwang [1992]). So (30.15) is converted into the following standard linear programming problem:

$$
\begin{aligned}
maximize \quad & \frac{4\mathbf{c}^0 + \mathbf{c}^- + \mathbf{c}^+}{6}\mathbf{x} \\
subject\ to \quad & A\mathbf{x} \le \mathbf{b} \\
& \mathbf{x} \ge 0
\end{aligned}
\tag{30.16}
$$

Other weighted-sum strategies also may be used.

The second approach starts with the observation that our goal is to maximize the triangular fuzzy number $(\mathbf{c}^-\mathbf{x}, \mathbf{c}^0\mathbf{x}, \mathbf{c}^+\mathbf{x})$. Therefore, instead of maximizing the three values $\mathbf{c}^-\mathbf{x}, \mathbf{c}^0\mathbf{x}$ and $\mathbf{c}^+\mathbf{x}$ simultaneously, we may maximize $\mathbf{c}^0\mathbf{x}$ (the center), minimize $\mathbf{c}^0\mathbf{x} - \mathbf{c}^-\mathbf{x}$ (the left leg), and maximize $\mathbf{c}^+\mathbf{x} - \mathbf{c}^0\mathbf{x}$ (the right leg). In this way, the triangular membership function is pushed to the right. Thus, the problem (30.15) is changed to another multiple objective linear programming problem, as follows:

$$
\begin{aligned}
minimize \quad & z_1 = (\mathbf{c}^0 - \mathbf{c}^-)\mathbf{x} \\
maximize \quad & z_2 = \mathbf{c}^0\mathbf{x} \\
maximize \quad & z_3 = (\mathbf{c}^+ - \mathbf{c}^0)\mathbf{x} \\
subject\ to \quad & A\mathbf{x} \le \mathbf{b} \\
& \mathbf{x} \ge 0
\end{aligned}
\tag{30.17}
$$

A method to solve this problem is to characterize the three objective functions by membership functions and then maximize their α-cuts. Specifically, we first get the solutions:

$$
\begin{aligned}
z_1^P &= \min_{\mathbf{x} \in X}(\mathbf{c}^0 - \mathbf{c}^-)\mathbf{x}, \quad z_1^N = \max_{\mathbf{x} \in X}(\mathbf{c}^0 - \mathbf{c}^-)\mathbf{x} \\
z_2^P &= \max_{\mathbf{x} \in X}\mathbf{c}^0\mathbf{x}, \quad z_2^N = \min_{\mathbf{x} \in X}\mathbf{c}^0\mathbf{x} \\
z_3^P &= \max_{\mathbf{x} \in X}(\mathbf{c}^+ - \mathbf{c}^0)\mathbf{x}, \quad z_3^N = \min_{\mathbf{x} \in X}(\mathbf{c}^+ - \mathbf{c}^0)\mathbf{x}
\end{aligned}
\tag{30.18}
$$

where $X = \{\mathbf{x} | A\mathbf{x} \leq \mathbf{b}, \mathbf{x} \geq 0\}$. The solutions z_i^P are called Positive Ideal Solution and z_i^N are called Negative Ideal Solution. Then, define the following three membership functions to characterize the three objectives:

$$\mu_{z_1}(\mathbf{x}) = \begin{cases} 1 & if \quad (\mathbf{c}^0 - \mathbf{c}^-)\mathbf{x} < z_1^P \\ \frac{z_1^N - (\mathbf{c}^0 - \mathbf{c}^-)\mathbf{x}}{z_1^N - z_1^P} & if \quad z_1^P \leq (\mathbf{c}^0 - \mathbf{c}^-)\mathbf{x} \leq z_1^N \\ 0 & if \quad (\mathbf{c}^0 - \mathbf{c}^-)\mathbf{x} > z_1^N \end{cases} \qquad (30.19)$$

$$\mu_{z_2}(\mathbf{x}) = \begin{cases} 1 & if \quad \mathbf{c}^0\mathbf{x} > z_2^P \\ \frac{\mathbf{c}^0\mathbf{x} - z_2^N}{z_2^P - z_2^N} & if \quad z_2^N \leq \mathbf{c}^0\mathbf{x} \leq z_2^P \\ 0 & if \quad \mathbf{c}^0\mathbf{x} < z_2^N \end{cases} \qquad (30.20)$$

$$\mu_{z_3}(\mathbf{x}) = \begin{cases} 1 & if \quad (\mathbf{c}^+ - \mathbf{c}^0)\mathbf{x} > z_3^P \\ \frac{(\mathbf{c}^+ - \mathbf{c}^0)\mathbf{x} - z_2^N}{z_2^P - z_2^N} & if \quad z_3^N \leq (\mathbf{c}^+ - \mathbf{c}^0)\mathbf{x} \leq z_3^P \\ 0 & if \quad (\mathbf{c}^+ - \mathbf{c}^0)\mathbf{x} < z_3^N \end{cases} \qquad (30.21)$$

Finally, the problem is solved by solving the following standard linear programming problem:

$$\begin{aligned} maximize \quad & \alpha \\ subject\ to \quad & \mu_{z_i}(\mathbf{x}) \geq \alpha, \ i = 1, 2, 3 \\ & A\mathbf{x} \leq \mathbf{b}, \ \mathbf{x} \geq 0 \end{aligned} \qquad (30.22)$$

30.4   Linear Programming with Fuzzy Constraint Coefficients

Consider the linear programming problem with fuzzy constraint coefficients (30.5). Again, for simplicity and without loss of much generality, we assume that $\tilde{A} = [\tilde{a_{ij}}]$ consists of triangular fuzzy numbers, that is, $\tilde{a_{ij}} = (a_{ij}^-, a_{ij}^0, a_{ij}^+)$ and $\tilde{A} = (A^-, A^0, A^+)$, where $A^- = [a_{ij}^-], A^0 = [a_{ij}^0]$ and $A^+ = [a_{ij}^+]$. Then the problem becomes

$$\begin{aligned} maximize \quad & \mathbf{cx} \\ subject\ to \quad & (A^-\mathbf{x}, A^0\mathbf{x}, A^+\mathbf{x}) \leq \mathbf{b} \\ & \mathbf{x} \geq 0 \end{aligned} \qquad (30.23)$$

Using the most-likely criterion as in (30.16), we convert (30.23) into the following standard linear programming problem:

$$\begin{aligned} maximize \quad & \mathbf{cx} \\ subject\ to \quad & \frac{4A^0 + A^- + A^+}{6}\mathbf{x} \leq \mathbf{b} \\ & \mathbf{x} \geq 0 \end{aligned} \qquad (30.24)$$

Up to this point, we have solved the three basic fuzzy linear programming problems (30.3)-(30.5). Other types of fuzzy linear programming problems are essentially the combination of the three basic problems and therefore can be solved using similar approaches. For example, consider the problem where all the coefficients are fuzzy numbers, that is,

$$
\begin{aligned}
maximize \quad & \tilde{\mathbf{c}}\mathbf{x} \\
subject\ to \quad & \tilde{A}\mathbf{x} \le \tilde{\mathbf{b}} \\
& \mathbf{x} \ge 0
\end{aligned}
\tag{30.25}
$$

Assume that $\tilde{\mathbf{c}}$, $\tilde{A}$ and $\tilde{\mathbf{b}}$ consist of triangular fuzzy numbers, that is, $\tilde{\mathbf{c}} = (\mathbf{c}^-, \mathbf{c}^0, \mathbf{c}^+)$, $\tilde{A} = (A^-, A^0, A^+)$ and $\tilde{\mathbf{b}} = (\mathbf{b}^-, \mathbf{b}^0, \mathbf{b}^+)$, then (30.25) can be converted into the following multiple objective linear programming problem:

$$
\begin{aligned}
minimize \quad & z_1 = (\mathbf{c}^0 - \mathbf{c}^-)\mathbf{x} \\
maximize \quad & z_2 = \mathbf{c}^0\mathbf{x} \\
maximize \quad & z_3 = (\mathbf{c}^+ - \mathbf{c}^0)\mathbf{x} \\
subject\ to \quad & A^-\mathbf{x} \le \mathbf{b}^-,\ A^0\mathbf{x} \le \mathbf{b}^0, A^+\mathbf{x} \le \mathbf{b}^+ \\
& \mathbf{x} \ge 0
\end{aligned}
\tag{30.26}
$$

We can use the method for (30.17) to solve this problem.

30.5   Comparison of Stochastic and Fuzzy Linear Programming

Stochastic linear programming deals with situations where the coefficients $\mathbf{c}$ and A and the resources $\mathbf{b}$ are imprecise and described by random variables. Although stochastic programming has been extensively studied since the late 1950s, its applications in solving real problems are limited. The main problems of stochastic programming, as pointed out by Lai and Hwang [1992], are: (i) lack of computational efficiency, and (ii) inflexible probabilistic doctrines that might not be able to model the real imprecise meaning of decision makers. Fuzzy programming overcomes these two problems to some extent. In this section, we compare a particular stochastic programming method, the chance-constrained programming model, with the corresponding fuzzy programming method.

The chance-constrained programming model, developed by Charnes and Cooper [1959], is defined as:

$$
\begin{aligned}
maximize \quad & z = \sum_j \bar{c}_j x_j \\
subject\ to \quad & P\left\{\sum_j \bar{a}_{ij} x_j \le \bar{b}_i\right\} \ge 1 - \alpha_i, \forall i \\
& x_j \ge 0,\ \forall j
\end{aligned}
\tag{30.27}
$$

where $\bar{c}_j, \bar{a}_{ij}$ and $\bar{b}_i$ are random variables, $P\{\}$ denotes probability measure, and α_i are small positive constants. The name "chance-constrained" follows from each constraint $\sum_j \bar{a}_{ij}x_j \le \bar{b}_i$ being realized with a minimum probability $1 - \alpha_i$. For illustrative purposes, we assume that $\bar{a}_{ij}$ and $\bar{b}_i$ are Gaussian distributed with known means and variances. Let $\bar{g}_i = \sum_j \bar{a}_{ij}x_j - \bar{b}_i$, then the probabilistic constraint in (30.27) can be treated as

$$P\{\bar{g}_i \le 0\} = P\left\{ \frac{\bar{g}_i - E[\bar{g}_i]}{Var[\bar{g}_i]^{1/2}} \le \frac{0 - E[\bar{g}_i]}{Var[\bar{g}_i]^{1/2}} \right\} \ge 1 - \alpha_i \qquad (30.28)$$

or equivalently,

$$P\{\bar{g}_i \le 0\} = \Phi\left(\frac{0 - E[\bar{g}_i]}{Var[\bar{g}_i]^{1/2}} \right) \ge 1 - \alpha_i \qquad (30.29)$$

where Φ is the CDF of the standard Gaussian distribution. Since $\bar{a}_{ij}$ and $\bar{b}_i$ are Gaussian random variables, $\bar{g}_i$ is also normally distributed with

$$E[\bar{g}_i] = \sum_j E[\bar{a}_{ij}]x_j - E[\bar{b}_i] \qquad (30.30)$$

$$Var[\bar{g}_i] = \mathbf{x}^T D_i \mathbf{x} \qquad (30.31)$$

where $\mathbf{x} = (x_1, ..., x_n)^T$ and

$$D_i = \begin{pmatrix} Var[\bar{a}_{i1}] & \cdots & Cov[\bar{a}_{i1}, \bar{a}_{in}] & Cov[\bar{a}_{i1}, \bar{b}_i] \\ \vdots & & \vdots & \vdots \\ Cov[\bar{a}_{in}, \bar{a}_{i1}] & \cdots & Var[\bar{a}_{in}] & Cov[\bar{a}_{in}, \bar{b}_i] \\ Cov[\bar{b}_i, \bar{a}_{i1}] & \cdots & Cov[\bar{b}_i, \bar{a}_{in}] & Var[\bar{b}_i] \end{pmatrix} \qquad (30.32)$$

Let S_{α_i} be the normal value such that $\Phi(S_{\alpha_i}) = 1 - \alpha_i$, then the constraint $P\{\bar{g}_i \le 0\} \ge 1 - \alpha_i$ is realized if and only if

$$\frac{0 - E[\bar{g}_i]}{Var[\bar{g}_i]^{1/2}} \ge S_{\alpha_i} \qquad (30.33)$$

Substituting (30.30)-(30.31) into (30.33) and replacing $\bar{c}_j$ by its mean c_j, we convert the stochastic linear programming problem into the following deterministic programming problem:

$$maximize \qquad z = \sum_j c_j x_j$$

$$subject\ to \qquad \sum_j E[\bar{a}_{ij}]x_j - E[\bar{b}_i] + S_{\alpha_i}\mathbf{x}^T D_i \mathbf{x} \le 0, \forall i \qquad (30.34)$$

$$x_j \ge 0, \ \forall j$$

Clearly, (30.34) is a nonlinear programming problem that is not easy to solve. In fuzzy linear programming, on the other hand, we replace the coefficients and resources by triangular fuzzy numbers. From Sections 30.2-30.4 we see that the fuzzy linear programming problems can be converted into the standard linear programming problem that is much easier to solve than the nonlinear programming problem (30.34). Therefore, fuzzy linear programming is more efficient than stochastic linear programming from a computational point of view.

30.6   Summary and Further Readings

In this chapter we have demonstrated the following:

- Where and how does fuzziness appear in linear programming and the classification of fuzzy linear programming problems.

- How to solve the linear programming problem with fuzzy recources.

- How to solve the linear programming problem with fuzzy objective coefficients.

- How to solve the linear programming problem with fuzzy constraint coefficients.

- The advantages of fuzzy linear programming over stochastic linear programming.

A very good book on fuzzy mathematical programming is Lai and Hwang [1992]. The books Chen and Hwang [1992] and Lai and Hwang [1994] also covered related topics. For classical linear programming, see Luenberger [1984].

30.7   Exercises

Exercise 30.1. Show that the linear programming problem (30.1) can be transformed into the following problem:

$$
\begin{aligned}
maximize \quad & \mathbf{c'x'} \\
subject\ to \quad & A'\mathbf{x'} = \mathbf{b'} \\
& \mathbf{x'} \geq 0
\end{aligned}
\tag{30.35}
$$

Exercise 30.2. Using graphics to find the solution to the linear programming problem (30.2).

Exercise 30.3. Using graphics to find the solution to the following linear programming problem:

$$
\begin{aligned}
minimize \quad & x_1 - 2x_2 \\
subject\ to \quad & 3x_1 - x_2 \geq 1 \\
& 2x_1 + x_2 \leq 6 \\
& x_1 \geq 0,\ 0 \leq x_2 \leq 2
\end{aligned}
\tag{30.36}
$$

Exercise 30.4. Show that the max-min optimization problem (30.10) is equivalent to the mathematical programming problem (30.11).

Exercise 30.5. A company has been producing a highly profitable decorative material that came in two versions, x and y. Five different component ingredients were needed: golden thread, silk, velvet, silver thread, and nylon. The prices of these inputs and their technological contributions to both x and y are given in Table 30.1. The profit margins are $400 per unit of x and $300 per unit of y. In order to maintain these margins, the company does not allow "substantially more than $2,600" to be spent on the purchase of the components. Formulate the problem into a linear programming problem with fuzzy resources and solve it using the method in Section 30.2.

Table 30.1. Inputs and technological coefficients

Resource	Technological coefficients		Price/unit
	x	y	
golden thread	4	0	30
silk	2	6	40
velvet	12	4	9.5
silver thread	0	3	20
nylon	4	4	10

Exercise 30.6. Solve the following fuzzy linear programming problems:

(a)

$$\begin{aligned} maximize \quad & 5x_1 + 4x_2 \\ subject\ to \quad & (4,2,1)x_1 + (5,3,1)x_2 \leq (24,5,8) \\ & (4,1,2)x_1 + (1,0.5,1)x_2 \leq (12,6,3) \\ & x_1, x_2 \geq 0 \end{aligned} \qquad (30.37)$$

(b)

$$\begin{aligned} maximize \quad & 6x_1 + 5x_2 \\ subject\ to \quad & (5,3,2)x_1 + (6,4,2)x_2 \leq (25,6,7) \\ & (5,2,3)x_1 + (2,1.5,1)x_2 \leq (13,7,4) \\ & x_1, x_2 \geq 0 \end{aligned} \qquad (30.38)$$

Exercise 30.7. Develop one or two methods that are different from those in Section 30.3 to solve the linear programming problem with fuzzy objective coefficients.

Exercise 30.8. Develop one or two methods that are different from those in Section 30.4 to solve the linear programming problem with fuzzy constraint coefficients.

Exercise 30.9. Can you find some disadvantages of fuzzy linear programming over stochastic linear programming? Explain your answer.

Chapter 31

Possibility Theory

31.1 Introduction

Possibility theory was initialized by Zadeh [1978] as a complement to probability theory to deal with uncertainty. The justification for possibility theory was best stated by Zadeh [1978]:

> The pioneering work of Wiener and Shannon on the statistical theory of communication has led to a universal acceptance of the belief that information is intrinsically statistical in nature and, as such, must be dealt with by the methods provided by probability theory.
>
> Unquestionably, the statistical point of view has contributed deep insights into the fundamental processes involved in the coding, transmission and reception of data and played a key role in the development of modern communication, detection and telemetering systems. In recent years, however, a number of other important applications have come to the fore in which the major issues center not on the transmission of information but on its meaning. In such applications, what matters is the ability to answer questions relating to information that is stored in a database as in natural language processing, knowledge representation, speech recognition, robotics, medical diagnosis, analysis of rare events, decision-making under uncertainty, picture analysis, information retrieval and related areas.
>
> ... our main concern is with the meaning of information -- rather than with its measure -- the proper framework for information analysis is possibilistic rather than probabilistic in nature, thus implying that what is needed for such an analysis is not probability theory but an analogous -- and yet different -- theory which might be called the theory of possibility.

Since Zadeh [1978], there has been much research on possibility theory. There are two approaches to possibility theory: one, proposed by Zadeh [1978], was to introduce possibility theory as an extension of fuzzy set theory; the other, taken by Klir and Folger [1988] and others, was to introduce possibility theory in the framework of Dempster-Shafer's theory of evidence. The first approach is intuitively plausible and is closely related to the original motivation for introducing possibility theory—representing the meaning of information. The second approach puts possibility theory on an axiomatic basis so that in-depth studies can be pursued. In the next two sections, we will introduce the basic ideas of these two approaches, respectively.

31.2   The Intuitive Approach to Possibility

31.2.1   Possibility Distributions and Possibility Measures

This approach starts with the concept of fuzzy restriction. Let x be a variable that takes values in the universe of discourse U and A be a fuzzy set in U. Then the proposition "x is A" can be interpreted as putting a *fuzzy restriction on x* and this restriction is characterized by the membership function μ_A. In other words, we can interpret $\mu_A(u)$ as the *degree of possibility* that $x = u$. For example, let x be a person's age and A be the fuzzy set "Young." Suppose we know that "the person is Young" (x is A), then $\mu_A(30)$ could be interpreted as the degree of possibility that the person's age is 30. Formally, we have the following definition.

Definition 31.1. Given fuzzy set A in U and the proposition "x is A," the *possibility distribution associated with x*, denoted by π_x, is defined to be numerically equal to the membership function of A, that is,

$$\pi_x(u) = \mu_A(u) \qquad (31.1)$$

for all $u \in U$.

As an example, consider the fuzzy set "small integer" defined as

$$small\ integer = 1/1 + 1/2 + 0.8/3 + 0.6/4 + 0.4/5 + 0.2/6 \qquad (31.2)$$

Then, the proposition "x is a small integer" associates x with the possibility distribution

$$\pi_x = 1/1 + 1/2 + 0.8/3 + 0.6/4 + 0.4/5 + 0.2/6 \qquad (31.3)$$

where the term such as 0.8/3 signifies that the possibility that x is 3, given that "x is a small integer," is 0.8.

Now let x be a person's age and A be the fuzzy set "Young." Given "x is A," we know that the possibility of $x = 30$ equals $\mu_A(30)$. Sometimes, we may ask the question: "What is the possibility that the person's age is between 25 and 35, given that the person is Young?" One eligible answer to this question is $\sup_{u \in [25,35]} \mu_A(u)$. The generalization of this example gives the concept of possibility measure.

Definition 31.2. Let C be a crisp subset of U and π_x be a possibility distribution associated with x. The *possibility measure* of x belonging to C, denoted by

$Pos_x(C)$, is defined by

$$Pos_x(C) = \sup_{u \in C} \pi_x(u) \tag{31.4}$$

As an illustration of possibility measure, consider the fuzzy set "small integer" defined by (31.2) and give the proposition "x is a small integer." Let $C = \{3, 4, 5\}$, then the possibility measure of x equals 3,4 or 5 is

$$Pos_x(C) = \max_{u \in \{3,4,5\}} \pi_x(u)$$
$$= \max[0.8, 0.6, 0.4] = 0.8 \tag{31.5}$$

31.2.2 Marginal Possibility Distribution and Noninteractiveness

Let $x = (x_1, ..., x_n)$ be a vector taking values in $U = U_1 \times \cdots \times U_n$ and A be a fuzzy set (relation) in U with membership function $\mu_A(u_1, ..., u_n)$. Then, given the proposition "x is A" we have from Definition 31.1 that the possibility distribution of x is $\pi_x = \mu_A$; we call this π_x the *basic distribution*.

Now let $q = (i_1, ..., i_k)$ be a subsequence of $(1, ..., n)$ and $x_{(q)} = (x_{i_1}, ..., x_{i_k})$. The *marginal possibility distribution associated with* $x_{(q)}$, denoted by $\pi_{x_{(q)}}$, is defined as the projection of π_x on $U_{(q)} = U_{i_1} \times \cdots \times U_{i_k}$, that is,

$$\pi_{x_{(q)}}(u_{(q)}) = \sup_{u_{(q')} \in U_{(q')}} \pi_x(u) \tag{31.6}$$

where $u_{(q)} = (u_{i_1}, ..., u_{i_k}), q' = (j_1, ..., j_m)$ is a subsequence of $(1, ..., n)$, which is complementary to q, $u_{(q')} = (u_{j_1}, ..., u_{j_m})$ and $U_{(q')} = U_{j_1} \times \cdots \times U_{j_m}$. For example, if $n = 5$ and $q = (i_1, i_2) = (2, 4)$, then $q' = (j_1, j_2, j_3) = (1, 3, 5)$.

As a simple illustration, assume that $U = U_1 \times U_2 \times U_3$ with $U_1 = U_2 = U_3 = \{a, b\}$ and

$$\pi_x(u_1, u_2, u_3) = 0.8/(a, a, a) + 1/(a, a, b) + 0.6/(b, a, a) + 0.2/(b, a, b) + 0.5/(b, b, b) \tag{31.7}$$

Then, the marginal possibility distribution of (x_1, x_2) is

$$\pi_{(x_1, x_2)}(u_1, u_2) = \max_{u_3 \in \{a,b\}} \pi_x(u_1, u_2, u_3)$$
$$= 1/(a, a) + 0.6/(b, a) + 0.5/(b, b) \tag{31.8}$$

By analogy with the concept of independence of random variables, the variables $x_{(q)}$ and $x_{(q')}$ are called *noninteractive* if and only if

$$\pi_x(u_1, ..., u_n) = \min[\pi_{x_{(q)}}(u_{i_1}, ..., u_{i_k}), \pi_{x_{(q')}}(u_{j_1}, ..., u_{j_m})] \tag{31.9}$$

where q and q' are defined as before. Because of the *min* operation, an increase in the possibility of u_1 cannot be compensated by a decrease in the possibility of u_2 and vice versa; this is the intuitive meaning of noninteractiveness.

31.2.3 Conditional Possibility Distribution

Again, by analogy with the concept of conditional probability distributions, we can define conditional possibility distributions in the theory of possibility. As in the previous subsection, let $q = (i_1, ..., i_k)$ and $q' = (j_1, ..., j_m)$ be complementary subsequences of $(1, ..., n)$ and $(a_{j_1}, ..., a_{j_m})$ be a given point in $U_{j_1} \times \cdots \times U_{j_m}$. Then, the *conditional possibility distribution* of $x_{(q)}$ given $x_{(q')} = (a_{j_1}, ..., a_{j_m})$ is defined as

$$\pi_{x_{(q)}}(u_{i_1}, ..., u_{i_k} | x_{j_1} = a_{j_1}, ..., x_{j_m} = a_{j_m}) = \pi_x(u_1, ..., u_n)|_{u_{j_1} = a_{j_1}, ..., u_{j_m} = a_{j_m}}$$
(31.10)

For the possibility distribution (31.7), as an example, we have

$$\pi_{(x_2, x_3)}(u_2, u_3 | x_1 = a) = 0.8/(a, a) + 1/(a, b)$$
(31.11)
$$\pi_{(x_2, x_3)}(u_2, u_3 | x_1 = b) = 0.6/(a, a) + 0.2/(a, b) + 0.5/(b, b)$$
(31.12)

In (31.10), the conditional possibility distribution is conditioned on a singleton value $x_{(q')} = (a_{j_1}, ..., a_{j_m})$. Now suppose $x_{(q')}$ is only fuzzily restricted by the possibility distribution $\pi_{x_{(q')}}$, what is the conditional possibility distribution of $x_{(q)}$ given $x_{(q')}$? In Zadeh [1978], this conditional possibility distribution is defined as follows. First, cylindrically extend $x_{(q')}$ from $U_{j_1} \times \cdots \times U_{j_m}$ to $U_1 \times \cdots \times U_n$, that is,

$$\pi_{\bar{x}_{(q')}}(u_1, ..., u_n) = \pi_{x_{(q')}}(u_{j_1}, ..., u_{j_m})$$
(31.13)

where $\bar{x}_{(q')}$ is the cylindrical extension of $x_{(q)}$. Then, the conditional possibility distribution of $x_{(q)}$ given $x_{(q')}$ is defined as

$$\pi_{x_{(q)}}(u_{i_1}, ..., u_{i_k} | x_{(q')}) = \sup_{u_{(q')} \in U_{j_1} \times \cdots \times U_{j_m}} \min[\pi_x(u_1, ..., u_n), \pi_{\bar{x}_{(q')}}(u_1, ..., u_n)]$$
(31.14)

where $u_{(q')} = (u_{j_1}, ..., u_{j_m})$. That is, we take the intersection of π_x and $\pi_{\bar{x}_{(q')}}$ and define the conditional possibility distribution as the projection of this intersection onto $U_{i_1} \times \cdots \times U_{i_k}$.

Consider, again, the possibility distribution π_x of (31.7) and let

$$\pi_{(x_1, x_2)}(u_1, u_2) = 0.4/(a, a) + 0.8/(b, a) + 1/(b, b)$$
(31.15)

Then, the cylindrical extension of (x_1, x_2) is

$$\pi_{\overline{(x_1, x_2)}}(u_1, u_2, u_3) = 0.4/(a, a, a) + 0.4/(a, a, b) + 0.8/(b, a, a) + 0.8/(b, a, b)$$
$$1/(b, b, a) + 1/(b, b, b)$$
(31.16)

Substituting (31.16) and (31.7) in (31.14), we obtain

$$\pi_{x_3}(u_3|x_1, x_2) = 0.6/a + 0.5/b \tag{31.17}$$

31.3   The Axiomatic Approach to Possibility

In Klir and Folger [1988] and Klir and Yuan [1995], possibility theory was developed within the framework of Dempster-Shafer's evidence theory. Since this approach was built on an axiomatic basis, a variety of mathematical properties can be derived. In evidence theory, there are two important measures: plausibility measure and belief measure; the basic concepts in possibility theory were defined through these two measures.

31.3.1   Plausibility and Belief Measures

In the definition of probability measures, there is a strong requirement—the additivity axiom, that is,

$$Pro(A \cup B) = Pro(A) + Pro(B) \tag{31.18}$$

where $Pro(\cdot)$ denotes probability measure, and A and B are subsets of the domain U such that $A \cap B = \phi$ (empty set). Plausibility and belief measures are defined by relaxing this additivity axiom in different ways. Specifically, we have the following definitions.

Definition 31.3. Given a universal set U and a nonempty family $\mathcal{F}$ of subsets of U, a *plausibility measure* is a function $Pl : \mathcal{F} \to [0,1]$ such that

- (p1) $Pl(\phi) = 0$ and $Pl(U) = 1$ (boundary conditions)

- (p2) for all $A_i \in \mathcal{F}$,

$$Pl(A_1 \cap \cdots \cap A_n) \leq \sum_j Pl(A_j) - \sum_{j<k} Pl(A_j \cup A_k) + \cdots + (-1)^{n+1} Pl(A_1 \cup \cdots \cup A_n) \tag{31.19}$$

  (subadditivity)

- (p3) for any increasing sequence $A_1 \subset A_2 \subset \cdots$ in $\mathcal{F}$, if $\bigcup_{i=1}^{\infty} A_i \in \mathcal{F}$, then

$$\lim_{i \to \infty} Pl(A_i) = Pl(\bigcup_{i=1}^{\infty} A_i) \tag{31.20}$$

  (continuity from below)

Definition 31.4. Given a universal set U and a nonempty family $\mathcal{F}$ of subsets of U, a *belief measure* is a function $Bel : \mathcal{F} \to [0,1]$ such that

- (p1) $Bel(\phi) = 0$ and $Bel(U) = 1$ (boundary conditions)

- (p2) for all $A_i \in \mathcal{F}$,

$$Bel(A_1\cup\cdots\cup A_n) \geq \sum_j Bel(A_j) - \sum_{j<k} Bel(A_j\cap A_k) + \cdots + (-1)^{n+1} Bel(A_1\cap\cdots\cap A_n)$$

(31.21)

(superadditivity)

- (p3) for any decreasing sequence $A_1 \supset A_2 \supset \cdots$ in $\mathcal{F}$, if $\bigcap_{i=1}^{\infty} A_i \in \mathcal{F}$, then

$$\lim_{i\to\infty} Bel(A_i) = Bel(\bigcap_{i=1}^{\infty} A_i)$$

(31.22)

(continuity from above)

Letting $n = 2, A_1 = A$, and $A_2 = \bar{A}$ in (31.19) and (31.21), we obtain the following basic inequalities of plausibility and belief measures:

$$Pl(A) + Pl(\bar{A}) \geq 1$$

(31.23)

$$Bel(A) + Bel(\bar{A}) \leq 1$$

(31.24)

Furthermore, it can be shown that plausibility and belief measures are related through

$$Pl(A) + Bel(\bar{A}) = 1$$

(31.25)

Hence, we can say that plausibility and belief measures are complementary. Plausibility and belief measures also are known as upper and lower probabilities, because it can be shown that

$$Bel(A) \leq Pro(A) \leq Pl(A)$$

(31.26)

Evidence theory is a very rich field and interesting readers are referred to Shafer [1976] and Guan and Bell [1991].

31.3.2   Possibility and Necessity Measures

In the intuitive approach in Section 31.2, possibility distributions are induced by membership functions of fuzzy sets and possibility measures are defined according to the possibility distributions. In the axiomatic approach here, possibility measures are defined as a special plausibility measure, and possibility distributions are derived from the possibility measures. Since plausibility measure has a complementary belief measure in the sense of (31.25), the so-called necessity measure was introduced to represent the complement of possibility measure. Specifically, we have the following definitions.

Definition 31.5. A *possibility measure Pos* : $\mathcal{F} \to [0,1]$ is defined as a special plausibility measure with the subadditivity condition (31.19) replaced by

$$Pos(\bigcup_{k \in K} A_k) = \sup_{k \in K} Pos(A_k) \qquad (31.27)$$

where K is an arbitrary index set, and $\bigcup_{k \in K} A_k \in \mathcal{F}$.

Definition 31.6. A *necessity measure Nec* : $\mathcal{F} \to [0,1]$ is defined as a special belief measure with the superadditivity condition (31.21) replaced by

$$Nec(\bigcap_{k \in K} A_k) = \inf_{k \in K} Nec(A_k) \qquad (31.28)$$

where K is an arbitrary index set, and $\bigcap_{k \in K} A_k \in \mathcal{F}$.

Since possibility and necessity measures are special plausibility and belief measures, respectively, (31.23)-(31.25) are satisfied, that is,

$$Pos(A) + Pos(\bar{A}) \geq 1 \qquad (31.29)$$
$$Nec(A) + Nec(\bar{A}) \leq 1 \qquad (31.30)$$
$$Pos(A) + Nec(\bar{A}) = 1 \qquad (31.31)$$

It can be shown that for every possibility measure Pos on $\mathcal{F}$ there exists a function $\pi : U \to [0, 1]$ such that

$$Pos(A) = \sup_{x \in A} \pi(x) \qquad (31.32)$$

for any $A \in \mathcal{F}$. This π is defined as the *possibility distribution*. That is, in this approach possibility distributions are induced by possibility measures.

In addition to (31.29)-(31.31), possibility and necessity measures have some interesting properties. For example,

$$\max[Pos(A), Pos(\bar{A})] = 1 \qquad (31.33)$$
$$\min[Nec(A), Nec(\bar{A})] = 0 \qquad (31.34)$$

where (31.33) is obtained from $1 = Pos(A \cup \bar{A}) = max[Pos(A), Pos(\bar{A})]$, and (31.34) is obtained from $0 = Nec(A \cap \bar{A}) = min[Nec(A), Nec(\bar{A})]$. Furthermore, from (31.31) and (31.33)-(31.34) we have that $Pos(A) < 1$ implies $Nec(A) = 0$ and that $Nec(A) > 0$ implies $Pos(A) = 1$. Indeed, if $Pos(A) < 1$, then from (31.33) we have $Pos(\bar{A}) = 1$ and hence, $Nec(A) = 1 - Pos(\bar{A}) = 0$ (using (31.31)). Similarly, $Nec(A) > 0$ implies $Nec(\bar{A}) = 0$ (from (31.34)) and therefore $Pos(A) = 1 - Nec(\bar{A}) = 1$ (from (31.31)).

Since the possibility and necessity measures are defined on an axiomatic basis, a variety of properties can be derived; see Klir and Yuan [1995] and Dobois and Prade [1988]. This gives possibility theory a rich content.

31.4   Possibility versus Probability

31.4.1   The Endless Debate

This book seems incomplete without a discussion on the possibility (fuzziness) versus probability issue. Indeed, this is the most controversial topic around fuzzy theory and has sparked very intensive debate in the past. To make the story more interesting, we classify the debate into three rounds.

The debate began with the possibilists'[1] argument that probability theory is not suitable to deal with uncertainties in natural languages. Probabilists do not agree. Probability theory, with its hundreds of years of history, is rich in theory and very successful in application. Probabilists argued that the membership function of a fuzzy set can be interpreted in terms of subjective probability (Loginov [1966]). In the 1980s, probabilists lunched a number of offences with the position papers Lindley [1982], Lindley [1987], and Cheeseman [1988], followed by discussions published in statistical journals. The conclusion was: "anything fuzzy can do, probability can do it equally well or better." Possibilists did not put an effective defense line against these strong offences, such that at the end of the first round of debate, probabilists had a total upper hand and fuzzy theory was badly treated or simply ignored in the following years.

After losing the first round of debate, possibilists, who were mostly engineers, switched their attention to engineering applications of fuzzy theory. Indeed, one could argue against the principles of a theory, but it would be very unconvincing if one tries to argue with solid applications. With the successes of fuzzy controllers in home electronics and industrial process control, possibilists lunched a major offense by publishing a special issue of the *IEEE Trans. on Fuzzy Systems* (Vol. 2, No. 2, 1994) on fuzziness versus probability. The conclusion was: "the two theories are complementary, and they deal with different types of uncertainties." Possibilists had an upper hand this time; the consequence is that the "ignoring" period of fuzzy theory was over and many scientists and engineers alike began to look at fuzzy theory seriously.

The third round of debate has just begun with the publication of a position paper by Laviolette, Seaman, Barrett and Woodall [1995] followed by six discussions written by mostly possibilists. The position paper laid out fuzzy and probability models side by side and compared the strengths and weaknesses of each. Although the two parties still had major differences, the atmosphere was much more cooperative. In the concluding remarks, Laviolette, Seaman, Barrett and Woodall [1995] wrote:

---

[1]We call proponents of fuzzy theory *possibilists* and proponents of probability theory *probabilists*; the boundary is, however, often fuzzy.

> Although we have serious reservations about some philosophical
> tenets of FST [fuzzy set theory], we do not claim that the theory has
> been useless. The well-documented successful applications,
> particularly in control theory, show that FST can work if carefully
> applied. ... Rather, we take a skeptical view, in that so far we have
> found no instances in which FST is uniquely useful -- that is, no
> solutions using FST that could not have been achieved at least as
> effectively as using probability and statistics.

This round of debate is currently going on and we will definitely see more debate in the future.

Since fuzziness versus probability is a very controversial issue, we think the best we can do is to list the major differences between the two theories and to let the reader judge the arguments.

31.4.2 Major Differences between the Two Theories

We avoid discussing philosophical differences between the two theories, because any arguments in this category will spark controversy; interested readers are referred to the articles mentioned in the previous subsection. The following differences are mainly technical differences.

Difference 1: Although both theories deal with uncertainty, the practical problems they intend to solve are quite different. Fuzzy theory, on one hand, tries to formulate human reasoning and perceptions and therefore targets problems in areas such as industrial process control, pattern recognition, group decision making, etc., where human factors have a major impact. Probability theory, on the other hand, concentrates on such areas as statistical mechanics, data analysis, communications systems, etc., where human reasoning and perceptions do not play a major role.

Although probabilists claimed that probability could be used to formulate human knowledge if one wants to, the details have never been developed to the scale that fuzzy theory provides. In fact, the mainstream probabilists try to avoid modeling human behavior, as stated in Bernardo and Smith [1994]:

> It is important to recognize that the axioms we shall present are
> prescriptive, not descriptive. Thus they do not purport to describe
> the ways in which individuals actually do behave in formulating
> problems or making choices, neither do they assert, on some presumed
> 'ethical' basis, the ways in which individuals should behave. The
> axioms simply prescribe constraints which it seems to us imperative to
> acknowledge in those situations where an individual aspires to choose
> among alternatives in such a way to avoid certain forms of behavioural
> inconsistency. (p.23)

Consequently, although the two parties claim to deal with the same issue—uncertainty, the problems they really work on are quite different.

Difference 2: Fuzzy set theory discards the law of the excluded middle, whereas probability theory is built on classical set theory, in which the law of the excluded middle is fundamental. Due to this fundamental difference, the technical contents of the two theories are quite different.

Difference 3: Possibility measures, as defined in Section 31.3, replace the additivity axiom of probability (31.18) with the weaker subadditivity condition (31.19). This fundamental difference results in a sequence of other differences between possibility and probability theories, some of which were shown in Section 31.3.

Difference 4: From an application point of view, the computational algorithms resulting from the two theories and the information required to implement them usually are quite different, even if the algorithms are developed to solve the same problem. For example, many problems with using probability in artificial intelligence models come from multiplying two probabilities that are not independent; in fuzzy theory, this dependency information is not required.

There are many more differences and endless arguments. Because this book is written mainly for engineers or future engineers, it might be helpful to comment on how to look at the debate from an engineer's perspective.

31.4.3   How to View the Debate from an Engineer's Perspective

Why has there been such intensive debate on fuzziness versus probability? Besides the philosophical and technical differences, it might be helpful to notice the professional differences between scientists (most probabilists are scientists) and engineers (most possibilists are engineers). In some sense, the business of a scientist is to be skeptical of all claims, whereas a good engineer should be open minded and keeps all options open. Indeed, an ultimate goal of science is to discover the fundamental principles governing the universe; whereas the task of engineering is to build up things that did not exist in the universe before. Consequently, scientists are usually critical and concentrate on finding the defects in claims or theories; whereas engineers are typically pragmatists and use whatever methods that can best solve the problem. Therefore, from an engineer's perspective, we should definitely use the techniques provided by fuzzy theory as long as they can help us to produce good products, no matter what probabilists say about fuzzy theory. If some day in the future probabilists provide design tools for problems that are now solved by fuzzy techniques, we should definitely try them and compare them with their fuzzy counterparts. We would like to conclude this section, and this book, by quoting from Zadeh [1995]:

```
    In many cases there is more to be gained from cooperation than from
    arguments over which methodology is best. A case in point is the
    concept of soft computing. Soft computing is not a methodology
    -- it is a partnership of methodologies that function effectively in
    an environment of imprecision and/or uncertainty and are aimed at
    exploiting the tolerance for imprecision, uncertainty, and partial
```

truth to achieve tractability, robustness, and low solution costs. At
this juncture, the principal constituents of soft computing are fuzzy
logic, neurocomputing, and probabilistic reasoning, with the latter
subsuming genetic algorithms, evidential reasoning, and parts of
learning and chaos theories.

31.5 Summary and Further Readings

In this chapter we have demonstrated the following:

- The induction of possibility distributions from membership functions of fuzzy sets.

- The concepts of possibility measures, marginal and conditional possibility distributions, and noninteractiveness, based on the possibility distributions induced from membership functions.

- The concepts of plausibility and belief measures and their relationship.

- The induction of possibility and necessity measures from the plausibility and belief measures and their properties.

- The similarities and differences between probability theory and fuzzy set theory.

A special book on possibility theory is Dubois and Prade [1988]. The intuitive approach to possibility was taken from Zadeh [1978]. The details of the axiomatic approach to possibility can be found in Klir and Folger [1988] and Klir and Yuan [1995]. The debate on fuzziness versus probability appeared in a number of special issues in statistics or fuzzy journals, for example, in International Statistical Review (1982), Statistical Science (1987), IEEE Trans. on Fuzzy Systems (1994) and Technometrics (1995).

31.6 Exercises

Exercise 31.1. Consider the possibility distribution π_x of (31.7) and find the conditional possibility distributions of π_x given:

(a) $\pi_{(x_1,x_3)}(u_1, u_3) = 0.2/(a, a) + 0.9/(a, b) + 0.7/(b, b)$

(b) $\pi_{(x_2,x_3)}(u_2, u_3) = 0.8/(a, a) + 0.2/(b, a)$

(c) $\pi_{x_3}(u_3) = 0.6/a + 0.5/b$

(d) $\pi_{x_1}(u_1) = 0.6/a$

Exercise 31.2. Given two noninteractive marginal possibility distributions $\pi_x = (1, .8, .5)$ and $\pi = (1, .7)$ on sets $X = \{a, b, c\}$ and $Y = \{\alpha, \beta\}$, respectively, determine the corresponding basic distribution.

Exercise 31.3. Repeat Exercise 31.2 for the following marginal possibility distributions:

(a) $\pi_x = (1, .7, .2)$ on $X = \{a, b, c\}$ and $\pi_y = (1, 1, .4)$ on $Y = \{\alpha, \beta, \gamma\}$

(b) $\pi_x = (1, .9, .6, .2)$ on $X = \{a, b, c, d\}$ and $\pi_y = (1, .6)$ on $Y = \{\alpha, \beta\}$

Exercise 31.4. Prove that plausibility and belief measures are complement, that is, prove (31.25).

Exercise 31.5. Prove that probability measure is bounded by plausibility and belief measures, that is, prove the truth of (31.26).

Exercise 31.6. Show that possibility and necessity measures are special cases of plausibility and belief measures, respectively.

Exercise 31.7. Let the possibility measure *Pos* be defined in Definition 31.5. Show that there exists a function $\pi : U \to [0, 1]$ such that (31.32) is true.

Exercise 31.8. Find an example that shows that possibility and objective probability are different.

Bibliography

1. Aliev, R., F. Aliev, and M. Babaev [1991], *Fuzzy Process Control and Knowledge Engineering in Petrochemical and Robotic Manufacturing*, Verlag TÜV Rheinland, Köln.

2. Altrock, C.V., B. Krause, and H.J. Zimmermann [1992], "Advanced fuzzy logic control of a model can in extreme situations," *Fuzzy Sets and Systems*, 48, no. 2, pp. 41-52.

3. Anderson, B.D.O. and J.B. Moore [1990], *Optimal Control: Linear Quadratic Methods*, Prentice-Hall, NJ.

4. Aracil, J., A. Ollero, and A. Garcia-Cerezo [1989], "Stability indices for the global analysis of expert control systems," *IEEE Trans. on Systems, Man, and Cybern.*, 19, no. 5, pp. 998-1007.

5. Åström, K.J., and B. Wittenmark [1995], *Adaptive Control*, Addison-Wesley Publishing Company, MA.

6. Baldwin, J.F. [1993], "Fuzzy reasoning in FRIL for fuzzy control and other knowledge-based applications," *Asia-Pacific Engineering J.*, 3, no. 1-2, pp. 59-82.

7. Bandler,W., and L.J. Kohout [1980], "Semantics of implication operators and fuzzy relational products," *International J. of Man-Machine Studies*, 12, no. 1, pp. 89-116.

8. Bellman, R.E., and M. Giertz [1973], "On the analytic formalism of the theory of fuzzy sets," *Information Sciences*, 5, pp. 149-156.

9. Bellman, R.E., and L.A. Zadeh [1970], "Decision-making in a fuzzy environment," *Management Science*, 17, no. 4, pp. 141-164.

10. _____ [1977], "Local and fuzzy logics," in *Modern Uses of Multiple-Valued Logic*, J. M. Dunn and G. Epstein, eds., Reidel Publ., Dordrecht, Netherlands: Reidel

405

Publ., pp. 103-165.

11. Berenji, H.R., and P. Khedkar [1992], "Learning and tuning fuzzy logic controllers through reinforcements," *IEEE Trans. on Neural Networks*, 3, no. 5, pp. 724-740.

12. Bernard, J. A. [1988], "Use of rule-based system for process control," *IEEE Contr. Syst. Mag.*, 8, no. 5, pp. 3-13.

13. Bernardo, J.M., and A.F.M. Smith [1994], *Bayesian Theory*, John Wiley, NY.

14. Bezdek, J. C. [1981], *Pattern Recognition with Fuzzy Objective Function Algorithms*, Plenum Press, NY.

15. _____ [1994], "Editorial: Fuzziness versus probability—again (!?)," *IEEE Trans. on Fuzzy Systems*, 2, no. 1, pp. 1-3.

16. Bezdek, J.C., R.J. Hathaway, M.J. Sabin, and W.T. Tucker [1987], "Convergence theory for fuzzy c-means: counterexamples and repairs," *IEEE Trans. on Systems, Man, and Cybern.*, 17, no. 5, pp. 873-877.

17. Bezdek, J.C., and S.K. Pal, eds. [1992], *Fuzzy Models for Pattern Recognition: Methods that Search for Patterns in Data*, IEEE Press, NY.

18. Biglieri, E., A. Gersho, R. D. Gitlin, and T. L. Lim [1984], "Adaptive cancellation of nonlinear intersymbol interference for voiceband data transmission," *IEEE J. on Selected Areas in Communications*, SAC-2, 5, pp. 765-777.

19. Bonissone, P. P., S. Dutta, and N.C. Wood [1994], "Merging strategic and tactical planning in dynamic and uncertain environments," *IEEE Trans. on Systems, Man, and Cybern.*, 24, no. 6, pp. 841-862.

20. Box, G. E. P., and G. M. Jenkins [1976], "*Time Series Analysis: Forecasting and Control*," Holden-Day Inc., 1976.

21. Braae, M., and D.A. Rutherford [1979], "Theoretical and linguistic aspects of the fuzzy logic controller," *Automatica*, 15, pp. 553-577.

22. Brown, M., and C. Harris [1994], *Neurofuzzy Adaptive Modeling and Control*, Prentice Hall, Englewood Cliffs, NJ.

23. Bryson, A.E., and Y.C. Ho [1975], *Applied Optimal Control*, Hemisphere, NY.

24. Buckley, J.J. [1992a], "Solving fuzzy equations," *Fuzzy Sets and Systems*, 50, no. 1, pp. 1-14.

25. _____ [1992b], "Universal fuzzy controllers," *Automatica*, 28, no. 6, pp. 1245-1248.

26. _____ [1993], "Sugeno type controllers are universal controllers," *Fuzzy Sets and Systems*, 53, no. 3, pp. 299-303.

27. Buckley, J.J., and Y. Hayashi [1994], "Can fuzzy neural nets approximate continuous fuzzy functions?" *Fuzzy Sets and Systems*, 61, no. 1, pp. 43-52.

28. Chang, C.L. [1968], "Fuzzy topological spaces," *J. of Math. Analysis and Applications*, 24, no. 1, pp. 182-190.

29. Charnes, A., and W.W. Cooper [1959], "Chance-constrainted programming," *Management Science*, 5, pp. 73-79.

30. Cheeseman, P. [1988], "An inquiry into computer understanding," (with discussions) *Computational Intelligence*, 4., pp. 58-142.

31. Chen, J.Q., and L.J. Chen [1993], "Study on stability of fuzzy closed-loop control systems," *Fuzzy Sets and Systems*, 57, no. 2, pp. 159-168.

32. Chen, S., C. F. N. Cowan, and P. M. Grant [1991], "Orthogonal least squares learning algorithm for radial basis function networks," *IEEE Trans. on Neural Networks*, 2, 2, pp. 302-309.

33. Chen, S., G. J. Gibson, C. F. N. Cowan, and P. M. Grant [1990], "Adaptive equalization of finite non-linear channels using multilayer perceptrons," *Signal Processing*, 20, pp. 107-119.

34. _____ [1991], "Reconstruction of binary signals using an adaptive radial-basis-function equalizer," *Signal Processing*, 22, pp. 77-93.

35. Chen, S.J., and C.L. Hwang [1992], *Fuzzy Multiple Attribute Decision Making: Methods and Applications*, Springer-Verlag, NY.

36. Chiu, S., S. Chand, D. Moore, and A. Chaudhary [1991], "Fuzzy logic for control of roll and moment for a flexible wing aircraft," *IEEE Control Systems Magazine*, 11, 4, pp. 42-48.

37. Chu, C.K., and J.M. Mendel [1994], "First break refraction event picking using fuzzy logic systems," *IEEE Trans. on Fuzzy Systems*, 2, no. 4, pp. 255-266.

38. Cox, E. [1994], *The Fuzzy Systems Handbook*, Academic Press, MA.

39. Cowan, C. F. N., and P. M. Grant (eds.) [1985], "*Adaptive Filters*," Prentice-Hall, Inc., Englewood Cliffs, NJ.

40. Cybenko, G. [1989], "Approximation by superpositions of a sigmoidal function," *Mathematics of Control, Signals, and Systems*.

41. Czogala, E., and W. Pedrycz [1982], "Fuzzy rule generation for fuzzy control," *Cybernetics and Systems*, 13, no. 3, pp. 275-293.

42. Di Nola, A., W. Pedrycz, S. Sessa, and E. Sanchez [1991], "Fuzzy relation equations theory as a basis for fuzzy modeling: an overview," *Fuzzy Sets and Systems*, 40, no. 3, pp. 415-429.

43. Di Nola, A., S. Sessa, W. Pedrycz, and E. Sanchez [1989], *Fuzzy Relation Equations and Their Applications to Knowledge Engineering*, Kluwer, Boston.

44. Dombi, J. [1982], "A general class of fuzzy operators, the De Morgan class of fuzzy operators and fuzziness measures induced by fuzzy operators," *Fuzzy Sets and Systems*, 8, no. 2, pp. 149-163.

45. Driankov, D., H. Hellendoorn, and M. Reinfrank [1993], *An Introduction to Fuzzy Control*, Springer-Verlag, Berlin.

46. Dubois, D., and H. Prade [1980], *Fuzzy Sets and Systems: Theory and Applications*, Academic Press, Inc., Orlando, Florida.

47. _____ [1985], "A review of fuzzy set aggregation connectives," *Information Sciences*, 36, no. 1-2, pp. 85-121.

48. _____ [1987a], "Fuzzy numbers: an overview," in *Analysis of Fuzzy Information – Vol. 1: Mathematics and Logic*, Bezdek, J.C., ed., CRC Press, FL, pp. 3-39.

49. _____ [1987b], "Necessity measures and the resolution principle," *IEEE Trans. on Systems, Man, and Cybern.*, 17, no. 3, pp. 474-478.

50. _____ [1988], *Possibility Theory*, Plenum Press, NY.

51. _____ [1991], "Fuzzy sets in approximate reasoning, Parts 1 and 2," *Fuzzy Sets and Systems*, 40, no. 1, pp. 143-244.

52. _____ [1993], "Fuzzy sets and probability: Misunderstanding, bridges and gaps," *Proc. 2nd IEEE Intern. Conf. on Fuzzy Systems*, San Francisco, pp. 1059-1068.

53. Duda R.O., and P.E. Hart [1973], *Pattern Classification and Scene Analysis*, John Wiley & Sons, NY.

54. Falconer, D. D. [1978], "Adaptive equalization of channel nonlinearities in QAM data transmission systems," *The Bell System Technical Journal*, 57, 7, pp. 2589-2611.

55. Filev, D.P., and R.R. Yager [1993], "Three models of fuzzy logic controllers," *Cybernetics and Systems*, 24, no. 2, pp. 91-114.

56. Gegov, A. [1994], "Multilevel intelligent fuzzy control of oversaturated urban traffic networks," *Intern. J. of Systems Science*, 25, no. 6, pp. 967-978.

57. Goguen, J.A. [1967], "L-fuzzy sets," *J. of Math. Analysis and Applications*, 18, no. 1, pp. 145-174.

58. Gottwald, S. [1979], "Set theory for fuzzy sets of higher level," *Fuzzy Sets and Systems*, 2, no. 2, pp. 125-151.

59. Green, M., and D.J.N. Limebeer [1995], *Linear Robust Control*, Prentice Hall, Englewood Cliffs, NJ.

60. Guan, J.W. and D.A. Bell [1991], *Evidence Theory and Its Applications, Vols. I and II*, North-Holland, NY.

61. Gupta, M.M., J.B. Kiszka, and G.M. Trojan [1986], "Multivariable structure of fuzzy control systems," *IEEE Trans. on Systems, Man, and Cybern.*, 16, no. 5, pp. 638-656.

62. Gupta, M.M., and D.H. Rao [1994], "On the principles of fuzzy neural networks," *Fuzzy Sets and Systems*, 61, no. 1, pp. 1-18.

63. Hansson, A., P. Gruber, and J. Tödtli [1994], "Fuzzy anti-reset windup for PID controllers," *Control Eng. Practice*, 2, no. 3, pp. 389-396.

64. Harris, C.J., C.C. Moore, and M. Brown [1993], *Intelligent Control: Aspects of Fuzzy Logic and Neural Nets*, World Scientific, Singapore.

65. Hauser, J., S. Sastry,, and P. Kokotovic [1992], "Nonlinear control via approximate input-output linearization: the ball and beam example," *IEEE Trans. on Automatic Control*, 37, 3, pp. 392-398.

66. Hellendoorn, H., and R. Palm [1994], "Fuzzy system technologies at Siemens R&D," *Fuzzy Sets and Systems*, 63, no. 3, pp. 245-269.

67. Hellendoorn H., and C. Thomas [1993], "Defuzzification in fuzzy controllers," *J. of Intelligent & Fuzzy Systems*, 1, no. 2, pp. 109-123.

68. Higashi, M., and G.J. Klir [1982], "On measure of fuzziness and fuzzy complements," *Intern. J. of General Systems*, 8 , no. 3, pp. 169-180.

69. Hirota, K., A. Arai, and S. Hachisu [1989], "Fuzzy controlled robot arm playing two-dimensional ping-pong game," *Fuzzy Sets and Systems*, 32, no. 2, pp. 149-159.

70. Holmblad, L.P., and J.J. Østergaard [1982], "Control of a cement kiln by fuzzy logic," In: Gupta, M.M., and E. Sanchez, eds., *Fuzzy Information and Decision Processes*, North-Holland, Amsterdam, pp. 398-409.

71. Hornik, K., M. Stinchcombe, and H. White [1989], "Multilayer feedforward networks are universal approximators," *Neural Networks*, 2, pp. 359-366.

72. Hwang, Y.R., and M. Tomizuka [1994], "Fuzzy smoothing algorithms for variable structure systems," *IEEE Trans. on Fuzzy Systems*, 2, no. 4, pp. 277-284.

73. Ioannou, P.A., and P.V. Kokotovic[1983], *Adaptive Systems with Reduced Models*, Springer-Verlag, NY.

74. Isidori, A. [1989], *Nonlinear Control Systems*, Springer-Verlag, Berlin.

75. Itoh, O., K. Gotoh, T. Nakayama, and S. Takamizawa [1987], "Application of fuzzy control to activated sludge process," *Proc. 2nd IFSA Congress*, Tokyo, Japan, pp. 282-285.

76. Jang, J.R. [1993], "ANFIS: Adaptive-network-based fuzzy inference system," *IEEE Trans. on Systems, Man, and Cybern.*, 23, no. 3, pp. 665-685.

77. Johansen, T.A. [1994], "Fuzzy model based control: stability, robustness, and performance issues," *IEEE Trans. on Fuzzy Systems*, 2, pp. 221-234.

78. Kandel, A., and G. Langholz, eds. [1994], *Fuzzy Control Systems*, CRC Press, FL.

79. Kaufmann, A., and M.M. Gupta [1985], *Introduction to Fuzzy Arithmetic: Theory and Applications*, Van Nostrand, NY.

80. Kickert, W. J. M., and H. R. Van Nauta Lemke [1976], "Application of a fuzzy controller in a warm water plant," *Automatica*, 12, 4, pp. 301-308.

81. Kiszka, J., M. Gupta, and P. Nikiforuk [1985], "Energetistic stability of fuzzy dynamic systems," *IEEE Trans. Systems, Man, and Cybern.*, 15, 5, pp. 783-792.

82. Klir, G.J., and T. Folger [1988], *Fuzzy Sets, Uncertainty, and Information*, Prentice Hall, Englewood Cliffs, NJ.

83. Klir, G.J., and B. Yuan [1995], *Fuzzy Sets and Fuzzy Logic: Theory and Applications*, Prentice Hall, Englewood Cliffs, NJ.

84. Kosko, B. [1991], *Neural Networks and Fuzzy Systems*, Prentice Hall, Englewood Cliffs, NJ.

85. Kosko, B. [1993], *Fuzzy Thinking: The New Science of Fuzzy Logic*, Hyperion, NY.

86. Kruse, R., J. Gebhardt, and F. Klawonn, *Foundations of Fuzzy Systems*, John Wiley & Sons, 1994.

87. Kuipers, B., and K. Åström [1994], "The composition and validation of heterogeneous control laws," *Automatica*, 30, no. 2, pp. 233-249.

88. Lai, Y.J., and C.L. Hwang [1992], *Fuzzy Mathematical Programming*, Springer-Verlag, NY.

89. Lai, Y.J., and C.L. Hwang [1994], *Fuzzy Multiple Objective Decision Making*, Springer-Verlag, NY.

90. Langari, G., and M. Tomizuka [1990], "Stability of fuzzy linguistic control systems," *Proc. IEEE Conf. on Decision and Control*, pp. 2185-2190.

91. Larkin, L. I. [1985], "A fuzzy logic controller for aircraft flight control," in *Industrial Applications of Fuzzy Control*, ed. M.Sugeno, North-Holland, Amsterdam, pp. 87-104.

92. Larsen, P. M. [1980], "Industrial application of fuzzy logic control," *Int. J. Man Mach. Studies*, 12, 1, pp. 3-10.

93. Laviolette, M., J.W. Seaman, Jr., J.D. Barrett, and W.H. Woodall, "A probabilistic and statistical view of fuzzy methods," *Technometrics*, 37, no. 3, pp. 249-261.

94. Lee, C. C. [1990], "Fuzzy logic in control systems: fuzzy logic controller, part I and II," *IEEE Trans. on Syst., Man, and Cybern.*, 20, 2, 404-435.

95. Lewis, F.L., K. Liu, and A. Yesildirek [1993], "Neural net robot controller with guaranteed tracking performance," *Proc. 1993 Int. Symp. on Intelligent Control*, pp. 225-231.

96. Lewis, F.L., S.Q. Zhu, and K. Liu [1995], "Function approximation by fuzzy systems," *Proc. 1995 American Control Conf.*, Seattle, pp. 3760-3764.

97. Lin, C.T. [1994], *Neural Fuzzy Control Systems with Structure and Parameter Learning*, World Scientific, Singapore.

98. Lindley, D. V. [1982], "Scoring rules and the inevitability of probability," (with discussions), *International Statistical Review*, 50, pp. 1-26.

99. _____ [1987], "The probability approach to the treatment of uncertainty in artificial intelligence and expert systems," *Statistical Science*, 2, 1, pp. 17-24.

100. Liu, Y.M. [1985], "Some properties of convex fuzzy sets," *J. of Math. Analysis and Applications*, 111, no. 1, pp. 119-129.

101. Ljung, L. [1991], "Issues in system identification," *IEEE Control Systems Magazine*, pp. 25-29.

102. Loginov, V.J. [1966], "Probability treatment of Zadeh membership functions and their use in pattern recognition," *Engineering Cybern.*, pp. 68-69.

103. Lowen, R. [1980], "Convex fuzzy sets," *Fuzzy Sets and Systems*, 3, no. 3, pp. 291-310.

104. Luenberger, D. G. [1969], *Optimization by Vector Space Methods*, John Wiley, NY.

105. _____ [1984], *Linear and Nonlinear Programming*, Addison-Wesley Publishing Company, Inc., Reading, MA.

106. Maiers, J., and Y. S. Sherif [1985], "Applications of fuzzy sets theory," *IEEE Trans. Syst. Man Cybern.*, 15, 6, 175-189.

107. Mamdani, E. H. [1974], "Applications of fuzzy algorithms for simple dynamic plant," *Proc. IEE*, 121, 12, pp. 1585-1588.

108. Mamdani, E. H., and S. Assilian [1975], "An experiment in linguistic synthesis with a fuzzy logic controller," *Int. J. Man Mach. Studies*, 7, 1, pp. 1-13.

109. Mendel, J.M. [1994], *Lessons in Digital Estimation Theory*, Prentice Hall, Englewood Cliffs, NJ.

110. McNeill, D., and P. Freiberger [1993], *Fuzzy Logic: The Discovery of a Revolutionary Computer Technology—and How It Is Changing Our World*, Simon & Schuster, NY.

111. Miller, W. T., R. S. Sutton, and P. J. Werbos eds. [1990], *Neural Networks for Control*, The MIT Press, Cambridge, MA.

112. Narendra, K. S., and A. M. Annaswamy [1987], "A new adaptive law for robust adaptation with persistent excitation," *IEEE Trans. on Automatic Control*, 32, no. 2, pp. 134-145.

113. _____ [1989], *Stable Adaptive Systems*, Prentice Hall, Englewood Cliffs, NJ.

114. Narendra, K. S., and K. Parthasarathy [1990], "Identification and control of dynamical systems using neural networks," *IEEE Trans. on Neural Networks*, 1, 1, pp. 4-27.

115. Nguyen, D., and B. Widrow [1990], "The truck backer-upper: an example of self-learning in neural networks," *IEEE Cont. Syst. Mag.*, 10, 3, pp. 18-23.

116. Nie, J., and D.A. Linkens [1994], *Fuzzy Neural Control: Principles, Algorithms, and Applications*, Prentice Hall, Englewood Cliffs, NJ.

117. Pal, S.K., and D.K. Majumder [1986], *Fuzzy Mathematical Approach to Pattern Recognition*, John Wiley, New York, NY.

118. Palm, R. [1992], "Sliding mode fuzzy control," *Proc. 1st IEEE Inter. Conf. on Fuzzy Systems*, pp. 519-526, San Diego.

119. Pappis, C. P., and E. H. Mamdani [1977], "A fuzzy logic controller for a traffic junction," *IEEE Trans. Syst. Man Cybern.*, 7, 10, pp. 707-717.

120. Pedrycz, W. [1991], "Processing in relational structures: fuzzy relational equations," *Fuzzy Sets and Systems*, 40, no. 1, pp. 77-106.

121. _____ [1992], "Fuzzy neural networks with reference neurons as pattern classifiers," *IEEE Trans. on Neural Networks*, 3, no. 5, pp. 770-775.

122. _____ [1993], *Fuzzy Control and Fuzzy Systems*, John Wiley & Sons, Inc., New York, NY.

123. Pitas, I., and A. N. Venetsanopoulos [1990], "*Nonlinear Digital Filters*," Kluwer Academic Publishers, Boston.

124. Polycarpou, M. M., and P. A. Ioannou [1991], "Identification and control of nonlinear systems using neural network models: design and stability analysis," USC EE-Report.

125. Powell, M. J. D. [1981], *Approximation Theory and Methods*, Cambridge University Press, Cambridge.

126. Raju, G.V.S., J. Zhou, and R.A. Kisner [1991], "Hierarchical fuzzy control," *Int. J. of Control*, 54, no. 5, pp. 1201-1216.

127. Raju, G.V.S., and J. Zhou [1993], "Adaptive hierarchical fuzzy controller," *IEEE Trans. on Systems, Man, and Cybern.*, 23, no. 4, pp. 973-980.

128. Rescher, N. [1969], *Many -Valued Logic*, McGraw-Hill, Inc., New York.

129. Rudin, W. [1976], *Principles of Mathematical Analysis*, McGraw-Hill, Inc., New York.

130. Ruspini, E.H. [1969], "A new approach to clustering," *Information and Control*, 15, no. 1, pp. 22-32.

131. Sanner, R. M., and J. E. Slotine [1991], "Gaussian networks for direct adaptive control," *Proc. American Control Conf.*, pp. 2153-2159.

132. Sastry, S., and M. Bodson [1989], *Adaptive Control: Stability, Convergence, and Robustness*, Prentice Hall, Englewood Cliffs, NJ.

133. Sastry, S., and Isidori, A. [1989], "Adaptive control of linearizable systems," *IEEE Trans. on Automatic Control*, 34, 11, pp. 1123-1131.

134. Schwartz, A. L. [1992], "Comments on "Fuzzy logic for control of roll and moment for a flexible wing aircraft,"" *IEEE Control Systems Magazine*.

135. Shafer, G. [1976], *A Mathematical Theory of Evidence*, Princeton University Press, Princeton, NJ.

136. Simpson, P.K. [1992], "Fuzzy min-max neural networks, Part 1: Classification," *IEEE Trans. on Neural Networks*, 3, no. 5, pp. 776-786.

137. Slotine, J. E., and W. Li [1991], *Applied Nonlinear Control*, Prentice Hall, Englewood Cliffs, NJ.

138. Spooner, J.T., and K.M. Passino [1995], "Stable adaptive control using fuzzy systems and neural networks," IVHS OSU Report 95-01.

139. Stallings, W. [1977], "Fuzzy set theory versus Bayesian statistics," *IEEE Trans. Systems, Man, and Cybern.*, 7, 3, pp. 216-219.

140. Sugeno, M. [1977], "Fuzzy measures and fuzzy integrals: A survey," In: Gupta, M., G.N. Saridis, and B.R. Gaines, eds., *Fuzzy Automata and Decision Processes*, North-Holland, NY, pp. 329-346.

141. Sugeno, M., ed. [1985], *Industrial Applications of Fuzzy Control*, North-Holland, NY.

142. Sugeno, M., and M. Nishida [1985], "Fuzzy control of model car," *Fuzzy Sets and Systems*, pp. 103-113.

143. Sugeno, M., T. Murofushi, T. Mori, T. Tatematsu, and J. Tanaka [1989], "Fuzzy algorithmic control of a model car by oral instructions," *Fuzzy Sets and Systems*, 32, pp. 207-219.

144. Sugeno, M., and G.T. Kang [1988], "Structure identification of fuzzy model," *Fuzzy Sets and Systems*, 28, no. 1, pp. 15-33.

145. Sugeno, M., and K. Tanaka [1991], "Successive identification of a fuzzy model and its applications to prediction of a complex system," *Fuzzy Sets and Systems*, 42, no. 3, pp. 315-334.

146. Takagi, T., and M. Sugeno [1985], "Fuzzy identification of systems and its applications to modeling and control," *IEEE Trans. on Systems, Man, and Cybern.*, 15, 1, pp. 116-132.

147. Tanaka, K., and M. Sano [1994], "A robust stabilization problem of fuzzy control systems and its application to backing up a truck-trailer," *IEEE Trans. on Fuzzy Systems*, 2, no. 2, pp. 107-118.

148. Tanaka, K., and M. Sugeno [1992], "Stability analysis and design of fuzzy control systems," *Fuzzy Sets and Systems*, 45, no. 2, pp. 135-156.

149. Terano, T., K. Asai, and M. Sugeno, eds. [1994], *Applied Fuzzy Systems*, Academic Press, MA.

150. Tong, R. M. [1980], "Some properties of fuzzy feedback systems," *IEEE Trans. Systems, Man, and Cybern.*, 10, 6, pp. 327-330.

151. _____ [1984], "A retrospective view of fuzzy control systems," *Fuzzy Sets and Systems*, 14, 3, pp. 199-210.

152. Tong, R. M., M. B. Beck, and A. Latten [1980], "Fuzzy control of the activated sludge wastewater treatment process," *Automatica*, 16, 6, pp. 695-701.

153. Togai, M., and H. Watanabe [1986], "Expert system on a chip: an engine for real-time approximate reasoning," *IEEE Expert Syst. Mag.*, 1, 55-62.

154. Turksen, I.B., and D.D.W. Yao [1984], "Representations of connectives in fuzzy reasoning: the view through normal form," *IEEE Trans. Systems, Man, and Cybern.*, 14, no. 1, pp. 146-151.

155. Tzafestas, S.G. (ed.) [1991], *Intelligent Robotic Systems*, Marcel Dekker, Inc., NY.

156. Umbers, I. G., and P. J. King [1980], "An analysis of human decision making in cement kiln control and the implications for automation," *Int. J. Man Mach. Studies*, 12, 1, pp. 11-23.

157. Uragami, M., M. Mizumoto, and K. Tananka [1976], "Fuzzy robot controls," *Cybern.*, 6, pp. 39-64.

158. Utkin, V.I. [1978], *Sliding Modes and Their Application to Variable Structure Systems*, MIR Publishers, Moscow.

159. Valavanis, K.P., and G.N. Saridis [1992], *Intelligent Robotic Systems: Theory, Design and Applications*, Kluwer Academic Publishers, Boston.

160. Vandegrift, M.W., F.L. Lewis, S. Jagannathan, and K. Liu [1995], "Adaptive fuzzy control of discrete-time dynamical systems," personal communication.

161. Vidyasagar, M. [1993], *Nonlinear Systems Analysis*, Prentice Hall, Englewood Cliffs, NJ.

162. Voxman, W., and R. Goetshel [1983], "A note on the characterization of max and min operations," *Information Sciences*, 30, no. 1, pp. 5-10.

163. Walsh, G., D. Tilbury, S. Sastry, and J.P. Laumond [1994], "Stabilization of trajectories for systems with nonholonomic constraints," *IEEE Trans. on Automatic Control*, 39, no. 1, pp. 216-222.

164. Wang, L. X. [1992], "Fuzzy systems are universal approximators," *Proc. IEEE International Conf. on Fuzzy Systems*, San Diego, pp. 1163-1170.

165. _____ [1993a], "Stable adaptive fuzzy control of nonlinear systems," *IEEE Trans. on Fuzzy Systems*, 1, no. 2, pp.146-155.

166. _____ [1993b], "Solving fuzzy relational equations through network training," *Proc. 2nd IEEE Intern. Conf. on Fuzzy Systems*, San Francisco, pp. 956-960.

167. _____ [1994a], *Adaptive Fuzzy Systems and Control: Design and Stability Analysis*, Prentice Hall, Englewood Cliffs, NJ.

168. _____ [1994b], "A supervisory controller for fuzzy control systems that guarantees stability," *IEEE Trans. on Automatic Control*, 39, no. 9, pp. 1845-1848.

169. _____ [1995a], "Analysis and design of fuzzy identifiers of nonlinear dynamic systems," *IEEE Trans. on Automatic Control*, 40, no. 1, pp. 11-23.

170. _____ [1996], "Universal approximation by hierarchical fuzzy systems," submitted for publication.

171. Wang, L.X., and J. M. Mendel [1992a], "Fuzzy basis functions, universal approximation, and orthogonal least squares learning," *IEEE Trans. on Neural Networks*, 3, no. 5, pp. 807-814.

172. _____ [1992b], "Generating fuzzy rules by learning from examples," *IEEE Trans. on Systems, Man, and Cybern.*, 22, no. 6, pp. 1414-1427.

173. _____ [1993], "Fuzzy adaptive filters, with application to nonlinear channel equalization," *IEEE Trans. on Fuzzy Systems*, 1, no. 3, pp. 161-170.

174. Wang, P.P., ed. [1993], *Advances in Fuzzy Theory and Technology, Vol. I*, Bookwrights Press, Durham, NC.

175. Wang, P.Z. [1990], "A factor spaces approach to knowledge representation," *Fuzzy Sets and Systems*, 36, no. 1, pp. 113-124.

176. Werbos, P. [1974], "New tools for prediction and analysis in the behavioral sciences," Ph.D. Thesis, Harvard University Committee on Applied Mathematics.

177. Werner, B. [1987], "An interactive fuzzy programming system," *Fuzzy Sets and Systems*, 23, pp. 131-147.

178. White, D.A., and D.A. Sofge, eds. [1992], *Handbook of Intelligent Control: Neuro-Fuzzy and Adaptive Approaches*, Van Nostrand, NY.

179. Yager, R.R. [1980], "On a general class of fuzzy connectives," *Fuzzy Sets and Systems*, 4, no. 3, pp. 235-242.

180. _____ [1986], "A characterization of the extension principle," *Fuzzy Sets and Systems*, 18, no. 3, pp. 205-217.

181. Yager, R.R., and D.P. Filev [1994], *Essentials of Fuzzy Modeling and Control*, John Wiley, New York, NY.

182. Yagishita, O., O. Itoh, and M. Sugeno [1985], "Application of fuzzy reasoning to the water purification process," in *Industrial Applications of Fuzzy Control*, ed. M. Sugeno, North-Holland, Amsterdam, pp. 19-40.

183. Yamakawa, T. [1989], "Stabilization of an inverted pendulum by a high-speed fuzzy logic controller hardware systems," *Fuzzy Sets and Systems*, 32, pp. 161-180.

184. Yasunobu, S., S. Miyamoto,, and H. Ihara [1983], "Fuzzy control for automatic train operation system," *Proc. 4th IFAC/IFIP/IFORS Int. Congress on Control in Transportation Systems*, Baden-Baden.

185. Yasunobu, S., and S. Miyamoto [1985], "Automatic train operation by predictive fuzzy control," in *Industrial Application of Fuzzy Control*, ed. M. Sugeno, Amsterdam: North-Holland, pp. 1-18.

186. Yasunobu, S., S. Sekino,, and T. Hasegawa [1987], "Automatic train operation and automatic crane operation systems based on predictive fuzzy control," *Proc. 2nd IFSA Congress*, Tokyo, Japan, pp. 835-838.

187. Yasunobu, S., and T. Hasegawa [1986], "Evaluation of an automatic container crane operation system based on predictive fuzzy control," *Control Theory Adv. Technol.*, 2, 3, pp. 419-432.

188. Ying, H. [1994], "Sufficient conditions on general fuzzy systems as function approximations," *Automatica*, 30, no. 3, pp. 521-525.

189. Zadeh, L. A. [1962], "From circuit theory to systems theory," *Proc. Institution of Radio Engineers*, 50, pp. 856-865.

190. _____ [1965], "Fuzzy sets," *Informat. Control*, 8, pp. 338-353.

191. _____ [1968], "Fuzzy algorithms," *Information and Control*, 12, no. 2, pp. 94-102.

192. _____ [1971a], "Toward a theory of fuzzy systems," in *Aspects of Network and System Theory*, eds. R. E. Kalman and N. DeClaris.

193. _____ [1971b], "Similarity relations and fuzzy ordering," *Information Sciences*, 3, no. 2, pp. 177-200.

194. _____ [1973], "Outline of a new approach to the analysis of complex systems and decision processes," *IEEE Trans. on Systems, Man, and Cybern.*, 3, 1, pp. 28-44.

195. _____ [1975], "The concept of a linguistic variable and its application to approx-imate reasoning I, II, III," *Information Sciences*, 8, pp. 199-251, pp. 301-357; 9, pp. 43-80.

196. _____ [1978], "Fuzzy sets as a basis for a theory of possibility," *Fuzzy Sets and Systems*, 1, no. 1, pp. 3-28.

197. _____ [1985], "Syllogistic reasoning in fuzzy logic and its application to usuality and reasoning with dispositions," *IEEE Trans. on Systems, Man, and Cybern.*, 15, 6 (1985), pp. 754-763.

198. _____ [1994], "Fuzzy logic, neural networks and soft computing," *Communications of the ACM*, 37, pp. 77-84.

199. _____ [1995], "Discussion: Probability theory and fuzzy logic are complementary rather than competitive," *Technometrics*, 37, no. 3, pp. 271-276.

200. Zeng, X.J., and M.G. Singh [1994], "Approximation theory for fuzzy systems—SISO case," *IEEE Trans. on Fuzzy Systems*, 2, no. 2, pp. 162-176.

201. _____ [1995], "Approximation accuracy analysis of fuzzy systems as function approximators," personal communication.

202. Zhao, Z.Y., M. Tomizuka, and S. Isaka [1993], "Fuzzy gain scheduling of PID controllers," *IEEE Trans. on Systems, Man, and Cybern.*, 23, no. 5, pp. 1392-1398.

203. Zimmermann, H.J. [1978], "Results of empirical studies in fuzzy set theory," In: Klir, G.J., ed., *Applied General Systems Research*, Plenum Press, NY, pp. 303-312.

204. Zimmermann, H.J. [1991], *Fuzzy Set Theory and Its Applications*, Kluwer Academic Publishers, Boston.

Index